College
Chemistry

the text of this book is printed
on 100% recycled paper

About the Author

John R. Lewis received the doctorate degree in physical chemistry from the University of Wisconsin in 1924. After teaching at Wisconsin University for an additional four years, he became an Associate Professor and, later, Professor of Chemistry at the University of Utah.

From 1942 to 1958 he was Head of the Department of Metallurgical Engineering, University of Utah. From 1958 to 1963 he taught metallurgy and did research in the same area. At present he is Professor Emeritus at the University of Utah.

Professor Lewis has published several papers and bulletins in the fields of physical chemistry, extractive metallurgy, and chemical education. He is a member of several professional, scientific, and honor societies, including the American Chemical Society, the Honor Society of Sigma Xi, and Phi Kappa Phi. He is listed in *American Men of Science*, *Who's Who in Engineering*, *Who Knows—And What*, and *Who's Who in America*.

COLLEGE CHEMISTRY

Ninth Edition

JOHN R. LEWIS
University of Utah

BARNES & NOBLE BOOKS

A DIVISION OF HARPER & ROW, PUBLISHERS

New York, Hagerstown, San Francisco, London

Preface

The ninth edition of College Chemistry reflects another thorough revision. To keep up with progress in modern chemistry, some of the less important chapters have been deleted and others have been rearranged or combined. The chapters on atomic structure and the periodic classification of the elements have been revised so completely that they are, essentially, new material. A new chapter on the chemical bond has also been added. The revised book contains a number of important new tables, graphs, and figures that will be very useful to students and others. The appendixes have been enlarged to include (1) an activity table of the metals and (2) a table containing the electronic populations of atoms.

This new edition, like the earlier ones, summarizes the chemistry found in the latest editions of the most widely used general chemistry texts and also new texts that have recently come on the market. And, like the earlier editions, this new College Chemistry may be used as a textbook.

The author and publishers are grateful to students and others who have written to them concerning earlier editions. It is to be hoped that comments and suggestions regarding the ninth edition will also be forthcoming.

The author wishes to thank Dr. Gladys Walterhouse, Editor, of Barnes and Noble, Inc. for her excellent work in editing this book. It has been markedly improved through her valuable suggestions.

I wish to thank Mrs. Iris Adams for the careful work she did in typing part of the manuscript. My thanks are also due to Mrs. Dawn Ann Bailey for stenographic work and especially for the many courtesies she extended to the author. They made his tasks easier.

Finally great credit and appreciation are due my wife, Hazel, for the careful work she did in preparing the Index.

Table of Contents

Periodic Table of the Elements

Key

Atomic number — $\rightarrow$ 11
Na
$3s^1$ — — Valence electron configuration
Atomic weight — $\rightarrow$ 22.9898

s Series of Representative Elements

p Series of Representative Elements

d Series of Transition Metals

f Series of Transition Metals

Noble Gases

Group	IA	IIA	IIIB	IVB	VB	VIB	VIIB	VIII	VIII	VIII	IB	IIB	IIIA	IVA	VA	VIA	VIIA	Noble Gases
1st period $n=1$	1 **H** $1s^1$ 1.00797																	2 **He** $1s^2$ 4.0026
2nd period $n=2$	3 **Li** $2s^1$ 6.939	4 **Be** $2s^2$ 9.0122											5 **B** $2s^2 2p^1$ 10.811	6 **C** $2s^2 2p^2$ 12.01115	7 **N** $2s^2 2p^3$ 14.0067	8 **O** $2s^2 2p^4$ 15.9994	9 **F** $2s^2 2p^5$ 18.9984	10 **Ne** $2s^2 2p^6$ 20.183
3rd period $n=3$	11 **Na** $3s^1$ 22.9898	12 **Mg** $3s^2$ 24.312											13 **Al** $3s^2 3p^1$ 26.9815	14 **Si** $3s^2 3p^2$ 28.086	15 **P** $3s^2 3p^3$ 30.9738	16 **S** $3s^2 3p^4$ 32.064	17 **Cl** $3s^2 3p^5$ 35.453	18 **Ar** $3s^2 3p^6$ 39.948
4th period $n=4$	19 **K** $4s^1$ 39.102	20 **Ca** $4s^2$ 40.08	21 **Sc** $3d^1 4s^2$ 44.956	22 **Ti** $3d^2 4s^2$ 47.90	23 **V** $3d^3 4s^2$ 50.942	24 **Cr** $3d^5 4s^1$ 51.996	25 **Mn** $3d^5 4s^2$ 54.938	26 **Fe** $3d^6 4s^2$ 55.847	27 **Co** $3d^7 4s^2$ 58.9332	28 **Ni** $3d^8 4s^2$ 58.71	29 **Cu** $3d^{10} 4s^1$ 63.54	30 **Zn** $3d^{10} 4s^2$ 65.37	31 **Ga** $4s^2 4p^1$ 69.72	32 **Ge** $4s^2 4p^2$ 72.59	33 **As** $4s^2 4p^3$ 74.9216	34 **Se** $4s^2 4p^4$ 78.96	35 **Br** $4s^2 4p^5$ 79.909	36 **Kr** $4s^2 4p^6$ 83.80
5th period $n=5$	37 **Rb** $5s^1$ 85.47	38 **Sr** $5s^2$ 87.62	39 **Y** $4d^1 5s^2$ 88.905	40 **Zr** $4d^2 5s^2$ 91.22	41 **Nb** $4d^4 5s^1$ 92.906	42 **Mo** $4d^5 5s^1$ 95.94	43 **Tc** $4d^5 5s^2$ (99)	44 **Ru** $4d^7 5s^1$ 101.07	45 **Rh** $4d^8 5s^1$ 102.905	46 **Pd** $4d^{10}$ 106.4	47 **Ag** $4d^{10} 5s^1$ 107.870	48 **Cd** $4d^{10} 5s^2$ 112.40	49 **In** $5s^2 5p^1$ 114.82	50 **Sn** $5s^2 5p^2$ 118.69	51 **Sb** $5s^2 5p^3$ 121.75	52 **Te** $5s^2 5p^4$ 127.60	53 **I** $5s^2 5p^5$ 126.9044	54 **Xe** $5s^2 5p^6$ 131.30
6th period $n=6$	55 **Cs** $6s^1$ 132.905	56 **Ba** $6s^2$ 137.34	57 **La** $5d^1 6s^2$ 138.91	72 **Hf** $4f^{14} 5d^2 6s^2$ 178.49	73 **Ta** $5d^3 6s^2$ 180.948	74 **W** $5d^4 6s^2$ 183.85	75 **Re** $5d^5 6s^2$ 186.2	76 **Os** $5d^6 6s^2$ 190.2	77 **Ir** $5d^7 6s^2$ 192.2	78 **Pt** $5d^9 6s^1$ 195.09	79 **Au** $5d^{10} 6s^1$ 196.967	80 **Hg** $5d^{10} 6s^2$ 200.59	81 **Tl** $6s^2 6p^1$ 204.37	82 **Pb** $6s^2 6p^2$ 207.19	83 **Bi** $6s^2 6p^3$ 208.980	84 **Po** $6s^2 6p^4$ (210)	85 **At** $6s^2 6p^5$ (210)	86 **Rn** $6s^2 6p^6$ (222)
7th period $n=7$	87 **Fr** $7s^1$ (223)	88 **Ra** $7s^2$ (226)	89 **Ac** $6d^1 7s^2$ (227)	104 **?**														

Lanthanides

58 **Ce** $4f^1 5d^1 6s^2$ 140.12	59 **Pr** $4f^3 5d^0 6s^2$ 140.907	60 **Nd** $4f^4 5d^0 6s^2$ 144.24	61 **Pm** $4f^5 5d^0 6s^2$ (145)	62 **Sm** $4f^6 5d^0 6s^2$ 150.35	63 **Eu** $4f^7 5d^0 6s^2$ 151.96	64 **Gd** $4f^7 5d^1 6s^2$ 157.25	65 **Tb** $4f^9 5d^0 6s^2$ 158.924	66 **Dy** $4f^{10} 5d^0 6s^2$ 162.50	67 **Ho** $4f^{11} 5d^0 6s^2$ 164.930	68 **Er** $4f^{12} 5d^0 6s^2$ 167.26	69 **Tm** $4f^{13} 5d^0 6s^2$ 168.934	70 **Yb** $4f^{14} 5d^0 6s^2$ 173.04	71 **Lu** $4f^{14} 5d^1 6s^2$ 174.97

Actinides

90 **Th** $5f^0 6d^1 7s^2$ 232.038	91 **Pa** $5f^2 6d^1 7s^2$ (231)	92 **U** $5f^3 6d^1 7s^2$ 238.03	93 **Np** $5f^4 6d^1 7s^2$ (237)	94 **Pu** $5f^6 6d^0 7s^2$ (242)	95 **Am** $5f^7 6d^0 7s^2$ (243)	96 **Cm** $5f^7 6d^1 7s^2$ (247)	97 **Bk** $5f^8 6d^1 7s^2$ (247)	98 **Cf** $5f^{10} 6d^0 7s^2$ (249)	99 **Es** $5f^{11} 6d^0 7s^2$ (254)	100 **Fm** $5f^{12} 6d^0 7s^2$ (253)	101 **Md** $5f^{13} 6d^0 7s^2$ (256)	102 **No** $5f^{14} 6d^0 7s^2$ (254)	103 **Lw** $5f^{14} 6d^1 7s^2$ (257)

Table of International Atomic Weights (1968)

Values in parentheses are estimated for isotopes of longest half-life in most cases.

Element	Symbol	Atomic No.	Atomic Weight	Element	Symbol	Atomic No.	Atomic Weight
Actinium	Ac	89	(227)	Mercury	Hg	80	200.59
Aluminum	Al	13	26.9815	Molybdenum	Mo	42	95.94
Americium	Am	95	(243)	Neodymium	Nd	60	144.24
Antimony	Sb	51	121.75	Neon	Ne	10	20.179
Argon	Ar	18	39.948	Neptunium	Np	93	(237)
Arsenic	As	33	74.9216	Nickel	Ni	28	58.71
Astatine	At	85	(210)	Niobium	Nb	41	92.906
Barium	Ba	56	137.34	Nitrogen	N	7	14.0067
Berkelium	Bk	97	(247)	Nobelium	No	102	(255)
Beryllium	Be	4	9.0122	Osmium	Os	76	190.2
Bismuth	Bi	83	208.980	Oxygen	O	8	15.9994^a
Boron	B	5	10.811^a	Palladium	Pd	46	106.4
Bromine	Br	35	79.904^b	Phosphorus	P	15	30.9738
Cadmium	Cd	48	112.40	Platinum	Pt	78	195.09
Calcium	Ca	20	40.08	Plutonium	Pu	94	(244)
Californium	Cf	98	(252)	Polonium	Po	84	(210)
Carbon	C	6	12.01115^a	Potassium	K	19	39.102
Cerium	Ce	58	140.12	Praseodymium	Pr	59	140.907
Cesium	Cs	55	132.905	Promethium	Pm	61	(147)
Chlorine	Cl	17	35.453^b	Protactinium	Pa	91	(231)
Chromium	Cr	24	51.996^b	Radium	Ra	88	(226)
Cobalt	Co	27	58.9332	Radon	Rn	86	(222)
Copper	Cu	29	63.546	Rhenium	Re	75	186.2
Curium	Cm	96	(247)	Rhodium	Rh	45	102.905
Dysprosium	Dy	66	162.50	Rubidium	Rb	37	85.47
Einsteinium	Es	99	(254)	Ruthenium	Ru	44	101.07
Erbium	Er	68	167.26	Samarium	Sm	62	150.35
Europium	Eu	63	151.96	Scandium	Sc	21	44.956
Fermium	Fm	100	(257)	Selenium	Se	34	78.96
Fluorine	F	9	18.9984	Silicon	Si	14	28.086^a
Francium	Fr	87	(223)	Silver	Ag	47	107.868^a
Gadolinium	Gd	64	157.25	Sodium	Na	11	22.9898
Gallium	Ga	31	69.72	Strontium	Sr	38	87.62
Germanium	Ge	32	72.59	Sulfur	S	16	32.064^a
Gold	Au	79	196.967	Tantalum	Ta	73	180.948
Hafnium	Hf	72	178.49	Technetium	Tc	43	(99)
Helium	He	2	4.0026	Tellurium	Te	52	127.60
Holmium	Ho	67	164.930	Terbium	Tb	65	158.924
Hydrogen	H	1	1.00797^a	Thallium	Tl	81	204.37
Indium	In	49	114.82	Thorium	Th	90	232.038
Iodine	I	53	126.9044	Thulium	Tm	69	168.934
Iridium	Ir	77	192.2	Tin	Sn	50	118.69
Iron	Fe	26	55.847^b	Titanium	Ti	22	47.90
Krypton	Kr	36	83.80	Tungsten	W	74	183.85
Lanthanum	La	57	138.91	Uranium	U	92	238.03
Lawrentium	Lw	103	(256)	Vanadium	V	23	50.942
Lead	Pb	82	207.19	Xenon	Xe	54	131.30
Lithium	Li	3	6.939	Ytterbium	Yb	70	173.04
Lutetium	Lu	71	174.97	Yttrium	Y	39	88.905
Magnesium	Mg	12	24.305	Zinc	Zn	30	65.37
Manganese	Mn	25	54.9380	Zirconium	Zr	40	91.22
Mendelevium	Md	101	(257)				

^{a} The atomic weight varies because of natural variations in the isotopic composition of the element. The observed ranges are boron, ±0.003; carbon, ±0.00005; hydrogen, ±0.00001; oxygen, ±0.0001; silicon, ±0.001; sulfur, ±0.003.

^{b} The atomic weight is believed to have an experimental uncertainty of the following magnitude; bromine, ±0.002; chlorine, ±0.001; chromium, ±0.001; iron, ±0.003; silver, ±0.003. For other elements the last digit given is believed to be reliable to ±0.5.

Chapter 1
Introduction

Chemistry is one of the physical sciences. It contributes to the other physical sciences as well as to the biological sciences. As a matter of fact, chemistry occupies such an important place in the affairs of mankind that all students should have a basic knowledge of elementary chemistry in order to better understand what is going on around them. Our country's triumph in placing men on the moon was made possible through the contributions of many people, but especially by those trained in the physical sciences, including chemistry. In addition mathematicians, the various types of engineers, computer scientists, and many technicians in various disciplines worked as a team to accomplish this great feat. As students go through this book they will become aware of the many contributions chemistry has made in the past and will be in a position to appreciate worthwhile contributions made in the future.

Basic Definitions.   *Chemistry* is the science that investigates the composition and structure of matter, the changes that matter undergoes, the amounts and kinds of energy necessary for these changes, and the laws that govern change. Let us now define the important terms used in this definition.

Science is classified or systematized knowledge. It is gained and verified by observation and correct scientific thinking. In many cases scientific relationships can be expressed by mathematical equations. Science may be divided into the physical sciences, such as physics, chemistry, and geology, and the biological sciences, such as botany and zoology.

Matter is anything that occupies space and has mass. Experience shows that in ordinary chemical changes (or reactions) matter can be neither created nor destroyed. This statement is not true for nuclear reactions (see Chapter 14). Matter that is homogeneous, such as sulfur, sugar, water, or silver, is called a *substance*.

Matter may exist in three different physical forms or states:

solid, liquid, and gaseous. Solids are rigid and have a definite form, usually crystalline. Liquids flow and assume the shape of the vessel in which they are stored. Gases diffuse and "fill" any container in which they are placed. It is possible to change a substance from one physical state to another by changing the conditions under which it is maintained. Changes of temperature or pressure are frequently used to bring about these transformations. For example, if the pressure remains constant, the physical state of water depends upon the temperature. Ice (water in the crystalline state) is stable below 0°C, water (as liquid) is stable from 0°C to 100°C, and steam (water vapor) is stable above 100°C.

Matter.   Ordinary matter is made up of elements and compounds. An element used to be thought of as a substance that had not yet been chemically decomposed to give two or more simpler substances. In terms of more recent classifications, an element is a substance the properties of which give it a definite place in the periodic table (see inside front cover), or a substance all the atoms of which have the same nuclear charge (atomic number).

Thus hydrogen and oxygen are elements because they have never been decomposed into simpler substances. Water is not an element because it can be decomposed into hydrogen and oxygen. An up-to-date periodic table provides for 103 elements. Some of the elements, such as uranium and radium, are radioactive. These elements are so interesting that they are given special consideration in Chapter 14.

A *compound* is a substance composed of two or more elements chemically combined in definite proportions by weight. When the term *weight* is used we assume the value used is the value obtained at sea level. Compounds are homogeneous. The constituent elements of a compound have lost their original identity and they can be separated only by chemical means. The energy stored within a compound is not equal to the sum of the energies possessed by the uncombined elements. Water, sodium chloride (table salt), and sucrose (cane or beet sugar) are examples of compounds. Water is made up of hydrogen and oxygen. The properties of water are quite different from the properties of its constituent elements. The same is true for sodium chloride and sucrose.

A body made up of two or more substances which retain their

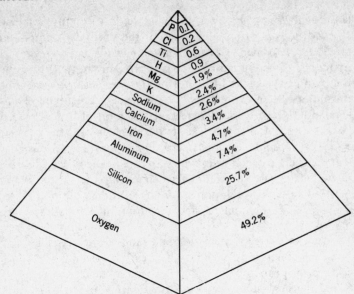

Fig. 1-1. Elemental composition of earth's crust.

own properties is called a *mixture*. Mixtures are heterogeneous, variable in composition, separable into their constituents by mechanical means, and composed of substances that retain their own energy contents. There is no gain or loss in chemical energy upon forming or destroying a mixture. Salt and pepper can be mixed together and then separated from each other without the gain or loss of chemical energy. A *solution* is a homogeneous body the composition of which can vary continuously within certain limits.

CHANGES IN MATTER. Matter undergoes physical and chemical changes. In a physical change the composition of the substance is not altered and the substance retains its identity. A rubber band will stretch and a copper wire will bend, but each object retains its identity. In a chemical change substances lose their identity, and the new substances formed have new physical and chemical properties. Thus, when a wax candle is burned in oxygen, the wax is changed into two new substances, carbon dioxide and water.

When elements or compounds combine, that is, react with each other to form more complex substances, the process may be called *chemical combination.* For example, if a mixture of iron filings and flowers of sulfur is heated, the chemical compound iron sulfide is formed. *Chemical decomposition* occurs when a compound such as mercuric oxide is thoroughly heated. Metallic mercury and oxygen are formed. In *simple replacement* reactions an uncombined element may replace an element already in chemical combination with another element. If a strip of iron is placed in a water solution of copper sulfate the iron replaces the copper. A *double replacement*, or *double decomposition*, as it is usually called, is a chemical change in which two substances react to give two new substances. For example, when a silver nitrate solution is added to a table salt solution (sodium chloride solution), silver chloride, a white precipitate, and a sodium nitrate solution are formed.

IDENTIFICATION OF MATTER. Substances are identified by enumerating their physical and chemical properties. The common physical properties include color, odor, taste, solubility in solvents, physical state, and density. For metals such as iron, gold, and silver, malleability, ductility, and conduction of heat and of electricity may also be included under physical properties. A physical description of hydrogen would entail stating that it is a colorless, odorless, tasteless gas that is slightly soluble in water and has the lowest density of any chemical substance. Chemical properties of substances are expressed in terms of the stability of the substance toward heat, light, and shock, and of its behavior when placed in contact with other substances at ordinary or elevated temperatures. See p. 184 for a list of the chemical properties of chlorine.

CLASSIFICATION OF ELEMENTS. Of the 103 known elements about 70 are metals and 33 are nonmetals. There is no sharp dividing line between the two groups; therefore, a few elements have properties of both.

With the exception of mercury (a liquid) *metals* are crystalline solids that are malleable and ductile and good conductors of heat and electricity. Excluding gold and copper, the metals are essentially silver-colored. The density of metals varies considerably, from lithium, sodium, and potassium, which are lighter than water, to platinum, which is about twenty times heavier than water.

The physical properties of *nonmetals* that are solids at ordinary temperatures can be summarized by saying that they have comparatively low densities, are poor conductors of heat and electricity, are brittle (if solid), and, in many cases, have characteristic colors. For example, sulfur is pale yellow, bromine is reddish brown, chlorine is greenish yellow, phosphorus is white or red, and carbon is transparent or black. It is interesting to note that of the known elements only two are liquids at ordinary temperatures, mercury, a metal, and bromine, a nonmetal.

CLASSIFICATION OF COMPOUNDS. In general, compounds are classified according to their composition and properties. Those of similar composition and properties are grouped together. This enables one to remember the chemistry of many more compounds than would be possible if each were considered separately. For example, *acids* and *bases* are the names of two groups of compounds, the constituent members of which have properties in common. Acids contain hydrogen that is replaceable by metals. Water solutions of acids have a sour taste, will change litmus paper from blue to red, and will neutralize bases. If, to a glass of water, one adds several drops of hydrochloric acid or a few drops of sulfuric acid, the addition of acid gives the water a sour taste. Further, if a strip of magnesium ribbon is added to the glass of water containing acid, it reacts with the water solution, liberating a gas, i.e., hydrogen. Bases are oxides or hydroxides of metals. A base will neutralize an acid. If the base is soluble it will change the color of litmus paper from red to blue.

Energy. We may define *energy* as the ability to do work. Matter always possesses energy in one form or another. Energy acts upon matter to produce chemical and physical changes. *Heat*, *light*, *kinetic* (*mechanical*), *electrical*, and *chemical* are the words used to describe different forms of energy. In ordinary chemical changes energy can neither be created nor be destroyed, but it can be changed from one form to another. This is a statement of the *law of conservation of energy*. When carbon combines chemically with oxygen, an oxide of carbon is formed. Chemical energy causes the reaction of carbon and oxygen to form, let us say, carbon dioxide. At the same time, however, heat is liberated and light may also be produced. In a lead storage battery chemical energy is changed to electrical energy. In 1905 Einstein developed an equation which indicated that matter can be converted into energy. We shall consider this in more detail in Chapter 14. We shall also give consideration to atomic energy in Chapter 2.

Divisions of Chemistry. The science of chemistry covers a broad field. Because of this fact it is generally divided into several areas of study. *General Chemistry* is a survey of the entire field of chemistry with particular emphasis on fundamental concepts and elementary laws. Textbooks in this branch of chemistry are given such titles as *General Chemistry*, *College Chemistry*, *Modern Chemistry*, *Introduction to Chemistry*, and *Essentials of Chemistry*. *Analytical Chemistry* is concerned with the detection, separation, and determination of substances and their constituents. Qualitative analysis identifies the constituents; quantitative analysis determines their amounts. *Physical Chemistry* is concerned with the laws underlying chemical changes. Where possible these laws are expressed mathematically. *Organic Chemistry* is the study of the compounds of carbon. There are more than a million compounds of carbon and so they are given special consideration. *Biochemistry* is the study of the compounds and chemical changes that are associated with living processes. Most of these compounds contain carbon. *Nuclear Chemistry* is treated in general chemistry texts and in modern texts in physical chemistry. Chapter 14 in this book is devoted to this subject and summarizes some of the results obtained when atomic nuclei are bombarded with suitable projectiles.

Other Important Definitions.   *Fact.*   An event or an occurrence is a fact. That which is true is a fact.

(1) It is a fact that two United States astronauts spent several hours on the moon. (2) It is a fact that water contains hydrogen and oxygen.

Hypothesis.   (1) A generalization based on a few facts. (2) A tentative explanation of experimental facts. Sometimes more than one hypothesis has been advanced to explain a given set of experimental facts. If this is the case, then the hypothesis which best explains the facts and which also explains newly discovered facts, in the same area, will be retained and the others discarded.

Theory.   A statement or statements based on many facts and on reason that explains facts and laws. A very satisfactory hypothesis may be advanced to become a theory. A well-established theory is not likely to be discarded. However, it may be revised to fit new facts.

Law.   A concise statement that summarizes a large number of facts. Often a law can be stated by a mathematical equation. In the statement of the law there is no attempt to explain why the law is true. Thus Boyle's gas law states that at constant temperature, the volume of a given mass of gas varies inversely with the pressure applied. Note that there is no attempt to explain why the gas expands or contracts with a pressure change. The kinetic theory accounts for this behavior of the gas.

Units of Measurement. The English system of measurement used in the United States, except for certain scientific work, is awkward and inconvenient. Table 1-1 illustrates this fact. Further, the conversion from one unit to another is always a nuisance.

Table 1-1

The English System of Measurement

Length Units

12 inches	=	1 foot
3 feet	=	1 yard
5280 feet	=	1 mile
1760 yards	=	1 mile
16.5 feet	=	1 rod
5.5 yards	=	1 rod

For lengths less than one inch, the units are common fractions such as:

1/2, 1/4, 1/8, 1/16, 1/32, etc.

Volume Units

4 gills	=	1 pint
2 pints	=	1 quart
4 quarts	=	1 gallon

Mass Units (Avoirdupois)

$437\frac{1}{2}$ grains	=	1 ounce
16 ounces	=	1 pound
2000 pounds	=	1 ton

The metric system used in western European countries and for certain scientific work in the United States is a decimal system. Therefore the conversion of one unit to another is simple and rapid.

The unit of *length* is the meter. Originally one meter was taken as one ten-millionth (.0000001) of the distance from the north pole to the equator. Recently the meter was more accurately defined as 1,650,763.73 wave lengths of the orange-red light emitted when krypton is excited by an electric discharge.

Table 1-2

The Metric System of Measurement

Units of:	Names of Units	Abbreviations	Numerical Value
Length	*Milli*meter	mm	0.001 m
	*Centi*meter	cm	0.01 m
	*Deci*meter	dm	0.1 m
	Meter	m	1.0 m
	*Kilo*meter	km	1000.0 m
Volume	*Milli*liter	ml	0.001 ml
	Liter	ℓ	1.0 ℓ
	Liter	1000 cc	1.0 ℓ
Mass	*Micro*gram	μ g	0.000001 g
	*Milli*gram	mg	0.001 g
	Gram	g	1.0 g
	*Kilo*gram	kg	1000.0 g

    Micro = one millionth
     Milli = one thousandth
    Centi = one hundredth
     Deci = one tenth
     Kilo = one thousand

Conversion Factors

Metric	*English*
1 meter	39.37 inches
1 kilometer	0.6214 mile
1 liter	1.06 quart
28.32 liters	1 cubic foot
29.57 cc	1 fluid ounce
453.6 g	1 pound

The unit of *mass* is the kilogram. The standard for this, kept in France, is a cylinder, made of a nonrusting platinum-iridium alloy. The mass of the cylinder is equal to the mass of a cubic decimeter of water at exactly 3.98°C.

The unit of *volume* is the liter. A liter is the volume of a 10 cm cube, which is equal to 1000 cubic centimeters.

MASS AND WEIGHT.   Mass refers to the quantity of matter in a given body (object). This quantity is constant and therefore does not depend on its location on the earth's surface. Thus an aluminum cube, one centimeter on an edge, contains the same mass whether it is at sea level or on top of a high mountain. However, the weight of the cube changes depending on its location. At sea level the cube weighs more than it does on the top of Pike's Peak in Colorado. On the moon its weight would be about one-sixth of its weight on the earth at sea level.

LARGE AND SMALL NUMBERS. Large and small numbers are conveniently expressed as powers of 10. Or, we may say they are expressed exponentially. Thus:

ORDINARY NUMBER	EXPONENTIAL FORM
1	1×10^0
10	1×10^1
100	1×10^2
1000	1×10^3
10000	1×10^4
100000	1×10^5
0.1	1×10^{-1}
0.01	1×10^{-2}
0.001	1×10^{-3}
0.0001	1×10^{-4}
0.00001	1×10^{-5}

The number to the right and above the ten (10), the superscript, is called the *exponent*. In multiplication, exponents are added; in division they are subtracted. Negative exponents are used for numbers less than one (1). It is customary to have only one digit before the decimal point. When the exponential form is used it is customary to use not more than three significant figures. Examples:

(a) Multiply 5×10^4 by $4 \times 10^3 = 20 \times 10^7$ or 2.0×10^8

(b) Divide 1.5×10^7 by 5×10^4 or

$$\frac{1.5 \times 10^7}{5 \times 10^4} \quad \text{or} \quad \frac{15 \times 10^6}{5 \times 10^4} = 3 \times 10^2$$

(c) Change 50,000 to the exponential form

$$50000 = 5 \times 10^4$$

(d) Change 0.000015 to the exponential form

$$0.000015 = 1.5 \times 10^{-5}$$

(e) Divide 80000 by 0.002 exponentially

$$\frac{8 \times 10^4}{2 \times 10^{-3}} = 4 \times 10^7$$

REVIEW QUESTIONS

1. Define each of the following:
   (a) chemistry, (b) science, (c) gas, (d) solid, (e) liquid.

2. (a) How can one change liquid water into ice?
   (b) How is water vapor changed into ice?

3. Distinguish between (a) an element and a compound; (b) a mixture of two solids, A and B, and a solid solution of A in B.

4. (a) Which of these three are physical changes? Why?
   (1) Splitting of a piece of wood.
   (2) Burning of a piece of coal.
   (3) Boiling of a kettle of water.
   (b) Which of these three are chemical changes? Why?
   (1) Heating of a piece of brass until it melts.
   (2) Heating of potassium chlorate until a gas is liberated.
   (3) Burning of a piece of magnesium ribbon.

5. How can one identify matter? List the common physical properties of matter.

6. What sort of tests could be made to determine the chemical properties of a piece of metal?

7. In terms of physical properties, how does sulfur, a nonmetal, differ from a piece of metallic lead?

8. Criticize this statement: A railroad-car wheel should be made of sulfur.

9. How is chemical energy converted into electrical energy?

10. How is electrical energy changed into heat energy?

11. Which is larger, a microgram or a kilogram? How many times larger?

12. How many cubic centimeters are in a cubic foot? Given: 1 inch = 2.54 cm.

13. Using the exponential form divide 27,000,000 by 9,000.

Chapter 2
Atomic Structure

In the first chapter of this book we pointed out that matter is made up of elements and compounds, and that compounds can be decomposed to give elements. On the other hand, elements cannot be decomposed to give simpler substances. (For the time being, we are ignoring the existence of subatomic particles such as protons, neutrons, and electrons.)

To go on from this point we can say that the chemistry of the 19th century and the first part of the 20th century was largely the study of elements and compounds. Chemists rejoiced in the discovery of new elements. They were interested in how compounds were made, what their physical and chemical properties were, and the formulation of the laws which summarized the various facts of chemistry known at that time.

In Part 1 of this chapter we shall state four of the basic, experimentally determined laws of chemical combination that have been obtained by the work of many investigators. After we have considered these laws we shall list the theories that have been developed to explain them. The importance of Dalton's atomic theory will then become evident. Certainly his theory explained a great many experimental facts known to him and to those who came after him.

Part 2 is concerned with the complexity of atoms and the evidence for this complexity.

Part 3 gives a somewhat simplified picture of the atom as we now think of it. The structures of atoms must account for the (experimental) properties they exhibit.

PART 1.   EXPERIMENTALLY DETERMINED BASIC LAWS

Law of Conservation of Matter.   This law states that in ordinary chemical reactions matter is neither created nor destroyed. When matter changes from one form to another, the new product or products possess the same mass (or for all practical purposes, the same weight) as the original substances. For example, place silver nitrate solution in one leg of an inverted Gothic Y tube (Figure 2-1) and a solution of sodium chloride in the other. Seal and weigh the tube. Finally invert the tube and weigh it again. Even though a chemical change has taken place, producing a curdy, white precipitate, there is no measurable change in weight.

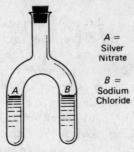

A = Silver Nitrate

B = Sodium Chloride

Fig. 2-1. Sealed tube in which a chemical reaction takes place without change in weight.

Law of Definite Composition. The composition of a pure compound is always the same. Thus, if 100 g of pure water is decomposed into oxygen and hydrogen we obtain 88.81 g of oxygen and 11.19 g of hydrogen. The same result is obtained each time the experiment is repeated. Or, if 100 g of calcium carbonate is decomposed by heating it, we always get 56 g of calcium oxide and 44 g of carbon dioxide.

Law of Multiple Proportions. When two elements A and B unite to form more than one compound, the weights of B which combine with a fixed weight of A stand to each other in the ratio of small whole numbers. Twelve grams of carbon combines with 16 g of oxygen to produce 28 g of carbon monoxide. Twelve grams of carbon also combines with 32 g of oxygen to produce 44 g of carbon dioxide. We see that the ratio of oxygen in carbon monoxide to the oxygen in carbon dioxide is 16:32 or 1:2.

Law of Combining Weights. Each element may have assigned to it a characteristic number or a simple multiple of that number, which represents the weight of the element that combines with other elements. Oxygen is taken as the standard because it is readily available and because it combines with most of the elements. For many years the equivalent or combining weight of oxygen was taken as 8 g. This choice made the combining weight of hydrogen, the lightest element, 1 g or slightly greater than 1 g. In 1961 carbon-12 was made the reference element for atomic weights. This makes the equivalent weight of oxygen 7.9997 g. However, for our purpose here, we shall continue to use 8 g as the combining weight of oxygen. Likewise we shall report the equivalent weights of the elements used in the illustration to four significant places only.

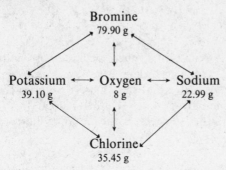

With the help of the accompanying diagram, we see that 8 g of oxygen combines with 22.99 g of sodium, 39.10 g of potassium, 35.45 g of chlorine, or 79.90 g of bromine. If 39.10 g of potassium or 35.45 g of chlorine combines with 8 g of oxygen, it follows that 39.10 is the equivalent weight of potassium and that this amount will combine with 35.45 g of chlorine. Other similar combinations take place in the same way.

Discontinuity of Matter.   Before we are ready to outline the theories that explain the laws listed above, we should point out that matter appears to be discontinuous. We shall cite experimental facts to indicate that matter is made up of discrete particles or units.

1. Substances in the liquid or solid states can be changed to the gaseous state. One volume of liquid water gives approximately 1700 volumes of steam measured at 100°C and 760 mm pressure. By a process called sublimation, crystals of iodine, upon heating in an open vessel, change directly to iodine vapor. Upon cooling the iodine vapor changes back to crystals.

2. Gases are highly compressible. Cylinders of compressed oxygen, hydrogen, and helium are available for industrial and scientific purposes.

3. Gases diffuse. A given mass of gas will "fill" any container

in which it is placed. If a small amount of ammonia gas is allowed to escape from a cylinder of ammonia, the odor of ammonia, through the process of diffusion, can be detected many feet from the cylinder.

4. Unlike gases, such as oxygen and methane, readily diffuse through each other. Since methane burns readily and since oxygen supports combustion, it is not difficult to form an explosive mixture of the two gases. A lighted match or an electric spark will initiate an explosion in such a mixture.

Theories of the Structure of Matter. As early as the 5th century B.C. Greek philosophers believed that matter was made up of small discrete particles. Democritus gave a rather reasonable picture of these particles but Aristotle succeeded in discrediting Democritus' ideas. Unfortunately, from the days of Aristotle until the 19th century the nature of matter was not given much thought.

DALTON'S ATOMIC THEORY.   Early in the 19th century John Dalton (1766–1844), an Englishman, gave us the atomic theory which bears his name. The theory states that each element is composed of very small indivisible units or particles called atoms. The atoms of each element are the same size and have the same mass; atoms of different elements have different sizes and masses. In chemical changes atoms combine, separate, or change places to form relatively simple and stable combinations (1:1, 1:2, 1:3, 2:3). These relatively stable combinations or units are now called molecules.

Dalton's atomic theory says nothing about the composition or structure of atoms. We know now that atoms are complex and are made up of smaller units which include the electron, the neutron, and the proton.

MOLECULAR THEORY.   This theory assumes the following: Substances are made up of small independent units called molecules. A molecule may be an atom or a group of atoms. All molecules of a substance are the same; molecules of different substances are different. The mass of a molecule is the sum of the masses of the atoms composing it. A molecule is the unit of a substance.

H_2, or H—H, is a molecule of hydrogen. H_2O, or H—O—H, is a molecule of water. A molecule of hydrogen contains two atoms of hydrogen. A molecule of water contains two atoms of hydrogen and one atom of oxygen. The word "atom" refers to elements.

It is correct to speak of an atom of hydrogen, but incorrect to speak of an atom of water. The word "molecule" may refer to either an element or a compound. It is correct to speak of a molecule of hydrogen or a molecule of water.

Use of Theories to Explain Experimental Laws. The four laws of chemical combination are explained in terms of the atomic and molecular theories.

LAW OF CONSERVATION OF MATTER. When sodium chloride (NaCl) solution and silver nitrate ($AgNO_3$) solution react, there is simply a rearrangement of atoms. The silver atoms combine with chlorine atoms to give silver chloride (AgCl) whereas sodium nitrate ($NaNO_3$) remains in solution. See Figure 2-1.

LAW OF DEFINITE COMPOSITION. Water molecules invariably contain the same number of hydrogen and oxygen atoms, and hence water will always contain the same percentage of the two elements. A calcium carbonate molecule contains one atom of calcium, one atom of carbon, and 3 atoms of oxygen. Consequently, pure calcium carbonate always contains 40% calcium, 12% carbon, and 48% oxygen.

LAW OF MULTIPLE PROPORTIONS. One atom of carbon combines with one atom of oxygen to give one molecule of carbon monoxide (CO). One atom of carbon combines with two atoms of oxygen to give one molecule of carbon dioxide (CO_2). It is obvious that the weight of the oxygen in a molecule of CO_2 is twice that of the oxygen in a molecule of carbon monoxide.

This relationship also holds for water and hydrogen peroxide. A molecule of water contains two atoms of hydrogen and one atom of oxygen. A molecule of hydrogen peroxide contains 2 atoms of hydrogen and 2 atoms of oxygen. It is apparent that the mass of oxygen in hydrogen peroxide is twice that in water.

LAW OF COMBINING WEIGHTS. The numbers which represent the combining weights of elements also represent the relative masses of the atoms of those elements. Sodium and chlorine combine in the ratio of 22.99 atomic weight units of sodium to 35.45 atomic weight units of chlorine. Sodium and bromine combine in the ratio of 22.99 units of sodium to 79.90 units of bromine. One atom of sodium combines with one atom of chlorine, and one atom of sodium combines with one atom of bromine. The difference, therefore, in the combining weights of the two elements represents the difference in their atomic masses. It should be pointed out here that the combining weight and the atomic weight are not the same for some elements. Thus, the combining weight for oxygen (8) is one-half its atomic weight, whereas the combining weight of aluminum is one-third its atomic weight.

PART 2.   COMPLEXITY OF ATOMS

Atoms, as John Dalton thought of them, were useful in explaining the laws of ordinary chemical combination and in accounting for the discontinuity of matter. However, we now know from experimental evidence that atoms are complex. We shall now consider some of the experimental evidence for atomic complexity and then describe the modern concept of the atom, partially at least, in terms of this evidence.

The Spectrum of an Element.   If an element is vaporized in a flame it produces light of characteristic wave lengths. This light may be analyzed with a spectroscope. Each element produces characteristic lines occurring at definite positions on the spectroscopic scale. The units of measurement on the scale are called

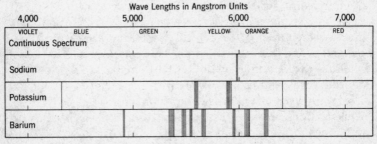

Fig. 2-2.   Spectral lines for three elements.

Ångstroms (Å). Each Ångstrom unit is 10^{-8} centimeter in length. Sodium vapor, for example, gives lines at about 5896 Å; potassium vapor gives lines near 7699 Å; barium vapor gives several lines. It is assumed that vibrations (or movements) by certain particles (electrons) within the atom produce these lines.

Electrons from Elements.   If an electric current of high potential is passed between two metallic electrodes in an evacuated tube, small particles are liberated from the cathode (negative electrode). These particles are electrons. Each weighs 1/1837 of a hydrogen atom. Here, then are particles appreciably smaller than the smallest and lightest atom. See Figure 2-3.

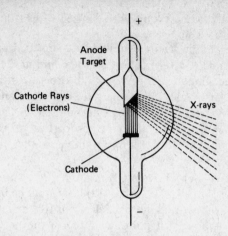

Fig. 2-3. An X-ray tube similar to the one used by Moseley to determine the atomic numbers of elements. The frequency of the X-rays depends on the material used in the anode target.

Spontaneous Decomposition of Certain Elements. Uranium, radium, thorium, and certain other elements undergo spontaneous decomposition; i.e., they are radioactive. Atoms of the radioelements decompose through the loss of alpha or beta particles. The latter are electrons; the former are charged helium atoms (helium ions).

Periodic Classification of Elements. Elements occurring in the same group in the periodic table (see Chapter 3 and inside the front cover) are similar in many of their physical and chemical properties. These similarities in properties are attributable to similarities in atomic structure. For example, sodium, potassium, and rubidium are in the same periodic group and have similar properties because their atomic structures are very similar. Likewise, chlorine, bromine, and iodine are in the same periodic group because their atomic structures are similar. Mendeleev in 1869 gave us the periodic table. He recognized there was a relationship between the atomic weights of elements and their physical and chemical properties. He expressed this relationship by means of his periodic law which states that the properties of elements are periodic functions of their atomic weights. However, Mendeleev and others were aware of the fact that there were a number of exceptions to the law. For example, if the law held, iodine would be placed in the same group with oxygen rather than with the halogens where it belongs.

At first the atomic scientists relied to a considerable extent on the periodic table for verification of their findings. However, as information accumulated concerning the structure of atoms, well-known defects in the periodic table were explained and corrected. Thus, it was Moseley (1887–1915) who pointed out, as a result of his classical experiments, that the properties of elements are periodic functions of their atomic numbers rather than of their atomic weights.

Present Conception of Atomic Structure.   The classical experiment is that of Lord Rutherford in 1911, wherein he bombarded a thin sheet of gold with alpha particles (positive helium ions). He found to his surprise that although most of the particles passed straight through the sheet of metal, about one in a hundred thousand was deflected, sometimes more than 90°. Accordingly, Rutherford suggested that an atom has a very small nucleus or center which is positively charged (and therefore repels positive alpha particles) and in which most of the mass of the atom is concentrated. See Figures 2-4 and 2-5. Figure 2-5 shows that all the α particles pass straight through the foil except those which are headed directly for the nuclei of atoms a and b. The positive nuclei of these two atoms repel the positive α particles, which are deflected.

The volume of an atom is many times greater than the volume of its nucleus. For example, the radius of the nucleus of the gold atom is 1.5×10^{-5} Å, whereas the radius of the gold atom (including the electrons outside the nucleus) is 1.5 Å.   This means

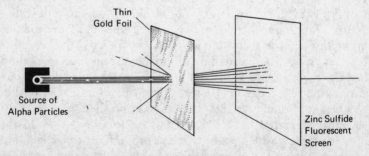

Fig. 2-4. Schematic drawing of Rutherford's classical experiment showing that atoms have very small nuclei. Note that most of the particles pass through the foil without being deflected.

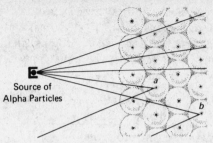

Fig. 2-5. A section of gold foil greatly magnified to show the paths of high-speed alpha particles through the gold atoms. But the alpha particles headed directly for the nucleus of atoms (*a*) and (*b*) are deflected sharply.

that the radius of the entire gold atom is 100,000 times greater than that of its nucleus.

ELECTRON. Through the work of J. J. Thomson, Robert A. Millikan, and their students and colleagues, the electron was discovered and its properties determined. It has been shown that an electron has a mass approximately 1/1837 that of a hydrogen atom and a negative charge of 1. Figure 2-3 shows cathode rays striking the anode. These rays are now known to be electrons. Further, they can be obtained from any element that is suitable for use as a cathode.

There is good evidence to support the assumption that there are no free electrons in the nucleus of the atom, but that there are as many electrons outside the nucleus as there are protons in the nucleus. This makes the atom electrically neutral.

PROTON. When a small amount of hydrogen is introduced into a suitable evacuated tube (see Figure 2-6) and the current turned on, the electrons from the perforated cathode speed

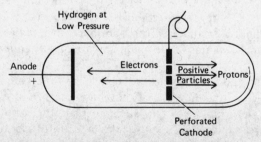

Fig. 2-6. Schematic drawing to suggest a method of preparing protons.

toward the anode. Some of these free electrons collide with hydrogen molecules and knock off their electrons. This process produces positive particles, now called protons, that pass through the perforations in the cathode where they may be detected. The equations may be written as follows:

$$H_2 + e \text{ (the projectile)} \longrightarrow H_2^+ + 2e$$
$$H_2^+ + e \text{ (the projectile)} \longrightarrow 2H^+ + 2e$$

$$\overline{H_2 + 2e \text{ (the projectile)} \longrightarrow 2H^+ + 4e}$$

The proton has a mass slightly less than that of the hydrogen atom and a positive charge of 1. It can be shown that an ordinary hydrogen atom is made up of one proton and one electron. This atom is electrically neutral, so we conclude that the charge of a proton and the charge of an electron are equal but of opposite sign.

NEUTRON.   The neutron has a mass slightly greater than that of the proton, but it has no charge. It was discovered by Sir James Chadwick in 1932. He bombarded beryllium atoms with high-energy alpha particles. A reaction took place in which beryllium atoms were converted into carbon atoms with the simultaneous liberation of high-energy particles of zero charge. These he called neutrons. The following equation (representing a nuclear reaction) is discussed in more detail in Chapter 14.

$$_4^9Be + \alpha \text{ (the projectile)} \longrightarrow {}_6^{12}C + {}_0^1n$$

POSITRON.   The positron is a particle of about the same mass as the electron but with a positive charge. It is sometimes called an antielectron or a positive electron. It was discovered in 1932 by the American physicist Carl D. Anderson. When an atom of radioactive silicon changes to an atom of aluminum, there is a simultaneous formation of a positron (e^+) or better ($_{+1}^0e$).

PHOTON.   There is well-established experimental evidence that indicates light is radiated from substances in small individual packets called quanta. These quanta are frequently called photons (see p. 24).

Positrons and photons are interesting and play an important role in science but they are of minor importance in elementary chemistry. As a matter of fact we can explain the essential chemical properties of atoms and molecules in terms of just three units (particles): protons, neutrons, and electrons. The following

Table 2-1 summarizes the essential information given here. Five "particles" are included in the table, the first three of which are most important to chemists.

Table 2-1

Fundamental Particles, Their Masses and Charges

Particle	Symbol	Mass (atomic mass units)	Electrical Charge
Proton	p	1.00758 (1)*....	+1
Neutron	n	1.00893 (1)	0
Electron	e^-	.000549 (0)	−1
Positron	e^+	.000549 (0)	+1
Alpha particle (Helium ion)	α	4.00380 (4)	+2

*The approximate mass is used in many calculations rather than the exact values.

ATOMIC NUMBER. As discussed on p. 32 Moseley pointed out that a periodic table based on atomic numbers is superior to one based on atomic weights. When he observed the X rays obtained when electrons were allowed to strike anode targets, in X-ray tubes, made of various elements, he discovered that as the nuclear charge (atomic number) increased the wave length of the X rays obtained decreased. This relationship is shown graphically in Figure 2-7.

The atomic nucleus, with the exception of an ordinary hydrogen atom (which contains 1 proton only), contains protons and neutrons. Thus a sodium atom contains 11 protons and 12

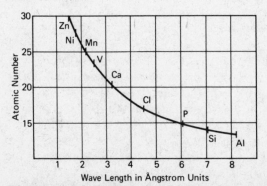

Fig. 2-7. Diagram showing the relation between the principal X-ray wave length and the atomic number.

neutrons. A fluorine atom contains 9 protons and 10 neutrons. The number of protons in the nucleus of an atom determines its total positive charge, which is, therefore, its atomic number. Thus, the atomic number of sodium is 11 and the atomic number of fluorine is 9. For all practical purposes the total mass of an atom is the sum of the protons and neutrons in its nucleus. In round numbers a sodium atom has a mass of 23 atomic mass units and a fluorine atom has a mass of 19.

Isotopes. We have a wealth of experimental evidence to prove that some atoms of an element weigh more than other atoms of the same element. Chlorine, for example, is made up of atoms that weigh 37 units and atoms that weigh 35 units.

Atoms of unequal weight (mass) but of the same nuclear charge (atomic number) are called isotopes. Aston, using the mass spectrograph, has shown that many of the elements are isotopic in character. In nature the percentage of each isotope in an element is nearly constant. Thus, of the two isotopes of chlorine we have 77.35 per cent of Cl-35 and 22.65 per cent of Cl-37. It is interesting to observe that some elements are made up of several isotopes (tin has ten) and others, such as sodium, appear to have only one isotope. See Figure 2-8.

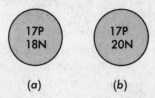

(a)   (b)

Fig. 2-8. The nuclei of the two isotopes of chlorine. (*a*) Atomic weight 35. (*b*) Atomic weight 37.

CHEMICAL PROPERTIES. The chemical properties of the isotopes of an element are similar but not identical. The slight difference in properties is accounted for by the difference in mass. The similarity is due to the fact that the atomic numbers, and consequently the configuration of the electrons, are the same.

ISOTOPIC MIXTURES. Most of the elements are mixtures of isotopes. However, elements such as sodium and iodine appear to have only one isotope each.

Although not correctly, we continue, as a matter of convenience, to write the isotopic weights as whole numbers. Thus, the three isotopes of oxygen are written 0-16, 0-17, and 0-18. Actually we should write 0-15.9950, 0-16.9992, and 0-17.992. Probably the only isotope whose weight is a whole number is C-12, the standard of atomic weights.

FRACTIONS OF ISOTOPES. The fraction of each isotope in chlorine can be found as follows: Let x equal the fraction of the atoms with isotopic weight 35. Then $(1 - x)$ is the fraction of the atoms with isotopic weight 37. (The answers will not be the same if the actual masses of the isotopes are used.) The total weight

Table 2-2
Selected Elements and Their Isotopes

Element	Atomic Number	Average Atomic Weight	Isotopes And Approximate Isotopic Weights
Hydrogen	1	1.00797	1, 2, 3
Boron	5	10.811	10, 11
Carbon	6	12.01115	12, 13, 14*
Oxygen	8	15.9994	16, 17, 18
Sodium	11	22.9898	23
Argon	18	39.948	40, 38, 36
Lead	82	207.19	206, 207, 208, 204

*Carbon-14 is radioactive.

of a unit quantity of chlorine, say a gram-atom, is 35.453 g. The fraction of this weight attributable to isotope 35 is $35x$; the fraction attributable to isotope 37 is $37(1 - x)$. Setting up an equation we have:

$$37(1 - x) + 35x = 35.453$$
$$37 - 37x + 35x = 35.453$$
$$37 - 2x = 35.453$$
$$2x = 1.547$$
$$x = .7735$$
$$(1 - x) = .2265$$

PART 3. ELECTRONIC ARRANGEMENTS IN ATOMS

The early idea regarding the positions occupied by electrons in atoms was arrived at by reasoning based on analogy. Just as the planets in the solar system revolve in circular or elliptical orbits around the sun, so likewise, it was thought, the electrons of an atom revolve at comparatively great distances about the nucleus of the atom. We know now that these interesting early ideas are inaccurate but we still think of electrons as occupying positions at great distances, relatively speaking, from the nucleus of the atom, in what we now call energy levels or energy shells. Before we present the current ideas concerning the electronic arrangements in atoms, we must introduce certain concepts as to the nature of light and its relations to electrons.

Quantum Theory. Experimental evidence indicates that light is dual in nature. As we shall presently see, this is important because it helped scientists work out the electronic structure of atoms. First, we have evidence that light consists of (electromagnetic) waves. Second, we also have evidence that light is made up of particles (quanta) now called photons. For example, when light of suitable energy strikes the surface of metallic cesium, electrons are emitted. This phenomenon, known as the *photoelectric effect*, could not happen unless the light striking the surface of the metal were made up of particles or photons. The concept of light being made up of photons is the basis of the quantum theory, which is attributed to Max Planck, who did his work near the beginning of the present century. The concepts formulated by Planck were used by Niels Bohr. He reasoned that electrons around the nucleus of an atom absorb or emit energy in packages (quanta) or photons. Bohr suggested that:

1. Electrons in atoms are located in energy levels at comparatively great distances from the nucleus.

2. The electrons in the energy level nearest the nucleus have lower energies than those in energy levels at greater distances from the nucleus.

3. The electrons in one energy level have only certain definite energy values. Those in other energy levels have definite but different energy values.

4. When an electron moves from one energy level to another it must receive (or give up) a discrete amount of energy (photons).

Quantum Numbers. Schrödinger developed an equation the solution of which gives the probability that the electron is present in a certain space or region in the atom but not in the nucleus. This equation is very difficult and does not belong in this book. However, the equation does supply us with 4 factors or parameters which are called quantum numbers. These are designated as n, l, m, s. If these quantum numbers are known for each electron in the atom, we have information to describe its behavior in the particular atom under consideration. The names and values for the 4 quantum numbers are:

n This is the principal quantum number and refers to the energy level (sometimes called a shell) in the atom where the electron in question is found. These energy levels are 1, 2, 3, 4, 5, 6, 7. Energy level 1 is nearest the nucleus; 7 is farthest away. Sometimes the energy levels are represented by capital letters. Thus energy level 1 is K, 2 is L, 3 is M, etc.

l This is the azimuthal quantum number. It is sometimes called the angular momentum quantum number. It relates to the spatial region formed as the electron moves in an elliptical orbit within the atom. These spatial regions within the atom, where electrons reside, are called *orbitals*. Sometimes l is called the orbital quantum number. The values for l are 0, 1, 2, 3, ..., $n - 1$.

m This letter refers to the magnetic quantum number and is related to the magnetic properties of the atom. The values for m are -1, 0, $+1$.

s This is the quantum number that refers to the spin of the electron and has a value of $-\frac{1}{2}$ or $+\frac{1}{2}$. Where there is a pair of electrons in a given space (an orbital) one electron has a clockwise spin and the other has a counterclockwise spin.

Energy Levels and Subenergy Levels. An energy level has as many subenergy levels as the number of the particular energy level. Thus energy level 1, the one nearest the nucleus, has one subenergy level. Energy level 2 has two subenergy levels or subshells. Energy level 3 has three subenergy levels and so on. See Table 2-3.

Table 2-3

Details for First Four Energy Levels

	1	2	3	4
Main energy levels	1	2	3	4
Subshells (n) in each level	1	2	3	4
Orbitals (n^2) in each level	1	4	9	16
Kinds of orbitals	s	s p	s p d	s p d f
	1	1 3	1 3 5	1 3 5 7
Electrons in each subshell	2	2 6	2 6 10	2 6 10 14
Total electrons each level ($2n^2$)	2	8	18	32

These subshells are designated as follows:

(*a*)   The lowest subshell (for any energy level) is called the *s* subshell.

(*b*)   The next higher subshell is the *p* subshell.

(*c*)   The third highest subshell is the *d* subshell.

(*d*)   The fourth highest subshell is the *f* subshell.

Pauli's Exclusion Principle.   Pauli pointed out that no two electrons in an atom can have the same set of quantum numbers. In other words, each electron must have its own set of quantum numbers. Table 2-4 gives the values for the quantum number needed.

Table 2-4
Values of Quantum Numbers Determined

Name of Quantum Number	Numerical Values
Principal	$n = 1, 2, 3, 4, 5, 6, 7$
Azimuthal	$l = 0, 1, 2, 3$
Magnetic	$m = -1, 0, +1$
Spin	$s = -\frac{1}{2}, +\frac{1}{2}$

The simplest method of satisfying Pauli's Exclusion Principle is to have electrons of opposite spin in each subshell. Thus in the helium atom there are 2 electrons in the 1*s* subshell. Therefore they must have opposite spins. In the lithium atom there are 2 electrons in the 1*s* subshell and only 1 electron in the 2*s* subshell. The two electrons in the 1*s* subshell must have opposite spins. Table 2-5 illustrates how Pauli's principle is obeyed.

Table 2-5
Pauli's Exclusion Principle Obeyed

Atom	Subshell	Electrons	Quantum Numbers			
He	1*s*	1st	$n = 1$	$l = 0$	$m = 0$	$s = +\frac{1}{2}$
		2nd	$n = 1$	$l = 0$	$m = 0$	$s = -\frac{1}{2}$
Li	1*s*	1st	$n = 1$	$l = 0$	$m = 0$	$s = +\frac{1}{2}$
		2nd	$n = 1$	$l = 0$	$m = 0$	$s = -\frac{1}{2}$
	2*s*	1st	$n = 2$	$l = 0$	$m = 0$	$s = +\frac{1}{2}$
Be	1*s*	1st	$n = 1$	$l = 0$	$m = 0$	$s = +\frac{1}{2}$
		2nd	$n = 1$	$l = 0$	$m = 0$	$s = -\frac{1}{2}$
	2*s*	1st	$n = 2$	$l = 0$	$m = 0$	$s = +\frac{1}{2}$
		2nd	$n = 2$	$l = 0$	$m = 0$	$s = -\frac{1}{2}$

Orbitals. An orbital is a region in a subshell of an atom in which one or two electrons are found. If two electrons are present they must have opposite spins (Pauli's Exclusion Principle), one clockwise and the other counterclockwise. Arrows may be used to represent the electrons in an orbital. Thus ↓ represents an incompletely filled orbital and ↓ ↑ represents a completely filled orbital. Orbitals may be further identified as *s* orbitals, *p* orbitals, *d* orbitals, and *f* orbitals. See Figures 2-9 and 2-10.

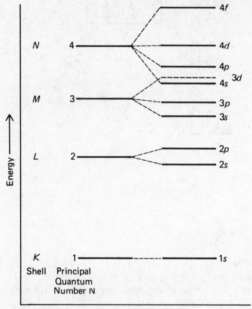

Fig. 2-9. Energy levels of atomic orbitals. Note that the 4*s* energy level is lower than the 3*d* energy level.

An *s* orbital is located in the subshell of lowest energy in each shell. Thus there is 1*s* orbital in each of the 7 energy levels. Likewise there may be 3*p* orbitals in each of the energy levels beginning with energy level 2, and continuing through 6. There can be 5*d* orbitals in each energy level beginning with level 3 and continuing through energy level 5. There are 7*f* orbitals possible in energy level 4 and also in energy level 5.

Building Electronic Configurations for Atoms. Suppose, for example, we wish to build an electronic configuration for the aluminum atom. The atomic number for this atom is 13 so we

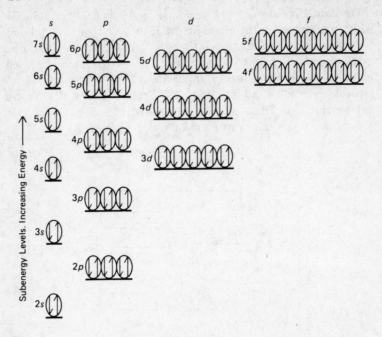

Fig. 2-10. Orbitals. A schematic subenergy level diagram for the s, p, d, and f atomic orbitals. Each orbital (circle) contains two arrows representing the two electrons of opposite spin in each orbital. Note that $4s$ is lower than $3d$, and $4d$ is higher than $5s$.

must have 13 electrons outside the nucleus. The electrons go into their places based on energy requirements. Energy level 1, which is nearest the nucleus, is filled first, then energy level 2 is filled, then 3 and so on. The conventional method of writing the electronic configuration of atoms is illustrated by the aluminum atom. Thus:

$$1s^2 2s^2 2p^6 3s^2 3p^1$$

In this notation the 1, 2, and 3 represent the energy levels; s (or p) the subshell in that particular energy level. The superscript 1, 2, or 6 represents the number of electrons present in the subshells occupied. In this atom there are six completed orbitals and 1 incomplete orbital. In each completed orbital there are two electrons present, one spinning clockwise, the other counterclockwise. The $1s$, $2s$, and $3s$ subshells each contain two electrons or one orbital. The $2p$ subshell contains six electrons or three orbitals and the $3p$ subshell has one electron or one incomplete orbital. Another example is the electronic configuration for chlorine, atomic number 17.

$$1s^2 2s^2 2p^6 3s^2 3p^5$$

Note that in the $3p$ subshell there are two completed orbitals and one incomplete orbital.

Hund's Rule. This rule, the multiplicity rule, states that if several orbitals of the same type are available, each orbital is first filled with only one electron before pairs of electrons are allowed to go into an orbital. Thus for vanadium, atomic number 23, we have:

$$1s^2 2s^2 2p^6 3s^2 3p^6 4s^2 3d^3$$

The $3d$ subshell contains only three electrons so these are present as single electrons in each of three unfilled orbitals.

Valence Electrons. The electrons in the highest energy levels of an atom are the ones that take part in chemical reactions. Thus in the aluminum atom the $3s^2$, $3p^1$ electrons are called *valence electrons*. If these three electrons are removed from the atom we have remaining an aluminum ion, written Al^{+3}. Likewise if a lithium atom loses its $2s^1$ electron we have a lithium ion (Li^{+1}). More on valence will be considered in Chapter 5.

Heisenberg's Uncertainty Principle. Before we leave the subject of atomic structure, we must consider the Uncertainty Principle of Heisenberg. He pointed out that it is not possible to know exactly the position and the velocity of an electron at the same moment. If the position of the electron is known then its velocity is not known precisely. This has forced us to think of atoms as containing electron clouds. Thus the ordinary hydrogen atom has a small nucleus surrounded by an electron cloud. This means that the electron is moving about the nucleus so rapidly that if we could pinpoint its position each microsecond we would have, for this atom, an area around the nucleus (the orbital) that resembled a spherical cloud. The electron clouds of more complex atoms, i.e., those of higher atomic number and with a corresponding higher number of electrons (and orbitals), have more complex shapes. See Figure 2-11 (*a*).

Schrödinger's quantum mechanics equation has been useful in giving us the shapes of the electron clouds that make up the orbitals in atoms. The hydrogen atom orbital with its 1*s* electron is, as stated above, spherical and therefore relatively simple. But atoms with *p*, *d*, or *f* orbitals in addition to the *s* orbitals have rather complex shapes and spatial arrangements. See Figure 2-11 (*b,c,d*).

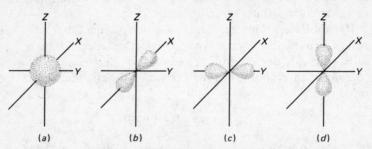

(*a*)   (*b*)   (*c*)   (*d*)

Fig. 2-11. Shapes of orbitals. (*a*) is the orbital for the hydrogen atom; (*b*), (*c*), and (*d*) represent *p* orbitals.

REVIEW QUESTIONS

1. State the law of conservation of matter (mass) as it applies to ordinary chemical reactions.
2. In what way does the Dalton atomic theory explain the law of definite composition and the law of multiple proportions?
3. Nitric oxide (NO) and nitrous oxide (N_2O) can be used as examples in explaining the law of multiple proportions. Show how to do this.
4. Why does pure water always contain 88.81% oxygen by weight?
5. Is Dalton's atomic theory of use in modern chemistry?
6. What evidence do we have that matter is discontinuous?
7. Explain this statement: All molecules of a substance are not the same both physically and chemically.
8. In what way does the spectrum of an element help in proving atoms are complex?
9. How can one remove electrons from an element?
10. How does a radioactive element, such as radium, change with time?
11. Give four reasons why we believe atoms are complex.
12. Define each of the following terms: (*a*) electron, (*b*) proton, (*c*) neutron, (*d*) photon.

13. What experimental evidence do we have that indicates the nucleus of an atom is very small compared with the whole atom?

14. Distinguish between atomic weight and atomic number of an element.

15. What experimental evidence did Chadwick have that suggested the existence of neutrons?

16. In an atom, where are the electrons located? Where are the protons and neutrons?

17. How do the isotopes of chlorine differ from each other?

18. How are the isotopes of chlorine separated from each other?

19. Explain the following statement: In atoms electrons occupy positions at comparatively great distances from the nuclei, in energy levels or energy shells.

20. What is the Bohr concept of atomic structure? List four postulates and explain them.

21. Schrödinger developed an equation which supplies us with four factors called quantum numbers. They are designated as n, l, m, and s. What does each of these four quantum numbers stand for?

22. What is an energy level? A subenergy level?

23. How do we designate the subshells in an atom?

24. What is an orbital? How many electrons are possible in an orbital?

25. Write the electronic structure of (a) krypton, (b) iodine.

Chapter 3
The Periodic Classification of the Elements

As pointed out in Chapter 2, there is a close relationship between the periodic classification of the elements and their atomic structures. This chapter is concerned with some of these relationships. First we shall give a brief historical review of how the table was developed and then point out how our present knowledge of atomic structure has been of great value in perfecting the present periodic table.

Historical Background. The early periodic classification of the elements was an outgrowth of the attempt to classify the elements in relation to their atomic weights. The "triads" of Döbereiner, the "octaves" of Newlands, and the "periodic law" of Mendeleev have all been fruitful in showing this relationship. However, since the work of Moseley, we know that the atomic number of an element is a more fundamental property than its atomic weight and, therefore, modern classifications (based on atomic numbers) are more accurate and more useful than those based on atomic weights.

TRIADS OF DÖBEREINER. In 1829, Döbereiner arranged many of the common elements in groups of three, called *triads*. He pointed out that the atomic weight of the intermediate element is equal to the arithmetical mean of the other two, and that the chemical and physical properties of the middle element are intermediate and may be predicted from the properties of the others. An example follows:

ELEMENT	ATOMIC WEIGHT
Calcium.	40.08
Strontium.	87.62
Barium	137.34

The arithmetical mean of the atomic weights of calcium and barium, 88.71, is very close to the atomic weight of strontium.

The chemical properties of strontium are intermediate between those of the other two.

OCTAVES OF NEWLANDS. In 1866, Newlands arranged the elements in the order of their increasing atomic weights. He found that every eighth element in the series is similar in chemical and physical properties. He called this the *law of octaves*. The table showing this relationship is:

H	Li	Be	B	C	N	O
F	Na	Mg	Al	Si	P	S
Cl	K	Ca	etc.			

PERIODIC LAW OF MENDELEEV. In 1869, Mendeleev announced the periodic law. It states that *the properties of elements*

Table 3-1

Periodic Table Based on Mendeleev's Original Table

PERIOD	GROUP I	GROUP II	GROUP III	GROUP IV	GROUP V	GROUP VI	GROUP VII	GROUP VIII		
1	1 **H** 1.0080							2 **He** 4.0026		
	A B	A B	A B	A B	A B	A B	A B			
2	3 **Li** 6.939	4 **Be** 9.012	5 **B** 10.81	6 **C** 12.011	7 **N** 14.007	8 **O** 15.999	9 **F** 18.998	10 **Ne** 20.183		
3	11 **Na** 22.99	12 **Mg** 24.31	13 **Al** 26.98	14 **Si** 28.09	15 **P** 30.974	16 **S** 32.06	17 **Cl** 35.453	18 **Ar** 39.948		
4	19 **K** 39.102	20 **Ca** 40.08	21 **Sc** 44.96	22 **Ti** 47.90	23 **V** 50.94	24 **Cr** 52.00	25 **Mn** 54.94	26 **Fe** 55.85	27 **Co** 58.93	28 **Ni** 58.71
	29 **Cu** 63.54	30 **Zn** 65.37	31 **Ga** 69.72	32 **Ge** 72.59	33 **As** 74.92	34 **Se** 78.96	35 **Br** 79.909	36 **Kr** 83.80		
5	37 **Rb** 85.47	38 **Sr** 87.62	39 **Y** 88.90	40 **Zr** 91.22	41 **Nb** 92.91	42 **Mo** 95.94	43 **Tc** (99)	44 **Ru** 101.07	45 **Rh** 102.90	46 **Pd** 106.4
	47 **Ag** 107.870	48 **Cd** 112.40	49 **In** 114.82	50 **Sn** 118.69	51 **Sb** 121.75	52 **Te** 127.60	53 **I** 126.90	54 **Xe** 131.30		
6	55 **Cs** 132.90	56 **Ba** 137.34	57 **La** 138.91	72 **Hf** 178.49	73 **Ta** 180.95	74 **W** 183.85	75 **Re** 186.2	76 **Os** 190.2	77 **Ir** 192.2	78 **Pt** 195.09
	79 **Au** 196.97	80 **Hg** 200.59	81 **Tl** 204.37	82 **Pb** 207.19	83 **Bi** 208.98	84 **Po** (210)	85 **At** (210)	86 **Rn** (222)		
7	87 **Fr** (223)	88 **Ra** (226)	89 **Ac** (227)							

are periodic functions of their atomic weights. That is, when all the elements are arranged (in table form) in the order of their increasing atomic weights, similar properties of the elements *recur* at regular intervals. This table consists of horizontal rows of elements (in order of their increasing atomic weights) called *series* or *periods* and vertical columns called *groups*. After passing through the first two series, the groups are divided into subgroups called A and B groups. The elements in these subgroups resemble each other enough in chemical and physical properties to be called *families* of elements. The length of a series is determined by the number of elements that must pass by (in review) before one appears that is similar in properties and structure to the first member of the series.

Table 3-1 is a periodic table based on Mendeleev's original table. However, in this table, unlike Mendeleev's table, the atomic numbers and the atomic weights are shown. The shaded areas indicate elements unknown to Mendeleev and his contemporaries. By means of his periodic law and the table he constructed, Mendeleev predicted elements would be discovered that had the proper physical and chemical properties to fill in the blank spaces in his table. For example, Mendeleev predicted that an element next to the zinc in period 4 would be discovered. His prediction was remarkably close to the actual values of gallium, which was in this case the element Mendeleev had in mind. The following values show this.

	PREDICTED		FOUND
Atomic weight...........	69	.....	69.72
Melting point............	low	.....	30.2°C
Specific gravity	5.9	.....	5.95
Formula of oxide........	X_2O_3	.....	Ga_2O_3
Action of air.............	none	.....	red heat necessary

Periodicity. Mendeleev and his contemporaries had no knowledge of the size of atoms, of their structures, or of the other properties associated with them. Not until the researches of Rutherford, Moseley, Bohr, Pauli, Schrödinger, Planck, and many others were available did chemists have a reasonable picture of atoms, their structures, and the relationship between structure and periodicity. Table 3-2 is an abbreviated periodic table in which the volume of atoms arranged in groups and in periods varies regularly, and in which there is a consistent change in the number of electrons present in the highest energy levels as we proceed across a period from left to right. Note, for example, that the

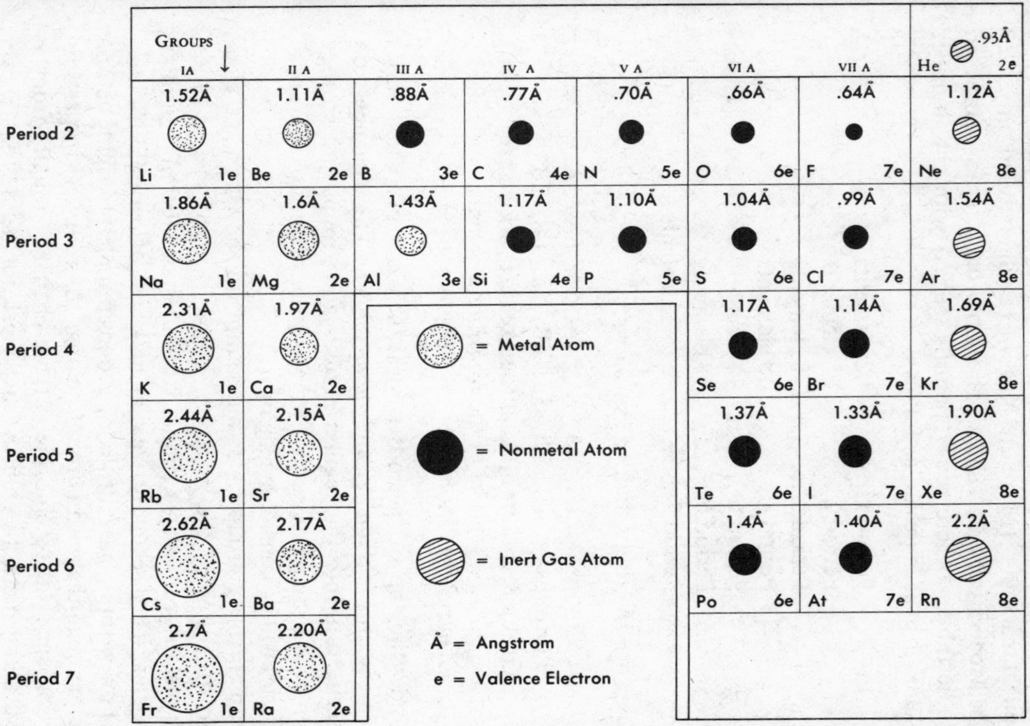

Groups	IA (1e)	IIA (2e)	IIIA (3e)	IVA (4e)	VA (5e)	VIA (6e)	VIIA (7e)	(8e)
Period 2	Li 1.52Å	Be 1.11Å	B .88Å	C .77Å	N .70Å	O .66Å	F .64Å	He .93Å
Period 3	Na 1.86Å	Mg 1.6Å	Al 1.43Å	Si 1.17Å	P 1.10Å	S 1.04Å	Cl .99Å	Ne 1.12Å
Period 4	K 2.31Å	Ca 1.97Å				Se 1.17Å	Br 1.14Å	Ar 1.54Å
Period 5	Rb 2.44Å	Sr 2.15Å				Te 1.37Å	I 1.33Å	Kr 1.69Å
Period 6	Cs 2.62Å	Ba 2.17Å				Po 1.4Å	At 1.40Å	Xe 1.90Å
Period 7	Fr 2.7Å	Ra 2.20Å						Rn 2.2Å

Legend:

= Metal Atom

= Nonmetal Atom

= Inert Gas Atom

Å = Angstrom

e = Valence Electron

Note the change in volume of atoms in group 1A from lithium (Li) to francium (Fr). Note also the change in the volume of atoms in the various periods. It is also apparent that the number of valence electrons in the periods increases from 1 to 8.

sodium atom has one electron in its highest energy level, magnesium has 2, aluminum has 3, silicon has 4, phosphorus has 5, sulfur has 6, chlorine has 7, and argon has 8. Frequently these electrons are called *valence electrons* because they take part in chemical reactions. (More on this subject will be found on p. 52.)

The Modern Periodic Table.   The table on p. ix is a modern periodic table.   There are 7 periods (series) with the number of elements in each period as follows:

Period 1 .... 2 elements
Period 2 .... 8 elements
Period 3 .... 8 elements
Period 4 .... 18 elements
Period 5 .... 18 elements
Period 6 .... 18 elements
Period 7 .... 3 (4?) elements

There are 16 groups divided as follows:

A Groups   1A, 2A, 3A, 4A, 5A, 6A, 7A
B Groups   1B, 2B, 3B, 4B, 5B, 6B, 7B
Group 8   iron, cobalt, nickel, etc., which are part of the *transition* elements.
Group Zero noble gases: He, Ne, etc.

Besides the above 7 periods (series) there are two other series of elements, the *lanthanides* and the *actinides*.   These elements do not seem to fit into the regular periods and groups and so they were placed in separate series at the bottom of the table.   These two series will be considered again.

Let us first use group IA as an example to show how the electronic structures of the members of this group, the so-called alkali metals group, have similar structures and therefore similar properties.   Then secondly, we shall make use of period 3 to illustrate how we must pass through 7 groups from left to right, before we come to another element that belongs to group IA. In passing from sodium (Na) to chlorine (Cl) we change gradually from a typical alkali metal to chlorine, a typical nonmetal. Finally we come to argon, which is an inert element because it has a stable structure. In passing from sodium to chlorine we see that sodium has 1 valence electron and chlorine has 7 valence electrons. Keep in mind that the valence electrons, i.e., those in the highest energy levels of each atom, are the ones that take part in chemical reactions.

To save time and space, the electronic structures of the inert elements, group 0, helium, neon, argon, etc., are used in writing the structures of other elements. Thus the complete electronic structure of, say, argon (atomic number 18) is: $1s^2 2s^2 2p^6 3s^2 3p^6$. In the abbreviated form this is written [Ar] and consequently, making use of this abbreviation, the structure of potassium (atomic number 19) is [Ar]$4s^1$. This is much shorter than the regular notation, which for potassium is $1s^2 2s^2 2p^6 3s^2 3p^6 4s^1$. See Tables 3-3 and 3-4. Review Chapter 2 to refresh your memory concerning details of atomic structure.

Table 3-3

Electronic Structure of Group IA

Element	Symbol	Atomic Number	Electrons in Shells
Hydrogen*	H	1	$1s^1$
Lithium	Li	3	$1s^2 2s^1$
Sodium	Na	11	[Ne]$3s^1$
Potassium	K	19	[Ar]$4s^1$
Rubidium	Rb	37	[Kr]$5s^1$
Cesium	Cs	55	[Xe]$6s^1$
Francium	Fr	87	[Rn]$7s^1$

Note that each element in this group has an s^1 electron in its highest energy level. These s^1 electrons, the so-called valence electrons, are the ones that take part in chemical reactions.

*Hydrogen is not an alkali metal but frequently it is listed with them.

Table 3-4

Electronic Structure, Period 3 Elements

Element	Symbol	Atomic Number	Electrons in Energy Shells
Sodium	Na	11	[Ne]$3s^1$
Magnesium	Mg	12	[Ne]$3s^2$
Aluminum	Al	13	[Ne]$3s^2 3p^1$
Silicon	Si	14	[Ne]$3s^2 3p^2$
Phosphorus	P	15	[Ne]$3s^2 3p^3$
Sulfur	S	16	[Ne]$3s^2 3p^4$
Chlorine	Cl	17	[Ne]$3s^2 3p^5$
Argon	Ar	18	[Ne]$3s^2 3p^6$

Metals and Nonmetals.   Of the 103 known elements, about two-thirds of them are metals. The others are nonmetals. However we sometimes say that certain elements are metalloids because their properties resemble those of both metals and nonmetals. Sodium, potassium, silver, and gold are typical metals; chlorine, sulfur, and phosphorus are typical nonmetals. Carbon in the form of graphite is sometimes thought of as a metalloid but in the form of diamond it is a typical nonmetal. In general metals have 1, 2, 3, or 4 valence electrons. Nonmetals have 4, 5, 6, or 7 valence electrons. The elements (metals) in group IA have 1 valence electron; those in group IIA have 2 valence electrons; those in group IIIA have 3 valence electrons; those in IVA have 4 valence electrons. The elements (nonmetals) in group VA have 5 valence electrons; those in VIA have 6 valence electrons; those in VIIA have 7 valence electrons.

Some periodic tables have a heavy diagonal line that begins at the left of boron B (atomic number 5) and ends at the left of astatine (At) atomic number 85. This line separates the metals from the nonmetals. To the right of the line are the nonmetals, to the left are the metals.

Other Transition Elements.   These are the elements found in the B subgroups of the periodic table. Consider the transition elements whose atomic numbers are 21–30 inclusive. These are: Sc, Ti, V, Cr, Mn, Fe, Co, Ni, Cu, and Zn. Refer to Appendix XII for the electronic structure of these elements. We see that the $4s$ subshells, with the exception of chromium and copper, are filled. However, as we increase in atomic number the $3d$ subshell obtains electrons, one for scandium up to ten for copper and zinc. Apparently these electrons are the ones that are responsible for the physical and chemical properties of these elements.

The lanthanide (and the actinide) series, placed at the bottom of the table because they do not seem to fit properly into periods 6 and 7, are sometimes referred to as the inner transition elements. For many years the lanthanides were called the *rare earths*. Those chemists who were trying to perfect the periodic table had great difficulty in finding a suitable place for the rare earths. Further, chemists who were interested in preparing samples of the rare earths had great difficulty in purifying them. This was because these elements were so much alike both physically and chemically and because they occurred together in nature. The most promising method was found to be fractional crystallization. We still have trouble in preparing pure samples of the lanthanides, but, thanks to our present knowledge of atomic structure, we know the reason why.

Beginning with element 58, cerium, we see that it differs from element 57, lanthanum, by the addition of an electron in the $4f$ subshell. No. 59, praseodymium, differs from No. 58 by the presence of three electrons in the $4f$ subshell. No 60, neodymium, has four electrons in the $4f$ subshell, and so on. For all of these elements the $6s$ subshell has two electrons. A generalized subshell formula may be written $4f^{0-14}5d^{0-1}6s^2$.

Value of Periodic Classification. The periodic table is valuable as an aid to memory. It is easier to remember the properties of an element if its position in the periodic table is known. Thus, radium occurs in group IIA and so its ordinary chemical properties will resemble the properties of calcium, strontium, and barium.

The experimental values obtained for melting points, density, electrical resistivity, and so on, for an element can be checked against neighboring elements in the periodic table. Unusual values will be detected and rechecked. For example, the density of cesium was reported as 2.4. This value seemed high when compared with the values for potassium and rubidium. Another determination gave the value 1.9, which is probably correct.

The periodic classification may serve as a guide in research work. For example, can sodium ethyl xanthate be used in place of potassium ethyl xanthate as a collector in flotation work? The periodic table seems to say "yes." Experimental work proves definitely that sodium ethyl xanthate can be used in place of the more expensive potassium ethyl xanthate.

REVIEW QUESTIONS

1. Outline the early attempts to make a periodic table of the elements.
2. In what way did Moseley's work on atomic numbers affect the periodic classification of the elements?
3. Suggest several ways in which the periodic table has been useful in solving chemical problems.
4. What problem is encountered in placing the rare earth elements (58–71) in the periodic table?
5. Distinguish between the lanthanide and actinide series.
6. What is the difference between a "period" and a "group" in the periodic table?

7. What are the transition elements and where are they located in the periodic table?

8. In terms of atomic structure explain why elements such as praseo-dymium (59) and neodymium (60) are very similar in chemical and physical properties. How does number 60 actually differ from number 59 in terms of electrons? Where is the electron located that makes the difference?

Chapter 4
The Chemical Bond

In Chapter 2 we were concerned with the structure of atoms, the make-up of the nucleus of the atom, and the arrangement of the electrons outside the nucleus. In Chapter 3 the periodic classification of the elements was discussed. We were concerned with the relationship between the structures of atoms and their physical and chemical properties. We saw that atoms of similar properties have similar structures. In this chapter we are concerned with the methods by which atoms combine to form molecules.

The Octet Rule. The inert gases, those in the zero group of the periodic table, with the exception of helium, have 8 electrons in their highest energy levels. This gives them a stable structure and as a consequence they are chemically inert. Thus neon, atomic number 10, has a structure $1s^2 2s^2 2p^6$, and argon (atomic number 18) is $1s^2 2s^2 2p^6 3s^2 3p^6$. G. N. Lewis in 1916 suggested that the other elements tend to interact to achieve an electronic structure similar to the 0 group elements. If they can do this they become more stable.

The Ionic Bond. Let us consider a reaction between a sodium atom and a chlorine atom. The electronic structure of the sodium atom is $1s^2 2s^2 2p^6 3s^1$. The electronic structure of the chlorine atom is $1s^2 2s^2 2p^6 3s^2 3p^5$. Now if these two atoms approach each other there will be, at the optimum distance apart, a rearrangement of the highest energy level electrons in which the $3s^1$ electron of the sodium atom becomes a $3p$ electron of the chlorine atom. The sodium atom now becomes a sodium ion with a charge of $+1$. Likewise the chlorine atom now becomes a chloride ion with a charge of -1. These two ions now have a more stable structure, i.e., they resemble a 0 group element structure. Certainly a sodium ion is more stable than a sodium atom. Likewise a chloride ion is more stable than a chlorine atom. There are now 8 electrons in the highest energy level of the sodium ion and there are now 8 electrons in the highest energy level of the chloride ion. This increase in stability is "explained" by the Lewis octet rule. In the following table (4-1) we give several examples. Each ion produced is more stable than the atom from which it was formed.

Table 4-1
Structure and Charge of Atoms and Their Ions

Substance	Symbol	Structure
Sodium atom	Na	$1s^2 2s^2 2p^6 3s^1 \ldots 0$
Sodium ion	Na^{+1}	$1s^2 2s^2 2p^6 \ldots \ldots +1$
Chlorine atom	Cl	$1s^2 2s^2 2p^6 3s^2 3p^5 \ldots 0$
Chloride ion	Cl^{-1}	$1s^2 2s^2 2p^6 3s^2 3p^6 \ldots -1$
Calcium atom	Ca	$1s^2 2s^2 2p^6 3s^2 3p^6 4s^2 \ldots 0$
Calcium ion	Ca^{+2}	$1s^2 2s^2 3p^6 3s^2 3p^6 \ldots +2$
Aluminum atom	Al	$1s^2 2s^2 2p^6 3s^2 3p^1 \ldots 0$
Aluminum ion	Al^{+3}	$1s^2 2s^2 2p^6 \ldots \ldots +3$

The following electronic equation shows the reaction between a sodium atom and a chlorine atom:

$$1s^2 2s^2 2p^6 3s^1 \; + \; 1s^2 2s^2 2p^6 3s^2 3p^5 \; = \quad 1s^2 2s^2 2p^6 \; + \; 1s^2 2s^2 2p^6 3s^2 3p^6$$
(sodium atom)   (chlorine atom)   (sodium ion) (chloride ion)

With symbols we have:

$$Na + Cl = Na^{+1} + Cl^{-1}$$

With dots as suggested by G. N. Lewis, we have:

$$Na \cdot + \cdot \overset{\cdot\cdot}{\underset{\cdot\cdot}{Cl}}: \; = \; Na^{+1} + \; : \overset{\cdot\cdot}{\underset{\cdot\cdot}{Cl}}:^{-1}$$

As shown by the electronic equation above this is a reaction involving two atoms giving a product made up of a positive ion and a negative ion. The bond between these two ions is an *ionic* bond or, as it is often called, an *electrovalent* bond. It is these electrovalent bonds that hold the sodium ions and chloride ions in place in a crystal of sodium chloride. Actually a crystal of sodium chloride is a macromolecule made up of sodium and chloride ions. See Chapter 8, p. 87 for more on this subject. Two other equations using dots to represent the electrons follow:

(a) $K \cdot + \cdot \overset{\cdot\cdot}{\underset{\cdot\cdot}{F}}: \; = \; K^{+1} + \; : \overset{\cdot\cdot}{\underset{\cdot\cdot}{F}}:^{-1}$

(b) $Ca \overset{\cdot\cdot}{} + 2 \cdot \overset{\cdot\cdot}{\underset{\cdot\cdot}{Br}}: \; = \; Ca^{+2} + 2 : \overset{\cdot\cdot}{\underset{\cdot\cdot}{Br}}:^{-1}$

In reaction (b) the calcium has 2 valence electrons so it can supply an electron to each of the two bromine atoms.

Covalent Bonds.   Unlike ionic bonds, a covalent bond is one where a pair of electrons, or more, are shared by at least two atoms. The simplest case is the sharing of two electrons in the hydrogen molecule. We can imagine that two hydrogen atoms approach each other until the two electrons, one from each atom, are shared. Thus

$$H^{\cdot} + H^{\cdot} = H:H$$

For fluorine we have

$$:\ddot{F}\cdot + \cdot\ddot{F}: = :\ddot{F}:\ddot{F}:$$

Covalent bonding also occurs between unlike atoms. Thus

(a) $H^{\cdot} + \cdot\ddot{Cl}: = H:\ddot{Cl}:$ (hydrogen chloride)

(b) $H^{\cdot} + \cdot\ddot{O}\cdot + H^{\cdot} = H:\ddot{O}:H$ (water)

(c) $4H^{\cdot} + \cdot\dot{\underset{\cdot\cdot}{C}}\cdot = H:\overset{H}{\underset{\ddot{H}}{C}}:H$ (methane)

(d) $\cdot\ddot{O}\cdot + \cdot\dot{C}\cdot + \cdot\ddot{O}\cdot = :\ddot{O}:C:\ddot{O}:$ (carbon dioxide)

COORDINATE COVALENT BOND.   This is a chemical bond in which at least one shared pair of electrons is contributed by the same atom. Thus, in changing ammonia (NH_3) to the ammonium ion (NH_4)$^{+1}$ by the addition of a hydrogen ion (H^{+1}) the nitrogen atom contributes the pair of electrons. If we use the dot formula of G. N. Lewis and represent the pair of electrons furnished by the nitrogen atom as (xx) we have:

$$H:\overset{XX}{\underset{\ddot{H}}{N}}:H + H^{+1} = [H:\overset{H}{\underset{\ddot{H}}{\overset{XX}{N}}}:H]^{+1}$$

Another example is the action of ammonia on boron trifluoride. Note that the nitrogen atom furnishes the pair of electrons.

$$H:\overset{H}{\underset{H}{N^X_X}} + \overset{:\ddot{F}:}{\underset{:\ddot{F}:}{\ddot{B}:\ddot{F}:}} = H:\overset{H}{\underset{H}{N^X_X}}\overset{F}{\underset{F}{\ddot{B}:\ddot{F}:}}$$

We should point out here that after a coordinate covalent bond is formed, it becomes like any other covalent bond.

The covalent bond between like atoms gives a symmetrical configuration. But covalent bonds between unlike atoms do not give a symmetrical configuration. See Figure 4-1. Further, these

H:H   :F:F: H:F:

(a)   (b)   (c)

Fig. 4-1. Covalent bonds between like and unlike atoms. The atoms are held together by covalent bonds; (a) and (b) are symmetrical but (c) is not. It is polar because the area near the fluorine atom has a greater negative charge than the area near the hydrogen atom.

molecules may be polar. This means that one area of the molecule may be negative with respect to another area. Polar molecules become oriented in an electric field. See Figure 4-2.

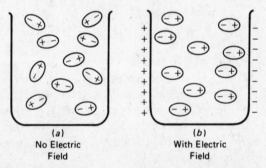

(a)   (b)
No Electric   With Electric
Field   Field

Fig. 4-2. The orientation of polar molecules in an electric field.

POLAR COVALENT BONDS. The bonding in certain molecules may be part ionic and part covalent. This will be more readily understood after we consider the phenomenon called electronegativity. This term is a measure of an atom's ability to attract and hold electrons, which is measured by determining the energy required to pull an electron from an atom. This is the ionization potential. It may be expressed in electron volts or in kilogram calories per mole (kcal/mole). There apparently is a close relationship between ionization potential and electronegativity. Thus fluorine has a very high ionization potential and so Linus Pauling assigned an arbitrary value of 4 to fluorine in his table of electronegativities. The other elements in the table have values less than 4. See the table of electronegativities in Table 4-2. As a matter of convenience this table is also given in the chapter on oxidation and reduction. See p. 138.

Table 4-2

Electronegativity Table

Most Electronegative

Fluorine	4.0
Oxygen	3.5
Nitrogen	3.0
Chlorine	3.0
Bromine	2.8
Carbon	2.5
Sulfur	2.5
Hydrogen	2.1
Copper	1.9
Manganese	1.6
Aluminum	1.5
Magnesium	1.2
Sodium	0.9
Potassium	0.8
Rubidium	0.8
Cesium	0.7

Least Electronegative
(Electropositive)

Now let us return to polar covalent bonds and give examples which will tell us something about the nature of these bonds. Consider the copper(I) chloride molecule (CuCl); using Table 4-2 we find that by subtracting the electronegativity value of copper from that of chlorine (3.0 − 1.9) we obtain a value of 1.1. This is interpreted to mean that the bond between copper and chlorine is partly ionic and partly covalent. The bond between fluorine and potassium in potassium fluoride (KF) is (4 − .8) or 3.2 so this particular bond is essentially ionic.

Molecular Orbitals. When atoms unite to form molecules there are new electronic distributions in space which are called *molecular orbitals*. The atomic orbitals have been modified considerably, in some cases, to give these molecular orbitals. If a molecular orbital is symmetrical about the axis connecting the two atoms it is called a sigma (σ) orbital. The hydrogen molecule is an example of a sigma (σ) s molecular orbital since it is formed by the interaction of two s orbitals. If there is a reaction between, say, a hydrogen atom and a fluorine atom to give hydrogen fluoride (HF) we have a sigma *sp* molecular orbital. The hydrogen molecule orbital is symmetrical but the hydrogen fluoride molecule orbital is not. See Figure 4-3.

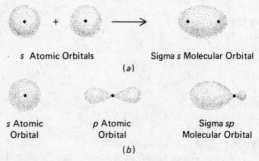

Fig. 4-3. Atomic and molecular orbitals. (*a*) Hydrogen molecular orbital. (*b*) Hydrogen fluoride molecular orbital.

Hydrogen Bonding. The highly electronegative atoms fluorine, oxygen, and nitrogen (see Table 4-2) when combined with hydrogen form molecules that are polar. The negative end of such molecules will attract the positive end of these molecules and form a bond. Since hydrogen is the positive end of these molecules the bond formed (not as strong as an ionic or covalent bond) is called a *hydrogen bond*. Liquid hydrogen fluoride (HF), and water (H_2O) are examples of molecules where the hydrogen bonding has a marked effect on the physical properties of these two compounds. For example the boiling point of liquid HF is much higher than that of liquid HCl. If there were no hydrogen bonding, HF should boil near −100°C. However, liquid HF boils at +19.4°C. Thus it takes more heat energy to break the hydrogen bonds. Likewise water would boil at a lower temperature if there were no hydrogen bonding present. See Figure 4-4. Proposed structures of clusters of HF and H_2O molecules are shown in Figure 4-5.

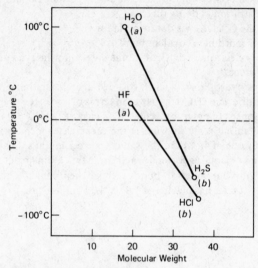

Fig. 4-4. Effect of hydrogen bonding on boiling points. (*a*) Boiling points of H_2O and HF are much higher than would be expected because of hydrogen bonding. (*b*) Boiling points of H_2S and HCl are much lower because of the absence of hydrogen bonding.

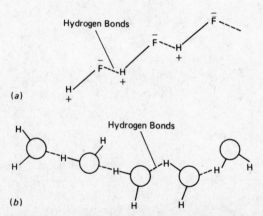

Fig. 4-5. Schematic representation of hydrogen bonding. (*a*) Hydrogen fluoride. (*b*) Water.

REVIEW QUESTIONS

1. State or explain G. N. Lewis' octet rule.
2. What makes a sodium ion more stable than a sodium atom?
3. Make an illustration to show an ionic bond.
4. Distinguish between an ionic bond and a covalent bond. Make use of an illustration in doing this.
5. What is the G. N. Lewis dot formula for CCl_4?
6. What is a coordinate covalent bond?
7. How does electronegativity of atoms enter into the type of chemical bonds formed?
8. Why does hydrofluoric acid (HF) have a higher boiling point than hydrochloric acid (HCl)? Answer in terms of hydrogen bonding.
9. What happens to polar molecules in an electric field?
10. On the Pauling scale fluorine is the most electronegative element, having a value of 4. (*a*) On the same scale cesium has a value of 0.7. What kind of bond is formed when these two atoms react with each other? (*b*) What kind of bond forms when fluorine reacts with copper? (Copper has a value of 1.9 on the Pauling scale.)

Chapter 5
Atomic Weights, Symbols, Formulas, Equations, and Valence

Students who have had courses in high school chemistry have been introduced to the subject matter presented in this chapter. But students who have not had such courses will find the material here to be basic to progress in elementary chemistry. It may be true that in studying the materials in Chapters 2, 3, and 4, the student came across terms and concepts that were not properly defined or not presented in sufficient detail for mastery. However, review of these three chapters in conjunction with a study of Chapter 5 should clear up the difficulties.

Atomic Weights. The basic postulates of the Dalton atomic theory have withstood the test of time in explaining the laws of chemical combination and other relationships. Scientists reasoned that since atoms of different elements have different masses it was apparent that the atom of one of the elements should be chosen as the standard of mass (or weight as it is commonly called). The choice of a standard has not been easy. At one time the atom of hydrogen, the lightest element, was taken as the standard. Later the oxygen atom became the standard. This was a good choice because oxygen combines with most of the elements. However when it was discovered that most elements, including oxygen, are isotopic mixtures (see p. 34 for a discussion of isotopes) it was decided to use an isotope of one of the elements as a standard. The chemists probably would have preferred O-16 but the physicists felt the need of another standard. Finally both groups agreed on an isotope of carbon, C-12. The value assigned to C-12 is exactly 12. Thus one atom of C-12 weighs exactly 12 *atomic weight units*. An atomic weight unit therefore is exactly one-twelfth the mass of a C-12 atom. See the atomic weight table on p. x .

Symbols. A chemical symbol is the first letter, or the first letter with another characteristic letter, of the name of an element. The symbol is more than an abbreviation. It represents one atom of the particular element and a definite weight of the element compared with the weight of an atom of the standard element. The symbol for phosphorus is P. This P represents one atom of phosphorus, the weight of which is 30.9738 atomic weight units. Or, Ca is the symbol for calcium. It represents one atom of calcium, the weight of which is 40.08 atomic weight units. See the atomic weight table.

Where a symbol has two letters, the first letter is capitalized and the second is a small letter. Thus, the symbols Ca, Cl, Cu, Co, and Cr represent one atom of calcium, chlorine, copper, cobalt, and chromium respectively.

Formulas and Computations Involving Formulas. A chemical formula consists of one symbol or a combination of symbols. If more than one atom of an element is present in the molecule, this is shown by a subscript. The formula shows the elements present and the number of atoms of each element in the molecule. Thus, $KClO_3$ is the formula for one molecule of potassium chlorate. This molecule contains one potassium atom, one chlorine atom, and three oxygen atoms.

FORMULAS AND MOLECULAR WEIGHTS. The molecular weight of a substance is the weight of one molecule of the substance compared with the weight of one atom of carbon-12.

The term *molecular weight* is not applicable to certain compounds such as the salts. For these compounds the simplest formula, corresponding to the percentage composition of the compound, is used. Here it is proper to use the term *formula weight*. Thus, the molecular weight of water (H_2O) is $2(1) + 16 = 18$ and the formula weight of potassium bromide (KBr) is $39.10 + 79.91 = 119.01$.

CALCULATION OF MOLECULAR OR FORMULA WEIGHT. The molecular weight (or formula weight) of a substance is obtained by adding the weights of its atoms. The weights of atoms are obtained from an atomic weight table. The formula of calcium carbonate is $CaCO_3$; the sum of its atomic weights is found as follows:

$$Ca = 40. \quad C = 12. \quad O = 3 \times 16 = 48.$$

$$40 + 12 + 48 = 100$$

100 is the formula weight of $CaCO_3$.

CALCULATION OF PERCENTAGE COMPOSITION. As a matter of convenience, for most calculations we round off the atomic weights as given in the table. Thus, we use 40 for calcium rather than 40.08, and 12 for carbon rather than 12.011. The percentage of each element in calcium carbonate may be obtained as follows:

$$\frac{Ca}{CaCO_3} = \frac{40}{100} \times 100 = 40\% \text{ calcium}$$

$$\frac{C}{CaCO_3} = \frac{12}{100} \times 100 = 12\% \text{ carbon}$$

$$\frac{3\,[O]}{CaCO_3} = \frac{48}{100} \times 100 = 48\% \text{ oxygen}$$

WEIGHT UNITS AND VOLUME UNITS. If a substance is a gas, the weight in grams of 22.4 liters of the gas measured at 0°C and 760 mm pressure is its molecular weight. The weight in grams of the carbon dioxide (CO_2) necessary to fill a 22.4-liter vessel at 0°C and 760 mm pressure is 44 g. Therefore, the molecular weight of carbon dioxide is 44. More on this subject is found in Chapter 7 of this book.

Chemical Equations. A chemical equation is a *qualitative* and *quantitative* expression of chemical changes. It shows the kind of molecules reacting, the products formed, the number of molecules entering the reaction, the number of molecules of products formed, and the proportions by weight (and volume in some reactions) in which substances react to give definite products. Thus,

$$2\,H_2 + O_2 \longrightarrow 2\,H_2O$$

2 molecules of hydrogen + 1 molecule of oxygen $\longrightarrow$ 2 molecules of water

$$2\,KClO_3 + \text{heat} \longrightarrow 2\,KCl + 3\,O_2$$

2 molecules of potassium chlorate $\longrightarrow$ + heat

2 molecules of potassium chloride + 3 molecules of oxygen

The steps in balancing equations are: On the left side of the equation mark (arrow) write the formulas of the substances taking part in the reaction. On the right side, write the formulas of the products formed. This is the *skeleton equation*. Adjust the number of molecules on the left and right sides so that there will be the same number of atoms of each kind on each side. In making the adjustments *do not change the formula of any substance.*

Examples.

$$CaCO_3 + heat \longrightarrow CaO + CO_2$$

This equation is balanced without taking more than one molecule of any of the three substances involved.

$$CaO + HCl \longrightarrow CaCl_2 + H_2O \text{ (skeleton equation)}$$

An inspection of this skeleton equation shows that we are short one hydrogen atom and one chlorine atom. The simple way to remedy this is to take two molecules of HCl. Thus,

$$CaO + 2\,HCl \longrightarrow CaCl_2 + H_2O \text{ (balanced equation)}$$

Let us examine an equation in which the amount of each substance must be adjusted.

$$KClO_3 + heat \longrightarrow KCl + O_2 \text{ (skeleton equation)}$$
$$2\,KClO_3 + heat \longrightarrow 2\,KCl + 3\,O_2 \text{ (balanced equation)}$$

This equation is balanced by taking two molecules of $KClO_3$. This gives two molecules of KCl and three molecules of oxygen. Notice that the correct formulas of the three substances involved were not changed, and that there are the same number of atoms of each element present on each side of the arrow ($\longrightarrow$): 2 K atoms, 2 Cl atoms, and 6 oxygen atoms.

Valence. The term valence is mentioned in Chapters 2, 3, and 4. In Chapters 2 and 3, it is stated that certain electrons of atoms are called valence electrons because they take part in chemical reactions. In Chapter 4 it is pointed out that *covalent* bonds contain valence electrons. Thus in the hydrogen molecule (H : H) the covalent bond is made up of 2 electrons, one from each atom.

It is possible however, to think of valence without associating it with atomic structure. This the early chemists did by building up their knowledge of valence upon experimental evidence and upon reasoning. Therefore, let us approach valence from the earlier point of view and then tie it in with atomic structure.

A valence is a number that represents the combining capacity of an atom. Let us choose the hydrogen atom as a standard with a valence of 1. It can be shown experimentally that two atoms of hydrogen combine with 1 atom of oxygen. Therefore, oxygen has a valence of 2. Likewise, three atoms of hydrogen combine with one atom of nitrogen so nitrogen must have a valence of 3.

| I | HCl | H_2O | NH_3 | CH_4 |
| II | NaCl | ZnO | AlN | $TiCl_4$ |

Following I, above, is a series of formulas each of which contains hydrogen plus one other element. We can tell by inspection what the valences of the other elements are. Chlorine has a valence of 1, oxygen a valence of 2, nitrogen 3, and carbon 4. Using the information gained from I, we learn from II that Na is 1, Zn is 2, Al is 3, and Ti is 4. It should not be difficult now to write the formulas of zinc chloride, aluminum chloride, aluminum oxide, carbon chloride, and titanium oxide. Aluminum chloride for example is $AlCl_3$.

RADICALS AND VALENCE. A radical* is a group of atoms, found in certain compounds, which react as a unit, i.e., as if it were a single atom. Radicals have valence. The radicals in the following compounds are enclosed in parentheses. The number below each radical gives its valence.

$$H(NO_3) \qquad H(C_2H_3O_2) \qquad H_2(SO_4) \qquad Na_2(C_2O_4) \qquad Na_3(PO_4)$$
$$1 \qquad\qquad 1 \qquad\qquad 2 \qquad\qquad 2 \qquad\qquad 3$$

In the following equation it is apparent that the radical is not changed. It remains (C_2O_4).

$$Na_2(C_2O_4) + CaCl_2 \longrightarrow 2\,NaCl + Ca(C_2O_4)$$

Note. What are called radicals in this chapter will be called ions in other chapters.

FORMULAS AND VALENCE. In writing the formula (usually the simplest formula) of any compound, all the valences of each atom should be used. To do this, adjust the number of atoms present in the molecule. The following rules are helpful:

1. Write the skeleton formula and indicate the valence of each atom or radical by Roman numerals.

2. Find the smallest multiple of these numbers and write it above the skeleton formula.

*Some textbook authors and research chemists refer to radicals as *functional groups*.

3. Divide this multiple by the valence number of each atom and write the quotient as a subscript of that atom.

The following examples will illustrate these three rules.

(1) What is the formula of aluminum oxide?

$$\overset{6}{\underset{\text{III} \quad \text{II}}{Al \quad O}} \longrightarrow Al_2O_3$$

(2) What is the formula of calcium phosphate?

$$\overset{6}{\underset{\text{II} \qquad \text{III}}{Ca \quad (PO_4)}} \longrightarrow Ca_3(PO_4)_2$$

(3) What is the formula of phosphoric pentoxide?

$$\overset{10}{\underset{\text{V} \quad \text{II}}{P \quad O}} \longrightarrow P_2O_5$$

SUGGESTIONS FOR REMEMBERING VALENCE. The hydrogen atom always has a valence of 1. Let it be the standard. Memorize the formulas of a few important compounds of hydrogen. The valence of the atom or radical combined with hydrogen can then be determined. Make a table similar to the one below.

FORMULA	NAME
H_2O	Hydrogen oxide (water)
HCl	Hydrogen chloride (hydrochloric acid)
$H_2(SO_4)$	Hydrogen sulfate (sulfuric acid)
$H(C_2H_3O_2)$	Hydrogen acetate (acetic acid)
$H(NO_3)$	Hydrogen nitrate (nitric acid)
$H_3(PO_4)$	Hydrogen phosphate (phosphoric acid)

Memorize the formulas of a few well-chosen compounds containing metal atoms of different valence.

UNIVALENT	BIVALENT	TRIVALENT	TETRAVALENT
$NaBr$	CaF_2	$AlCl_3$	$TiCl_4$
KCl	$Cu(NO_3)_2$	$Bi(NO_3)_3$	CH_4
$Ag(NO_3)$	$Ba(SO_4)$	$Fe(PO_4)$	SiF_4

Practice writing the formulas of the oxides, nitrates, sulfates, and chlorides of the metals in the memorized lists.

Example. What is the formula of copper phosphate? We see from the table that copper (Cu) is a bivalent metal and phosphate (PO_4) has a valence of 3. Accordingly,

$$
\overset{6}{\underset{II}{Cu}} \quad \underset{III}{(PO_4)} \longrightarrow Cu_3(PO_4)_2
$$

POSITIVE AND NEGATIVE VALENCE. Positive and negative valences are explained from the standpoint of atomic structure. Atoms with extra electrons have *negative valence*, and atoms that have lost electrons have *positive valence*. The following facts are important: The positive and negative charges (valences) of the atoms in a compound should total zero when added algebraically. Hydrogen and metals have positive valence. Nonmetals and nonmetal groups (acid radicals or ions) have negative valence. An element, either metal or nonmetal, in the free state has *zero valence*. An atom or radical that is positive or negative in charge is called an *ion*. In the compound NaCl the Na ion is $+1$ ("positive one") and the Cl ion is -1 ("negative one"). In the compound NH_4NO_3, the ammonium ion $(NH_4)^{+1}$ is $+1$ and the nitrate radical or ion $(NO_3)^{-1}$ is -1. In the compound $CaHPO_4$, the Ca is $+2$, the H is $+1$, and the PO_4 is -3.

VALENCE OF ATOMS IN A RADICAL OR ION. Consider each oxygen atom as having a valence of -2, each hydrogen atom as having a valence of $+1$, and the metals as having a valence of $+1$ or more.

Example. The valence of phosphorus in $K_3(PO_4)$ is $+5$, derived as follows:

$$3 K (+3) + P(?) + 4 O (-8) = 0. \qquad P = +5.$$

Many textbooks use the term *oxidation number* in place of *valence*. We now say that hydrogen, the metals, and certain radicals (ions) have *positive oxidation numbers* whereas oxygen, the nonmetals, and certain radicals (ions) have *negative oxidation numbers*. The following table shows the relationship of oxidation number to valence. (More is given on oxidation numbers in Chapter 13.

Table 5-1

The Relation between Valence and Oxidation Number

Atom or Ion	Valence	Oxidation Number
Hydrogen, H	+1	+1
Sodium, Na	+1	+1
Ammonium, NH_4^{+1}	+1	+1
Calcium, Ca	+2	+2
Aluminum, Al	+3	+3
Chlorine, Cl	−1	−1
Bromine, Br	−1	−1
Hydroxide, OH^{-1}	−1	−1
Nitrate, NO_3^{-1}	−1	−1
Oxygen, O	−2	−2
Sulfur, S	−2	−2
Carbonate, CO_3^{-2}	−2	−2
Phosphate, PO_4^{-3}	−3	−3
Phosphorus, P	−3	−3

Types of Chemical Reactions. In *direct combination*, two or more substances combine to form a more complex substance.

$$Cu + Cl_2 \longrightarrow CuCl_2$$
$$Fe + S \longrightarrow FeS$$

In a *chemical decomposition*, a substance is broken down into two or more simpler substances.

$$2\,KClO_3 + heat \longrightarrow 2\,KCl + 3\,O_2$$
$$2\,H_2O_2 + (catalyst) \longrightarrow 2\,H_2O + O_2 + (catalyst)*$$
$$2\,H_2O + electric\ current \longrightarrow 2\,H_2 + O_2$$

In a *single displacement* (*replacement*), an element replaces another element in a compound. For example, zinc replaces the copper in a copper sulfate solution.

$$Zn + CuSO_4 \longrightarrow ZnSO_4 + Cu$$

*A catalyst is a substance that aids in a chemical reaction without itself being permanently changed. It may be used more than once. See Chapter 12 for more on this subject.

Below, iron replaces the hydrogen in hydrochloric acid.

$$Fe + 2\,HCl \longrightarrow FeCl_2 + H_2$$

Double decomposition reactions are reactions in which the metal atoms or positive radicals (positive ions) of two compounds exchange places.

$$NaCl + AgNO_3 \longrightarrow AgCl + NaNO_3$$
$$NaOH + HCl \longrightarrow NaCl + H_2O$$
$$C_2H_5OH + HC_2H_3O_2 \rightleftharpoons C_2H_5C_2H_3O_2 + H_2O$$

A single arrow ($\longrightarrow$) indicates that the reaction is essentially complete. Reverse arrows ($\rightleftharpoons$) indicate the reaction is incomplete. See Chapter 12.

REVIEW QUESTIONS

1. What is a chemical symbol? What does it represent?
2. Distinguish between (a) a molecule of an element and a molecule of a compound, (b) a monoatomic and a diatomic molecule, and (c) a formula weight and a molecular weight.
3. How can one determine the molecular weight of a gas such as carbon dioxide?
4. What is a chemical equation?
5. Distinguish between a skeleton equation and a balanced equation.
6. Write the formulas of (a) sodium sulfate, (b) zinc sulfate, (c) titanium oxide, and (d) sodium ammonium hydrogen phosphate.
7. What other name can be ascribed to (a) sulfuric acid, (b) hydrogen acetate, (c) hydrochloric acid, and (d) water? Write the formulas of these compounds.
8. What is the valence (oxidation number) of phosphorus in (a) K_3PO_4, (b) P_2O_5, and (c) H_3PO_3?
9. Write an equation to show (a) a single replacement, (b) a double decomposition, and (c) a molecular decomposition.

Chapter 6
Acids, Bases, Salts, Nomenclature

Acids. *Earlier Concepts.* (1) An acid is a compound that contains hydrogen, replaceable by a metal, and the water solution of which changes the color of litmus paper (an indicator) from blue to red. (2) An acid is a substance that is dissociated in water solutions, furnishing hydrogen ions (H^+). Many compounds containing hydrogen, such as CH_4, C_2H_5OH, and C_6H_6, are not acids. Acetic acid ($H \cdot C_2H_3O_2$) has only one replaceable hydrogen. *New Concept.* An acid is a substance which gives up protons (H^+) to another substance. According to this point of view, acids may be molecules, positive ions, or negative ions. Thus, H_2O, HCl, NH_4^+, and HSO_4^- are acids. In water solutions the proton does not exist as such. It combines with a water molecule to form the hydronium ion (H_3O^+). In ammonia solutions the proton combines with NH_3 to give the ammounium ion NH_4^+. This point of view was first proposed by Brönsted and Lowry.

Classification of Acids. Acids are classified as *monobasic*, *dibasic*, or *tribasic*, depending on whether one, two, or three protons are furnished by the molecule or ion. Thus, HCl, $H \cdot C_2H_3O_2$, and HNO_3 are monobasic acids; sulfuric (H_2SO_4) and carbonic (H_2CO_3) are dibasic acids; and phosphoric (H_3PO_4) is a tribasic acid. Sometimes the terms used are *monoprotic*, *diprotic*, and *triprotic* acids, depending on the number of protons furnished per molecule.

Preparation of Acids. Acids may be prepared by several methods. The following are important. It should be noted that many acids are water solutions.

1. General Method. Treat a salt of the acid with a nonvolatile acid. Sulfuric acid is commonly used for this purpose.

$$NaCl + H_2SO_4 \longrightarrow NaHSO_4 + HCl$$
$$KNO_3 + H_2SO_4 \longrightarrow KHSO_4 + HNO_3$$

2. Dissolve an acid anhydride (an oxide of a nonmetal) in water.

$$SO_2 + H_2O \longrightarrow H_2SO_3 \text{ (sulfurous acid)}$$
$$CO_2 + H_2O \longrightarrow H_2CO_3 \text{ (carbonic acid)}$$

3. Direct combination of hydrogen with certain nonmetals.

$$H_2 + Cl_2 \longrightarrow 2\,HCl \left.\begin{array}{l} \\ \end{array}\right\} \text{ Dissolve the product}$$
$$H_2 + Br_2 \longrightarrow 2\,HBr \quad \text{ in water.}$$

Strength of Acids.   Strong acids give up protons readily. Weak acids give up their protons with difficulty. The strength of acids may be determined: (1) By measuring the rate at which a metal above hydrogen in the activity series (see Appendix XI) will replace the hydrogen. Strong acids react rapidly; weak acids react slowly. (2) By measuring the electrical conductivity of a solution of the acid. The strong acids are the best conductors.

In comparing the strength of acids, it is necessary that equivalent concentrations be used. These are obtained when unit volumes of each acid solution contain the same weight of replaceable hydrogen (available proton). The following table gives information for several acids. See Chapters 11 and 12 for information on ionization.

Table 6-1
Strength of Acids

Acid	Formula	Ions	Ionization Constant	Strength
Hydrochloric	HCl	H^+, Cl^-	None	Strong
Hydrobromic	HBr	H^+, Br^-	None	Strong
Nitric	HNO_3	H^+, NO_3^-	None	Strong
Sulfuric	H_2SO_4	H^+, HSO_4^-	None	Strong
Bisulfate Ion	HSO_4^-	H^+, SO_4^{--}	1.2×10^{-2}	Medium
Phosphoric	H_3PO_4	$H^+, H_2PO_4^-$	0.75×10^{-2}	Medium
Nitrous	HNO_2	H^+, NO_2^-	0.45×10^{-3}	Medium
Acetic	$H \cdot C_2H_3O_2$	$H^+, C_2H_3O_2^-$	1.8×10^{-5}	Weak
Hydrocyanic	HCN	H^+, CN^-	4×10^{-10}	Weak

Note. In water solutions of acids the proton (H^+) is always associated with a water molecule, thus, H_3O^+. The ionization constant is considered in Chapter 12.

Properties of Acids.   Acids turn blue litmus red. They neutralize bases to form salts. Nonoxidizing acids react with metals above hydrogen in the activity series, forming salts and liberating hydrogen. Oxidizing acids react with metals above or below hydrogen in the activity series, but pure hydrogen is not produced. Water solutions conduct an electric current. Acid solutions have a sour taste.

$$2 HNO_3 + Ca(OH)_2 \longrightarrow Ca(NO_3)_2 + 2 H_2O \text{ (neutralization)}$$
$$Zn + 2 HCl \longrightarrow ZnCl_2 + H_2 \text{ (replacement)}$$
$$3 Ag + 4 HNO_3 \longrightarrow 3 AgNO_3 + 2 H_2O + NO \text{ (oxidation)}$$

Bases. *Earlier Concepts.* (1) A base is an oxide or a hydroxide of a metal. If soluble in water, it turns red litmus paper blue. (2) In water solutions a base furnishes hydroxyl ions (OH^-) and some positive ion other than hydrated proton. Water-soluble bases are called *alkalies*. *New Concept.* A base is a molecule or ion that will combine with a proton. According to this point of view, molecules and negative ions (anions) are bases. Thus, H_2O, NH_3, OH^-, and SO_4^{--} are bases.

Classification of Bases. Bases are classified as *monoacid*, *diacid*, or *triacid*, depending on whether the molecule (or ion) is capable of accepting one, two, or three protons. This classification applies more particularly to the metallic hydroxides.

Potassium hydroxide (KOH) is a monoacid base; $Ca(OH)_2$ is a diacid base; and $Al(OH)_3$ is a triacid base.

Preparation of Bases. Bases may be prepared by the following methods.

1. Action of water on a basic anhydride (oxide of a metal).

$$Na_2O + H_2O \longrightarrow 2 NaOH$$

2. Dissolve an alkali metal in water.

$$2 Na + 2 H_2O \longrightarrow 2 NaOH + H_2$$

3. Treat a salt with an alkali.

$$FeCl_3 + 3 NaOH \longrightarrow Fe(OH)_3 + 3 NaCl$$

4. Dissolve ammonia in water.

$$NH_3 + H_2O \longrightarrow NH_4OH$$

Properties of Bases. Water-soluble bases form OH^- ions and turn red litmus blue. Bases neutralize acids to form salt solutions. Water-soluble bases conduct an electric current. They have a bitter taste. Solutions of sodium or potassium hydroxide will react with certain metals to liberate hydrogen. For example,

$$6 \text{ NaOH} + 2 \text{ Al} \longrightarrow 2 \text{ Na}_3\text{AlO}_3 + 3 \text{ H}_2 \text{ (sodium aluminate)}$$
$$2 \text{ NaOH} + 2 \text{ Al} + 2 \text{ H}_2\text{O} \longrightarrow 2 \text{ NaAlO}_2 + 3 \text{ H}_2$$

Relation between Acids and Bases. Since according to modern theory acids are proton donors and bases are proton acceptors, it is obvious that for every acid there is a corresponding base.

$$\text{Acid} \rightleftharpoons \text{Proton} + \text{Base}$$
$$\text{HCl} \rightleftharpoons \text{H}^+ + \text{Cl}^-$$
$$\text{H}_2\text{SO}_4 \rightleftharpoons \text{H}^+ + \text{HSO}_4{}^-$$
$$\text{H}_3\text{O}^+ \rightleftharpoons \text{H}^+ + \text{H}_2\text{O}$$
$$\text{NH}_4^+ \rightleftharpoons \text{H}^+ + \text{NH}_3$$

Strong bases accept protons readily but weak bases do not.

The following equations show the reactions between acids and bases according to the Brönsted-Lowry theory. Note that this theory considers water (H_2O) to be a base because it combines with a proton (H^+) to give (H_3O^{+1}).

First Acid		First Base		Second Acid		Second Base
HCl	+	H_2O	$\rightleftharpoons$	H_3O^{+1}	+	Cl^{-1}
H_2SO_4	+	H_2O	$\rightleftharpoons$	H_3O^{+1}	+	$HSO_4{}^{-1}$
HCl	+	NH_3	$\rightleftharpoons$	NH_4^{+1}	+	Cl^{-1}
H_2O	+	NH_3	$\rightleftharpoons$	NH_4^{+1}	+	OH^{-1}
$H \cdot C_2H_3O_2$	+	NH_3	$\rightleftharpoons$	NH_4^{+1}	+	$C_2H_3O_2{}^{-1}$

If we arrange some of the common acids and bases in the order of their strengths we have:

	Strong		Weak
	HCl	....	Cl^-
	H_2SO_4	....	HSO_4^-
	HSO_4^-	....	SO_4^{--}
Acids	$H \cdot C_2H_3O_2$	....	$C_2H_3O_2^-$	Bases
	H_2CO_3	....	HCO_3^-
	HCO_3^-	....	CO_3^{--}
	H_2O	....	OH^-
	Weak		Strong

Electron-Pair Concept. In 1923, G. N. Lewis of the University of California proposed an acid-base theory more general in application than the Brönsted and Lowry theory. According to the Lewis theory, *an acid is an electron-pair acceptor and a base is an electron-pair donor.* This means we can have acids and bases without H^+ (protons) and OH^- (hydroxyl ions). Thus BaO is a base, and SO_3 is an acid.

$$Ba\!:\!\ddot{O}\!: + \quad \overset{\displaystyle :\!\ddot{O}\!:}{\underset{\displaystyle :\!\ddot{O}\!:}{S\!:\!\ddot{O}\!:}} \longrightarrow Ba\!:\!\ddot{O}\!:\!\underset{\displaystyle :\!\ddot{O}\!:}{\overset{\displaystyle :\!\ddot{O}\!:}{S}}\!:\!\ddot{O}\!:$$

Base   Acid   Salt

Also, this theory will explain reactions such as

$$(a) \qquad H^{+1} + \quad \overset{\displaystyle H}{\underset{\displaystyle H}{:\!\ddot{N}\!:\!H}} \longrightarrow \left[\overset{\displaystyle H}{\underset{\displaystyle \ddot{H}}{H\!:\!N\!:\!H}} \right]^{+1}$$

Acid   Base

$$(b) \qquad H^{+1} + H\!:\!\ddot{O}\!:\!H \longrightarrow \left[H\!:\!\overset{\displaystyle H}{\ddot{O}}\!:\!H \right]^{+1}$$

Acid   Base

Comment Concerning Salts. Many modern general chemistry textbooks give little or no space to substances known as salts. Some authors point out that in terms of the Brönsted-Lowry theory of acids and bases, common salts such as table salt (NaCl), potassium iodide (KI), sodium nitrate ($NaNO_3$), sodium carbonate (Na_2CO_3), and a great many others contain or furnish anions that are proton acceptors and therefore are bases. From an industrial point of view salts are very important and the chemical literature is replete with information about them. Consequently some basic information about salts will be included in this book.

Salts. *Earlier Concepts.* (1) A salt is a substance containing a metal (or positive ion) and an acid radical (negative ion). (2) A salt is the compound formed when a metal replaces the replaceable hydrogen of an acid. (3) A salt is a substance which in solution furnishes a cation other than the hydrogen ion, and an anion other than the hydroxyl ion. *Modern Concepts.* Salts are electrovalent or ionic compounds that are crystalline. Their crystal lattices are made up of positive and negative ions. When these substances are melted or when they are dissolved in water, the ions can move freely and thus they are good conductors of an electric current.

A sodium chloride crystal is composed of sodium ions and chloride ions arranged in a lattice such that each sodium ion is surrounded by 6 chloride ions and each chloride ion is surrounded by 6 sodium ions. This means that simple molecules of NaCl do not exist in the crystal. Experiment shows that the molecular weight of vaporized sodium chloride is 58.44. Hence the simple formula, NaCl, is correct for the vapor state.

Classification of Salts.   Salts are classified according to their composition.

NORMAL SALTS. Normal salts contain neither replaceable hydrogen atoms (ions) nor hydroxyl groups (ions). Sodium carbonate (Na_2CO_3) and bismuth chloride ($BiCl_3$) are examples.

ACID SALTS.   (1) Acid salts contain replaceable hydrogen. (2) Acid salts contain a negative ion capable of dissociating a proton. Thus, sodium acid carbonate ($NaHCO_3$) and sodium acid phosphate (Na_2HPO_4) contain replaceable hydrogen or proton.

BASIC SALTS.   (1) Basic salts contain replaceable hydroxyl. (2) Basic salts contain OH^- ions (radicals) capable of combining with protons. Basic bismuth nitrate [$Bi(OH)_2(NO_3)$] and basic ferric acetate [$Fe(OH)_2(C_2H_3O_2)$] contain replaceable hydroxyl groups.

Neutralization.   Neutralization in water is a process in which an acid reacts with a base to form water and a solution of a salt. In general, neutralization is the transfer of a proton from an acid to a base. For example, when a solution of hydrogen chloride is added to a sodium hydroxide solution, the ion H_3O^+ reacts with OH^- to form water.

$$Cl^- + H_3O^+ + Na^+ + OH^- \longrightarrow 2\,H_2O + Na^+ + Cl^-$$

It is not necessary to write the chloride ion and the sodium ion because they have not reacted. In other words, they appear on both sides of the equation and may be canceled out. Thus, the above equation can be written

$$H_3O^+ + OH^- \longrightarrow 2\,H_2O$$

Preparation of Salts.   The following are common methods of preparing salts:

NEUTRALIZATION.   An acid reacts with a base to form a salt solution. If the water is evaporated, a salt remains. Salts made by the action of strong acids on strong bases are *neutral* to litmus or other indicators. Those made by the action of weak acids on strong bases (or vice versa), or by the action of weak acids on weak bases, react *acidic* or *basic* to indicators. This is due to *hydrolysis*.

$$HNO_3 + NaOH \longrightarrow NaNO_3 + H_2O \text{ (no hydrolysis)}$$
$$H \cdot C_2H_3O_2 + KOH \longrightarrow K \cdot C_2H_3O_2 + H_2O \text{ (hydrolysis)}$$

See Chapter 11 for a discussion of hydrolysis.

DIRECT COMBINATION OF A METAL AND A NONMETAL.   Under suitable conditions, metals and nonmetals combine directly.

$$Cu + Cl_2 \longrightarrow CuCl_2$$
$$Mg + Br_2 \longrightarrow MgBr_2$$

REPLACEMENT OF HYDROGEN.   Metals above hydrogen in the activity series replace hydrogen from acids to form salts.

$$Zn + H_2SO_4 \longrightarrow ZnSO_4 + H_2$$
$$Fe + 2HCl \longrightarrow FeCl_2 + H_2$$

REPLACEMENT OF HYDROGEN FROM AN ALKALI.   Zinc or aluminum reacts with sodium hydroxide to liberate hydrogen and form a salt.

$$2NaOH + Zn \longrightarrow Na_2ZnO_2 + H_2 \text{ (sodium zincate)}$$
$$6NaOH + 2Al \longrightarrow 2Na_3AlO_3 + 3H_2 \text{ (sodium aluminate)}$$

REPLACEMENT OF METAL.   An active metal will replace a less active metal from a salt to form another salt. In most cases, the salts are dissolved in water, but it does not appear in the equations.

$$2AgNO_3 + Cu \longrightarrow Cu(NO_3)_2 + 2Ag$$

INTERACTION OF TWO SALTS.   Two salts will react to form two new salts. This method is useful if one of the products is insoluble.

$$AgNO_3 + NaCl \longrightarrow AgCl + NaNO_3$$

ACTION OF A BASE ON A SALT.   A salt and a base will react to form a new salt and a new base. This method is useful if the base formed is insoluble.

$$CuSO_4 + 2NaOH \longrightarrow Cu(OH)_2 + Na_2SO_4$$

ACTION OF A NONVOLATILE ACID ON A SALT.   A nonvolatile acid will react on a salt to form a volatile new acid and a new salt. This is an important laboratory method.

$$H_2SO_4 + 2KNO_3 + heat \longrightarrow K_2SO_4 + 2HNO_3$$

Properties of Salts.   The properties of salts depend on their composition.

NORMAL SALTS.   Normal salts react neutral to indicators (if there is no appreciable hydrolysis). They have a salty taste. Many are white crystalline substances. Their water solutions conduct an electric current. The salts of copper, iron, cobalt, and certain other metals are colored.

ACID SALTS.   Acid salts react acid to litmus or other indicators. In solution they conduct an electric current. Sodium bicarbonate ($NaHCO_3$), an exception, is slightly basic to an indicator.

BASIC SALTS.   If soluble, basic salts react basic to indicators, and they conduct an electric current.

Nomenclature.   The systematic naming of chemical compounds is essential. If there were no rules to follow the student would be completely confused. Fortunately the name given to a substance is related to its composition and to its properties. The use of prefixes and suffixes helps to distinguish between compounds that contain the same elements and have similar properties, but do not have the same percentage composition. Thus, two compounds containing hydrogen, sulfur, and oxygen have acidic properties but one has the formula H_2SO_4, and the other's formula is H_2SO_3. The first is called sulfuric acid; the second is sulfurous acid.

ACIDS.   The naming of acids follows the rules given here. There are a number of exceptions.

Acids containing only two elements are *binary* acids. Thus, HCl and H_2F_2 are binary acids. Acids containing three elements are *ternary* acids. Many contain oxygen. Thus, HNO_3 and H_2CO_3 are ternary acids.

The prefix *hydro* and the suffix *ic* denote binary acids.

> HCl   *hydro*chlor*ic* acid
> H_2S *hydro*sulfur*ic* acid

The suffix *ic* denotes most of the common acids of a ternary series.

> H_2CO_3 carbon*ic* acid
> HNO_3 nitr*ic* acid

The suffix *ous* denotes ternary acids containing less oxygen than the *ic* acids.

HNO_2 nitr*ous* acid
H_2SO_3 sulfur*ous* acid

The prefix *hypo* and the suffix *ous* denote ternary acids containing the lowest percentage of oxygen.

HClO   *hypo*chlor*ous* acid

The prefix *per* and suffix *ic* denote ternary acids containing more oxygen than is present in *ic* acids

$HClO_4$ *per*chlor*ic* acid
$HMnO_4$ *per*mangan*ic* acid

SALTS.   Salts are named by changing the suffix of the corresponding acids.

The suffix *ide* denotes binary salts (salts containing two elements). The *ic* of the acid is changed to *ide*. Thus, NaCl is sodium chlor*ide* because it is the salt of hydrochlor*ic* acid.

The suffix *ate* denotes ternary salts corresponding to ternary acids ending in *ic*. Thus, K_2SO_4, a salt of sulfur*ic* acid, is potassium sulf*ate*.

The suffix *ite* denotes ternary salts corresponding to ternary acids ending in *ous*. Thus, NaClO, a salt of hypochlor*ous* acid, is sodium hypochlor*ite*.

The ternary salts of per—ic acids are named by changing the suffix *ic* to *ate*.

$KClO_4$ potassium *per*chlor*ate*
$KMnO_4$ potassium *per*mangan*ate*

BASES.   The naming of bases is not difficult since most bases contain the OH group and are called hydroxides. Thus, NaOH is sodium hydroxide and $Ca(OH)_2$ is calcium hydroxide. See Table 6-2 for the new method of naming compounds where the metal atom has more than one oxidation number (valence).

OXIDATION NUMBER AND NOMENCLATURE.   Some elements, such as some of the metals, have more than one oxidation number (valence). Thus, iron may have an oxidation number of a $+2$ or a $+3$. In the compound $FeSO_4$ the iron atom is a $+2$. In $Fe_2(SO_4)_3$ the iron atom is a $+3$. In terms of the older nomenclature the one with the $+2$ iron is called ferrous sulfate and the compound with the $+3$ iron is called ferric sulfate. In terms of the new nomenclature the former is called iron(II) sulfate and the latter is called iron(III) sulfate. The following table (Table 6-2) illustrates these relationships.

Table 6-2
Old and New Nomenclature

Formula	Old Name	New Name
$FeSO_4$	Ferrous sulfate	Iron(II) sulfate
$Fe_2(SO_4)_3$	Ferric sulfate	Iron(III) sulfate
Cu_2O	Cuprous oxide	Copper(I) oxide
CuO	Cupric oxide	Copper(II) oxide
$HgCl$	Mercurous chloride	Mercury(I) chloride
$HgCl_2$	Mercuric chloride	Mercury(II) chloride
$Cr(OH)_2$	Chromous hydroxide	Chromium(II) hydroxide
$Cr(OH)_3$	Chromic hydroxide	Chromium(III) hydroxide

The following table summarizes the naming of acids and their corresponding salts.

Table 6-3
The Names of Acids and Their Corresponding Salts

Formula of Acid		Name of Acid	Formula of Corresponding Salt	Name of Salt
Binary Acids	HCl	*Hydro* chlor *ic*	$NaCl$	Sodium chlor *ide*
	HBr	*Hydro* brom *ic*	$NaBr$	Sodium brom *ide*
	H_2S	*Hydro* sulfur *ic*	Na_2S	Sodium sulf *ide*
Ternary Acids	$HClO$	*Hypo* chlor *ous*	$NaClO$	Sodium *hypo* chlor *ite*
	$HClO_2$	—Chlor *ous*	$NaClO_2$	Sodium—chlor *ite*
	$HClO_3$	—Chlor *ic*	$NaClO_3$	Sodium—chlor *ate*
	$HClO_4$	*Per* chlor *ic*	$NaClO_4$	Sodium *per* chlor *ate*
	H_2SO_4	Sulfur *ic* (common)	Na_2SO_4	Sodium sulf *ate*
	H_2SO_3	Sulfur *ous*	Na_2SO_3	Sodium sulf *ite*
	$H_2S_2O_8$	*Per* sulfur *ic*	$Na_2S_2O_8$	Sodium *per* sulf *ate*
	$H_2N_2O_2$	*Hypo* nitr *ous*	$Na_2N_2O_2$	Sodium *hypo* nitr *ite*
	HNO_2	Nitr *ous*	$NaNO_2$	Sodium nitr *ite*
	HNO_3	Nitr *ic* (common acid)	$NaNO_3$	Sodium nitr *ate*

REVIEW QUESTIONS

1. What is an acid? Do all acids contain hydrogen? Oxygen? Give examples to illustrate your answer.
2. Give equations for the neutralization of potassium hydroxide (KOH) with: (*a*) a monobasic acid, (*b*) a dibasic acid, and (*c*) a tribasic acid.
3. Give two methods that will show that acetic acid is weaker than hydrochloric acid. Assume the acids are of equivalent concentrations.
4. Silver is soluble in nitric acid but not soluble in hydrochloric acid. Explain.
5. Distinguish between a base and an alkali.
6. (*a*) In what sense is NH_4^+ an acid?
   (*b*) HSO_4^- is also an acid. Why?
7. What is a diprotic acid? What happens to protons in water solutions?
8. What is the new concept of base?
9. Explain what is meant by the statement, "For every acid there is a corresponding base."
10. Explain G. N. Lewis' concept concerning an acid and a base.
11. In terms of the new nomenclature, what is the name of each of the following compounds? (*a*) FeO, (*b*) Fe_2O_3, (*c*) MnO_2, (*d*) $NiSO_4$, (*e*) CO, (*f*) $HgCl$.
12. What is the modern concept of a salt?
13. Distinguish between an acidic and a basic salt.
14. Write equations illustrating 4 methods of preparing salts.

Chapter 7
Gases and the Kinetic Theory

Matter can exist in three states. These are (1) the gaseous, (2) the liquid, and (3) the solid. As pointed out in Chapter 1 it is possible to change matter from one physical state to another. Such changes can be brought about by varying the temperature, or the pressure, or a combination of both.

Gases. Matter in that physical state in which it completely fills any vessel in which it is placed is called a *gas*. The following are characteristics of a gas:

1. *Expansibility.* This property is apparent with an *increase* of temperature or a *decrease* in pressure.

2. *Compressibility.* This property is apparent when the pressure is *increased* or the temperature *decreased*.

3. *Diffusibility.* The molecules of a gas tend to distribute themselves uniformly in any given space.

4. *Liquefiability.* All gases can be liquefied if cooled and compressed. For each gas there is a characteristic temperature to

Table 7-1

The Critical Temperature and Pressure for Various Substances

Substance	Critical Temperature	Critical Pressure (in atmospheres)*
Hydrogen	−239.9°C	12.8
Nitrogen	−147.1°C	33.5
Oxygen	−118.8°C	49.7
Carbon dioxide	+ 31.1°C	73
Ammonia	+132.4°C	111.5
Chlorine	+144.0°C	76.1
Water	+374.0°C	217.7
Ethyl alcohol	+243.1°C	63.1

*At sea level the atmospheric pressure is 760 mm. Often we refer to this pressure as *one atmosphere*.

69

which it must be cooled before it can be liquefied by the application of pressure. This is the *critical temperature*. The pressure required to liquefy the gas (at the critical temperature) is called the *critical pressure*.

The critical temperature and the critical pressure are characteristic for each gas. Table 7-1, p. 69 is included to show this for various substances.

From Table 7-1 we see why it is impossible to liquefy hydrogen, say at −100°C. At this temperature we are more than 100 degrees above the critical temperature and consequently it makes no difference what pressure is applied, the gas will not liquefy. On the other hand if the gas temperature is −245°C it can be liquefied by applying a pressure of 12.8 atmospheres or less. If we choose a gas such as carbon dioxide we can liquefy it by applying pressure to the gas at room temperature. The reason is that the critical temperature of CO_2 is +31.1°C.

The Kinetic Theory of Gases.   This theory explains why gases expand, contract, diffuse, exert a pressure, and liquefy. The postulates are:

1. All gases consist of discrete particles called molecules.

2. These particles are in rapid, ceaseless motion.

3. Collisions between molecules are perfectly elastic; i.e., there is no loss of energy when gas molecules collide with each other or with the walls of a container.

4. The distance between molecules is great when compared to their diameters.

Table 7-2 summarizes this information:

The quantitative behavior of gases can be summarized by means of the gas laws. These laws will now be considered and illustrative examples (calculations) given.

Boyle's Law.   If the temperature remains constant, the volume of a given mass of gas is *inversely* proportional to the pressure. This means that by doubling the pressure on 2 liters of gas, for example, the volume will decrease until 1 liter remains. Mathematically, the law is written:

$$\frac{V_1}{V_2} = \frac{P_2}{P_1} \quad \text{or} \quad P_1 V_1 = P_2 V_2$$

Here, V_1 and P_1 are the original volume and pressure, and V_2 and P_2 are the new volume and pressure.

Table 7-2

Facts Explained by Kinetic Theory

1. Gases are compressible.	Molecules are widely separated and therefore can be forced closer together.
2. Gases diffuse.	Molecules are in rapid, haphazard motion. There is plenty of room for gas molecules to intermingle.
3. Gases expand on heating.	The kinetic energy of the molecules increases and the speed of molecules thus increases also.
4. Gases exert pressure proportional to their concentration.	Molecules strike each other and the walls of the container without loss of energy. The more molecules there are per unit volume the greater the pressure on the walls of the container.
5. Gases can be liquefied.	Molecules attract each other at close range. Lowering the temperature and increasing the pressure forces the molecules near each other.

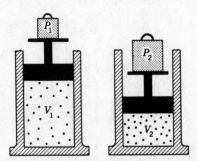

Fig. 7-1. The effect of pressure on the volume of a gas.

Problem 1. A given mass of gas, at constant temperature, occupies 500 ml at 700 mm* pressure. What volume will it occupy at 750 mm pressure?

Solution: Substitute the data into the formula and solve for V_2.

---

*In scientific work atmospheric pressure is usually expressed in millimeters of mercury. Recently the term *torr* has been adopted (in honor of Torricelli) as the unit of pressure and represents that pressure that will support a column of mercury 1 mm in height. However, we shall continue to use mm in expressing atmospheric pressure.

$$\frac{500 \text{ ml}}{V_2} = \frac{750 \text{ mm}}{700 \text{ mm}} \quad \text{or} \quad V_2 = 500 \text{ ml} \times \frac{700 \text{ mm}}{750 \text{ mm}} = 466.7 \text{ ml}$$

Problem 2. A given mass of gas, at constant temperature, occupies 600 ml at a pressure of 800 mm. What is its volume at 500 mm?
Solution:

$$\frac{600 \text{ ml}}{V_2} = \frac{500 \text{ mm}}{800 \text{ mm}} \quad \text{or} \quad V_2 = 600 \text{ ml} \times \frac{800 \text{ mm}}{500 \text{ mm}} = 960 \text{ ml}$$

Problem 3. At constant temperature a given mass of gas occupies a volume of 500 ml at a pressure of 700 mm. At what pressure will its volume be 250 ml?
Solution:

$$\frac{500 \text{ ml}}{250 \text{ ml}} = \frac{P_2}{700 \text{ mm}} \quad P_2 = 700 \text{ mm} \times \frac{500 \text{ ml}}{250 \text{ ml}} = 1400 \text{ mm}$$

The pressure of a gas is measured with a *barometer*. Figure 7-2 shows two forms of the common laboratory barometer based on

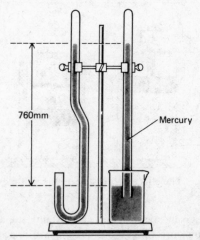

760mm

Mercury

Fig. 7-2. Common forms of mercury barometers.

the one invented by Torricelli about 300 years ago. The average pressure of the atmosphere at sea level is 760 mm. Frequently, as a matter of convenience, gas pressures are given in *atmospheres*. One atmosphere is 76 cm or 760 mm. Thus a pressure of 2.5 atmospheres equals 190 cm or 1900 mm.

Charles's Law. If the pressure remains constant, the volume of a gas varies *directly* with the Absolute temperature. This means that if the Absolute temperature of a gas is doubled, the volume is doubled also. Mathematically expressed, the law is:

$$\frac{T_1}{T_2} = \frac{V_1}{V_2}$$

Here, T_1 and V_1 are the original temperature and volume, and T_2 and V_2 are the new temperature and volume.

The Absolute (Kelvin) temperature is obtained by adding 273 to the Centigrade temperature. Thus, $20°C = 293°A$, or $0°A = -273°C$. The value 273 is used because a given mass of gas at $0°C$ will expand 1/273 of its volume for each degree rise in temperature. Likewise, it will contract 1/273 of its volume if the temperature is lowered one degree. If this relationship held through the entire temperature range, the volume of gas would be zero at $-273°C$. Lord Kelvin did pioneering work in this area and so we now use the term *Kelvin degrees* (°K) in place of Absolute degrees. Thus $0°C = 273°K$. (Keep in mind that many chemists still use the term Absolute degrees.)

Problem 1. A given mass of gas occupies 500 ml at $20°C$. What will its volume be at $40°C$ if the pressure remains constant?
Solution:

$$\frac{500 \text{ ml}}{V_2} = \frac{293°K}{313°K} \quad \text{or} \quad V_2 = 500 \text{ ml} \times \frac{313°K}{293°K} = 534.1 \text{ ml}$$

Note that it was necessary to convert Centigrade degrees to Kelvin degrees because the volume varies directly with the Kelvin (Absolute) temperature.

Problem 2. Pressure remaining constant, the volume of a given mass of gas at $35°C$ is 200 ml. At what temperature (Centigrade) will the volume be 250 ml?
Solution:

$$\frac{T_2}{308°K} = \frac{250 \text{ ml}}{200 \text{ ml}} \quad \text{or} \quad T_2 = \frac{250 \text{ ml}}{200 \text{ ml}} \times 308°K = 385°K$$

$$385°K - 273 = 112°C$$

Boyle's and Charles's Laws Combined. When these two gas laws are combined it is possible to solve problems where both the temperature and the pressure change. The combined laws may be written:

$$\frac{P_1 V_1}{T_1} = \frac{P_2 V_2}{T_2}$$

where P_1, V_1, T_1 and P_2, V_2, T_2 have the same significance as in the above discussion.

Problem 1. At 27°C and 700 mm pressure, a given mass of gas occupies 400 ml. What volume will it occupy at 7°C and 600 mm pressure?

Solution. Substitute in the formula and solve for V_2.

$$\frac{700 \text{ mm} \times 400 \text{ ml}}{300°\text{K}} = \frac{600 \text{ mm} \times V_2}{280°\text{K}}$$

$$V_2 \text{ ml} = \frac{700 \text{ mm} \times 400 \text{ ml} \times 280°\text{K}}{300°\text{K} \times 600 \text{ mm}} = 435.5 \text{ ml}$$

Note that mm and °K cancel out.

Problem 2. At 0°C and 600 mm pressure, a given mass of gas occupies 500 ml. If the pressure is increased to 800 mm, what must the Centigrade temperature be to give a volume of 600 ml?

Solution. Substitute the data in the formula and solve for T_2. The mm and ml cancel out.

$$\frac{600 \text{ mm} \times 500 \text{ ml}}{273°\text{K}} = \frac{800 \text{ mm} \times 600 \text{ ml}}{T_2}$$

$$T_2 = \frac{800 \text{ mm} \times 600 \text{ ml} \times 273°\text{K}}{600 \text{ mm} \times 500 \text{ ml}} = 436.8°\text{K}$$

$$436.8°\text{K} - 273 = 163.8°\text{C}$$

The "Common Sense" Method of Solving Gas Laws Problems. For Charles's law a fraction is made up of the two numbers which represent the original Absolute (Kelvin) temperature and the new Absolute (Kelvin) temperature to which the gas is to be changed. By inspection one decides, knowing the law, whether the gas is to expand or to contract. If the gas is to expand, the fraction will be greater than unity (one). If the gas is to contract, the fraction will be less than unity. The original volume multiplied by this fraction gives the volume the gas will occupy under the new conditions.

Example 1. A given mass of gas occupies 75 ml at 30°C. What volume will it occupy at 20°C? Inspection tells us that since the gas is to be cooled, the new volume will be less than 75 ml. Our fraction, then, will be less than unity.

Solution.   New volume = 75 ml × $\dfrac{293°K}{303°K}$ = 72.5 ml

For Boyle's law problems the fraction, made up of the two pressures, is multiplied by the original volume of the gas. This gives the new volume. It is obvious that the fraction will be greater than unity if the gas is to expand, and less than unity if the gas is to contract.

Example 2.   A given mass of gas occupies 90 ml at a pressure of 740 mm. What volume will the gas occupy when the pressure is changed to 760 mm? Inspection shows us that a contraction is to take place; therefore, the fraction will be less than unity. The fraction, multiplied by the given volume, gives the volume at the new pressure.

Solution.   New volume = 90 ml × $\dfrac{740 \text{ mm}}{760 \text{ mm}}$ = 87.6 ml

For Charles's and Boyle's laws combined, two fractions are formed: first, a fraction for the calculation of the temperature change and second, a fraction for the calculation of the pressure change. These two fractions are multiplied by the original gas volume.

Example 3.   A given mass of gas occupies 85 ml at 0°C and 740 mm pressure. What volume will the gas occupy at 25°C and 760 mm pressure? Inspection shows that (1) the temperature fraction is greater than unity, and (2) the pressure fraction is less than unity. Therefore, the equation will be:

Solution.   New volume = 85 ml × $\dfrac{298°K}{273°K}$ × $\dfrac{740 \text{ mm}}{760 \text{ mm}}$ = 90.3 ml

SUMMARY

1. *Charles's law:*

New volume = original volume × $\dfrac{\text{temperature}}{\text{fraction}}$.

2. *Boyle's law:* New volume = original volume × $\dfrac{\text{pressure}}{\text{fraction}}$.

3. *Charles's and Boyle's laws combined:*

New volume = original volume × $\dfrac{\text{temperature}}{\text{fraction}}$ × $\dfrac{\text{pressure}}{\text{fraction}}$.

Graham's Law of Diffusion.   Temperature and pressure of the gases being the same, the rates of diffusion of two gases are inversely proportional to the square roots of their densities. Mathematically this law can be expressed:

$$\frac{R_1}{R_2} = \frac{\sqrt{D_2}}{\sqrt{D_1}}$$

where R_1 and D_1 represent the rate of diffusion and density of the first gas, and R_2 and D_2 are the corresponding values for the second gas. Figure 7-3 is a sketch of an apparatus which demon-

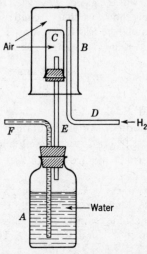

Fig. 7-3. The diffusion of hydrogen. Apparatus to show that hydrogen diffuses more rapidly than air. The inverted beaker B and the unglazed ceramic tube C are filled with air. When hydrogen is introduced through D it diffuses into C much more rapidly than air can diffuse out. This creates a pressure which forces water out of tube F.

strates that a light (low density) gas diffuses more rapidly than a gas of higher density. The following are given here to serve as examples of how problems based on Graham's law are solved.

Problem 1.   If the density of hydrogen is 1 and the density of oxygen is 16 which gas diffuses faster? How many times faster?

Solution 1.

$$\frac{R_{(hyd)}}{R_{(oxy)}} = \frac{\sqrt{D_{oxy}}}{\sqrt{D_{hyd}}} = \frac{\sqrt{16}}{\sqrt{1}} = \frac{4}{1} \text{ Hydrogen. 4 times faster}$$

Solution 2.   Reasoning from Avogadro's law (see p. 77) we know that the molecular weight of a gas can be substituted for its density in

Graham's law of diffusion. Thus for the above problem we have:

$$\frac{R_{(hyd)}}{R_{(oxy)}} = \frac{\sqrt{Molec.\ wt.\ of\ O_2}}{\sqrt{Molec.\ wt.\ of\ H_2}} = \frac{\sqrt{32}}{\sqrt{2}} = \frac{5.65}{1.4} = \frac{4}{1}\ Hydrogen.\ 4\ times\ faster$$

Problem 2. The molecular weight of SO_2 is 64. The molecular weight of He (helium) is 4. Which gas diffuses more rapidly? How many times more rapidly?

Solution

$$\frac{R_{(He)}}{R_{(SO_2)}} = \frac{\sqrt{64}}{\sqrt{4}} = \frac{8}{2} = \frac{4}{1}\ Helium.\ 4\ times\ faster$$

Law of Gay-Lussac. In any reaction in which any of the reacting substances or any of the products are gases, there is a *small-integer ratio* between the volumes of these gases. Thus:

One volume of hydrogen + one volume of chlorine = two volumes of hydrogen chloride.

$$H_2 + Cl_2 \longrightarrow 2\,HCl$$

One volume of nitrogen + three volumes of hydrogen = two volumes of ammonia.

$$N_2 + 3\,H_2 \longrightarrow 2\,NH_3$$

Avogadro's Law. Equal volumes of gases under the same conditions of temperature and pressure contain the *same number of molecules*. For example, a liter of hydrogen contains the same number of molecules as a liter of chlorine at the same temperature and pressure. By making use of Gay-Lussac's law and Avogadro's law it is possible to determine the number of atoms per molecule of the common gases. The reasoning is as follows: *One volume* of hydrogen reacts with *one volume* of chlorine to give *two volumes* of hydrogen chloride. Each molecule of the compound must contain at least one hydrogen atom and one chlorine atom. Since two volumes of hydrogen chloride were obtained from one volume of hydrogen (or one volume of chlorine) the simplest assumption is that each hydrogen molecule (or chlorine molecule) contains two atoms. No reaction is known where one volume of hydrogen gives more than two volumes of a gaseous compound containing hydrogen.

Molecular Weight of Gas. The molecular weight of oxygen is 32. This is the standard. One gram-molecular weight [a mole of oxygen (32 g) at 0°C and 760 mm pressure] occupies 22.4 liters. The temperature and pressure referred to here are called the Standard Temperature and Pressure (STP) or sometimes called Standard Conditions. The standard temperature is 0°C or 273°K; the standard pressure is 760 mm or 1 atmosphere. Because of Avogadro's law the weight of 22.4 liters of any gas (at STP) is the weight of one mole of that gas. This is an interesting and important relationship because we have a simple relation between the weight and volume of a gas. See Table 7-3.

Table 7-3

Weight-Volume Relationship of Gases

Gas	Formula	Molecular (Formula) Weight	Weight of 22.4 liters
Hydrogen	H_2	2	2 g
Helium	He	4	4 g
Oxygen	O_2	32	32 g
Nitrogen	N_2	28	28 g
Chlorine	Cl_2	71	71 g
Hydrogen chloride	HCl	36.5	36.5 g
Ammonia	NH_3	17	17 g

Problem 1. At STP, 1 g of methane (CH_4) occupies a volume of 1.4 liters. What is the molecular weight of methane?

Solution.

Let x represent the weight of 22.4 ℓ of methane.

$$\frac{x}{22.4\,\ell} = \frac{1\,g}{1.4\,\ell} \quad \text{or} \quad x = \frac{1\,g \times 22.4\,\ell}{1.4\,\ell}$$

1 g × 16 = 16 g per mole or the molecular weight is 16.

Problem 2. At STP, 1.144 g of oxygen occupies a volume of 800 ml. What is the weight of one mole of oxygen?

Solution: Let x = weight of 22.4 ℓ of oxygen.

$$\frac{1.144\,g\,O_2}{800\,ml} = \frac{x}{22,400\,ml}$$

$$x = \frac{1.144\,g \times 22,400\,ml}{800\,ml} = 32\,g$$

Dalton's Law of Partial Pressure. In a mixture of two or more gases, each gas exerts its pressure *independently*. In other words, the pressure exerted by a gas, in a mixture, is the same as if it alone occupied the volume. The total pressure is equal to the sum of the partial pressures.

The specific example of a gas confined over water will be used to outline the method. See Figure 7-4. (1) Find the total pressure of the system (to do this use a manometer or a barometer). (2)

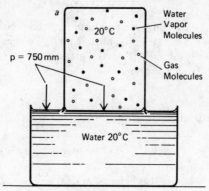

Fig. 7-4. Dalton's law of partial pressure is used in determining the volume of dry gas in container (*a*).

Determine the temperature of the confined gas. (3) Consult a water-vapor pressure table for the pressure of water at the given temperature. (4) Subtract the water-vapor pressure from the total pressure.

Problem 1. 250 ml of oxygen is held in a container over water. The total pressure is 730 mm and the temperature is 25°C. What is the pressure exerted by the oxygen? What is the volume of the dry oxygen in the mixture? What is the volume of the water vapor?

Solution: From the table (Appendix IV) we find that the vapor pressure of water at 25°C is 23.76 mm.

Oxygen pressure = 730 mm − 23.76 mm = 706.24 mm

Volume of oxygen = 250 ml × $\dfrac{706.24 \text{ mm}}{730 \text{ mm}}$ = 241.9 ml

Volume of water vapor = 250 ml − 241.9 ml = 8.1 ml

Problem 2. A 22.4-liter tank contains 2 g of hydrogen, 8 g of oxygen, and 22 g of carbon dioxide, at 0°C. What is the pressure exerted by each gas? What is the total pressure in the tank?

Solution: If the tank contained only 2 g of hydrogen (1 mole) the pressure would be one atmosphere (760 mm). If the tank contained only 8 g of oxygen (1/4 mole) the pressure would be 1/4 an atmosphere. If the tank contained only 22 g of carbon dioxide (1/2 mole) the pressure would be 1/2 an atmosphere. The total pressure, then, is equal to the sum of the partial pressures. Thus:

$$1 \text{ atm } + \tfrac{1}{4} \text{ atm } + \tfrac{1}{2} \text{ atm } = 1\tfrac{3}{4} \text{ atmospheres.}$$

or

$$1\tfrac{3}{4} \text{ atmospheres} \times \frac{760 \text{ mm}}{1 \text{ atm}} = 1330 \text{ mm of mercury}$$

Problem 3. At constant temperature, a certain container is filled with a mixture of gases as follows: nitrogen 60%, oxygen 20%, hydrogen 10%, and carbon dioxide 10%. The total pressure is 760 mm. What is the partial pressure of each gas present?

Solution.

$$\text{Partial pressure of N}_2 \ 760 \text{ mm } \times \frac{60}{100} = 456 \text{ mm}$$

$$\text{Partial pressure of O}_2 \ 760 \text{ mm } \times \frac{20}{100} = 152 \text{ mm}$$

$$\text{Partial pressure of H}_2 \ 760 \text{ mm } \times \frac{10}{100} = \ \ 76 \text{ mm}$$

$$\text{Partial pressure of CO}_2 \ 760 \text{ mm } \times \frac{10}{100} = \ \ \underline{76 \text{ mm}}$$

$$\text{Total pressure } = 760 \text{ mm}$$

Ideal Gas vs. Real Gas. The gas laws are obeyed perfectly only by "ideal" gases. Real gases such as hydrogen, nitrogen, helium, and oxygen obey the gas laws fairly well, especially over certain temperature and pressure ranges, but for other gases such as carbon dioxide and ammonia the deviation is quite marked. For example at 0°C 2 liters of oxygen at one atmosphere pressure will occupy 0.999 ℓ at 2 atmospheres pressure. This means that for oxygen molecules, there is a definite attraction for each other when they get close together. These attractive forces are now called *van der Waals forces*.

In terms of atomic structure, van der Waals attractive forces may be due to the attraction of the positive neuclei of atoms for the electrons of other atoms. It is apparent from Table 7-4 and Figure 7-5 that the value PV is less than 1 up to 350 atmospheres pressure. Above this pressure the PV values are greater than 1. At these higher pressures the volume actually occupied by the gas molecules becomes an important factor.

Table 7-4

Relation between Pressure and Volume of Gases (Boyle's Law)

Gas	$P_{(atm)}$	$V_{(liters)}$	$P \times V$
Ideal (perfect)	1	1	1.0
Oxygen	1	1	1.0
Ideal	100	0.01	1.0
Oxygen	100	0.0093	0.93
Ideal	200	0.005	1.0
Oxygen	200	0.0046	0.92
Ideal	300	0.0033	1.0
Oxygen	300	0.0032	0.96

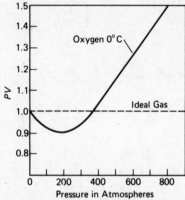

Fig. 7-5. The change in the value of PV for oxygen at various pressures. Note that PV for an ideal gas remains constant for all pressures but for oxygen PV is 1, then less than 1, and, above 350 atmospheres, becomes greater than 1.

REVIEW QUESTIONS

1. How would you prove that a gas expands on heating and contracts on cooling?
2. Distinguish between a gas, a liquid, and a solid.
3. State Boyle's law in words. State it by a mathematical equation.
4. What are the postulates of the kinetic theory of gases?
5. There are no perfect gases. How can one prove this experimentally?
6. (a) What is meant by the critical temperature of a gas? (b) the critical pressure of a gas?

7. In what way does the kinetic theory explain the compressibility of gases?

8. At constant temperature, a given mass of gas occupies a volume of 2 liters at 700 mm pressure. If temperature remains constant, at what pressure will the volume be 1 liter? *Ans.* 1400 mm.

9. State in words Graham's law of diffusion. What is the mathematical statement of Graham's law?

10. (*a*) What is Dalton's law of partial pressure?
    (*b*) Make up a problem that will make use of this law in the solution of the problem.

Chapter 8
Liquids and Solids

LIQUIDS

In gases, the molecules have little attraction for each other until they are forced, by increased pressure (and by lowering the temperature) to be near each other. Then van der Waals attractive forces between molecules come into action. On the other hand, molecules of a liquid do attract each other but they still have sufficient freedom of motion so there is random arrangement of molecules in the liquid. These forces of attraction are weak compared with chemical bonds. Liquids flow and assume the shape of the vessel in which they are stored. All molecules are in a state of constant motion, some moving slowly, others moving rapidly enough for them to escape from the surface of the liquid. Sealed in a partially filled vessel at constant temperature, a state of equilibrium is established where the number of molecules returning to the liquid surface just equals the number leaving. The pressure exerted at this point is the vapor pressure of the liquid at the temperature chosen. See Figure 8-1a. See also Appendix IV for the vapor pressures of water from 0°C to 100°C.

Boiling Point.   When the temperature of a liquid reaches that point where its vapor pressure is equal to the applied pressure (atmospheric pressure), the liquid boils. At this point bubbles of vapor form at the bottom of the vessel, rise through the liquid, and escape from its surface. The average atmospheric pressure at sea level is 760 mm. At this pressure water boils at 100°C. Usually it is customary for boiling points of liquids to be expressed for exactly 760 mm of pressure. At 650 mm pressure water boils at 96°C but at 800 mm pressure it boils at 101.4°C.

Surface Tension.   The forces on molecules at the surface of a liquid are not balanced. Because of these unbalanced forces there is a tendency for particles of water and other liquids to assume a spherical shape. Also the surface of a liquid in a container acts as though the liquid had a thin tough film or membrane covering

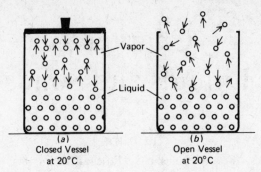

Fig. 8-1. The vapor pressure of a liquid. In (a) at constant temperature, an equilibrium exists between molecules in vapor and liquid states. In (b) there is no equilibrium. The liquid evaporates.

it. For example, if a dry steel sewing needle is carefully placed on the surface of some water, the needle floats, in spite of the fact that it has a much higher density than water.

Evaporation.   Due to the high (kinetic) energy of some molecules near the surface of a liquid, they escape from the surface into the surrounding atmosphere. If the liquid is confined in an open vessel it will continue to evaporate until all the liquid disappears. See Figure 8-1b.

Distillation.   This is a process in which water (or another

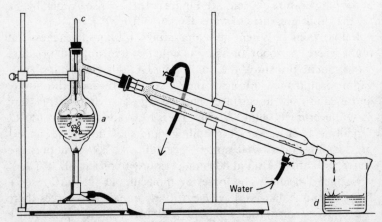

Fig. 8-2.   Distillation apparatus. (a) Distillation flask. (b) Condenser. (c) Thermometer. (d) Distillate in beaker.

liquid) is vaporized and the vapor condensed in a separate clean container. Figure 8-2 shows an apparatus used in distillation. The liquid is placed in the distillation flask (a) and then heated to boiling by means of a Bunsen (gas) burner. The boiling liquid is converted into a vapor and is subsequently liquefied in the condenser (b). It is then collected in the beaker (d). The purpose of distillation is the purification of the liquid. Distilled water is quite pure especially if the impurities in the water before distillation were salts and other nonvolatile substances.

Heating Curve.   Figure 8-3 shows graphically the effect of time on the changes in temperature (while applying heat at constant

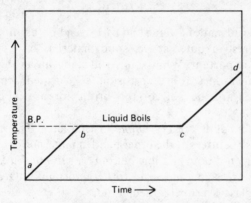

Fig. 8-3.   A heating curve.

pressure) of a chemically pure liquid. From (a) to (b) the temperature of the liquid rises continuously, say from room temperature to the boiling point (b). From (b) to (c), the boiling period, the temperature remains constant. At point (c) all the liquid is changed to vapor (gas) and then the temperature rises as shown by curve (c)–(d). The temperature remains constant during boiling because the excess heat energy applied is used up in (1) separating molecules in the liquid from each other and (2) changing liquid molecules to gaseous molecules.

Heat of Vaporization.   This term means the quantity of heat (in calories) required to change 1 g of the liquid, at the boiling temperature, to 1 g of vapor. The following table gives this information for several substances.

Table 8-1

Heat of Vaporization

Liquid	Heat of Vaporization Calories/g
Carbon tetrachloride (CCl_4)	46.4
Mercury (Hg)	68.0
Benzene (C_6H_6)	94.3
Ethyl alcohol (C_2H_5OH)	204.0
Acetone [$(CH_3)_2CO$]	263.0
Water (H_2O)	539.6

SOLIDS

Since the advent of atomic and hydrogen bombs and because of man's interest in outer space, solid materials have become increasingly important. What solids will stand up at high temperatures? Which ones resist corrosion at elevated temperatures? Which ones are too soft or too brittle for use in outer-space experiments?

As a rule, solids are *crystalline* in character. This means that they have structure. On the other hand, a material that appears to be solid may not be crystalline but only a *supercooled liquid*. Such substances are sometimes said to be amorphous. Ordinary soft (window) glass is a supercooled liquid.

Structure of Crystals. Crystalline substances that are pure have sharp melting (or freezing) points. The determination of the melting point of a crystalline substance is an excellent method of determining its purity.

Crystals are made up of atoms, ions, or molecules. These are arranged in a pattern of points called a *space lattice*. The arrangement of atoms, ions, or molecules in a space lattice is determined by X rays. When X rays pass through a crystal, they are diffracted by the particles (atoms, ions, or molecules) in the crystal. The diffracted rays are then allowed to strike a photographic plate or film. A pattern of spots is obtained on the developed plate from which it is possible to determine the arrangement of the particles in three-dimensional space. See Figure 8-4.

Most substances crystallize in one of six different systems based on their symmetry. These are isometric (cubic), hexagonal, tetragonal, orthorhombic, monoclinic, and triclinic. These six crystal-

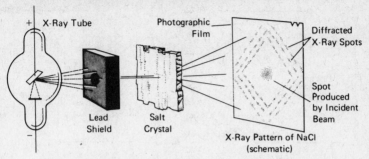

Fig. 8-4. Laue X-ray determination of crystal structure.

line systems are referred to imaginary axes that pass through the crystals. For example, cubic crystals (which are probably the simplest) have three axes, all the same length and at right angles to each other. See Figure 8-5.

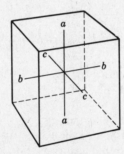

Fig. 8-5. A cube with its three axes, all of the same length and at right angles to each other.

Units of a Crystal. The points in a space lattice may be occupied by atoms, ions, or molecules. For example, the element argon forms cubic crystals with the units being argon atoms. Metallic sodium (and other metals) forms crystals in which the points in the space lattice are occupied by sodium ions (Na^+). These positive sodium ions are presumably immersed in an electron cloud or, as it is sometimes called, an "electron gas." On the other hand, table salt (sodium chloride) forms cubes where the individual units are sodium ions (Na^+) and chloride ions (Cl^-) arranged in such a way that each chloride ion is surrounded by 6

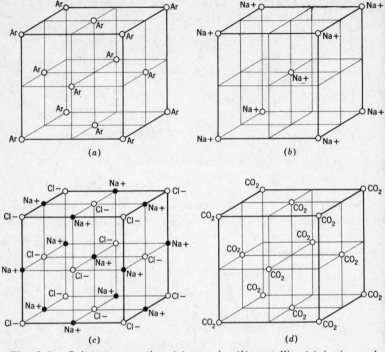

Fig. 8-6. Cubes representing (*a*) atomic, (*b*) metallic, (*c*) ionic, and (*d*) molecular types of crystals.

sodium ions and each sodium ion is surrounded by 6 chloride ions. Here, there are no free electrons, no electron cloud, because the electrons from sodium atoms are tied up to give chloride ions. Finally, the points in the space lattice may be occupied by molecules. This is true for solid carbon dioxide (dry ice). See Figure 8-6. Also see Table 8-2 for information as to the physical properties of crystals in terms of their lattice points.

Polymorphism. A substance may crystallize in more than one crystal form. For example, elemental sulfur can crystallize in the rhombic system and also in the monoclinic system. What happens in a given situation depends on the temperature. Likewise, the particles in a crystal may arrange themselves in more than one way without shifting into a new crystal system. For example, in the cubic system we can have a simple lattice, a face-centered lattice, or a body-centered lattice. Alpha iron is made up of body-

Table 8-2

Physical Properties of Crystals as Related to Their Lattice Points Composition

Type of Solid	Lattice Points	Bonding Forces	Hardness	Melting Point	Electrical Conductivity	Examples
Molecular	Molecules	van der Waals	Rather soft	Rather low	Very poor	H_2O, CO_2
Ionic	+ ions − ions	Ionic or electrostatic	Relatively high	Rather high	Very poor	NaCl, KNO_3
Covalent	Atoms	Electron pairs	Very high	Very high	Nonconductor or very low	C (Diamond) SiC (Carborundum)
Metallic	+ ions	Electrostatic between + ions and electrons	From soft to hard	Rather high	Good	Sodium Gold Silver Iron

centered cubic crystals; gamma iron is made up of face-centered cubic crystals. See Figure 8-7.

Alpha iron can be changed to gamma iron by heating the alpha iron to 910°C or above. It is interesting to note that gamma iron (face-centered cubic) is a good solvent for carbon, whereas alpha iron is a poor solvent for carbon. This one fact makes the science and art of heat treating of steels possible.

Lattice Defects. Most crystals are imperfect. Imperfection may be due to unoccupied lattice points, to atoms or ions not in their proper places, or to foreign bodies (atoms, ions, or molecules) present in the crystal. The physical properties of imperfect

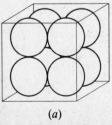 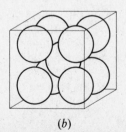 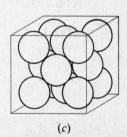

(a)   (b)   (c)

Fig. 8-7. Three different arrangments of particles in a cubic lattice: (a) simple cubic, (b) body centered, (c) face centered.

crystals, say of a metal such as iron, are much lower than would be the case if iron crystals were perfect. It has been estimated that the tensile strength of perfect iron crystals would be near 1,000,000 pounds per square inch, whereas the actual tensile strength of ordinary iron crystals is very much less than the value given. Some imperfections in crystals are called *line imperfections* or *dislocations*. Figure 8-8 illustrates an edge dislocation.

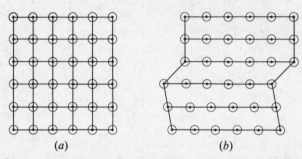

(a)   (b)

Fig. 8-8. A schematic representation of a space lattice (a) with no dislocation and (b) with an edge dislocation.

REVIEW QUESTIONS

1. What is meant by (a) the boiling point of a liquid? (b) the vapor pressure of a liquid?
2. Explain what is meant by the surface tension of a liquid.
3. What happens, in terms of the molecules, when a liquid evaporates?
4. Define distillation.
5. Why does a pure liquid such as distilled water boil at a constant temperature even though heat is being supplied constantly?
6. What is meant by the term heat of vaporization?
7. Why is a knowledge of the solid state a matter of importance in modern chemistry?
8. (a) What are the names of the 6 major crystalline systems? (b) What are the units of which crystals are built?
9. What is the difference between an atomic crystal and an ionic crystal?
10. Distinguish between simple cubic body-centered and face-centered cubic crystals.
11. What is a crystal lattice? What are lattice defects?
12. How are X rays useful in studying crystals?

Chapter 9
Oxygen and Hydrogen

Oxygen (8) $1s^2 2s^2 2p^4$*
Hydrogen (1) $1s^1$*

There are a number of reasons why the study of oxygen and hydrogen should come early in an elementary chemistry course. Some of these reasons (not necessarily in the order of their importance) are:

1. Oxygen and hydrogen are easily prepared in the laboratory. Many times the instructor demonstrates their preparation, but frequently the student is required to prepare them.

2. Oxygen is the most abundant element on earth. It occurs free (uncombined) in nature and also in chemical combination with most of the other elements.

3. Traces only of hydrogen occur free in nature, but it is a constituent of thousands of compounds.

Table 9-1
Composition of Earth's Crust

Weight Per Cent		Atomic Per Cent	
1. Oxygen	46.7	1. Oxygen	60.43
2. Silicon	27.7	2. Silicon	20.47
3. Aluminum	8.13	3. Aluminum	6.25
4. Iron	5.00	4. Hydrogen	2.88
5. Calcium	3.63	5. Sodium	2.55
6. Sodium	2.83	6. Calcium	1.88
7. Potassium	2.59	7. Iron	1.86
8. Magnesium	2.09	8. Magnesium	1.78
9. Titanium	0.44	9. Potassium	1.33
10. Hydrogen	0.14	10. Titanium	0.19

There seems to be some variation as to the actual percentages of each element present in the earth's crust. However, oxygen is the most abundant element and hydrogen is tenth or fourth, depending on which basis of comparison is used.

*The electronic structures of elements are important in considering their preparation and properties. Consequently they are given, as above, where they are appropriate.

4. Most of the compounds studied in elementary chemistry contain hydrogen or oxygen or both of them.

5. Oxygen is especially important from an industrial point of view as well as from the view of scientists. Its physical and chemical properties should be familiar to all students.

6. In terms of atomic percentage, hydrogen is the fourth most abundant element on earth. Considering the entire solar system, hydrogen is the most abundant element. See Table 9-1.

OXYGEN

Oxygen is an element occurring in nature both in the free state and in chemical combination. It was known to the Chinese centuries before it was prepared by Europeans. Scheele, a Swedish chemist, prepared oxygen in 1771; but John Priestley, an Englishman who prepared the gas in 1774, was given credit for its discovery because his account of its preparation was published first. Lavoisier, the great French chemist, noticed that the element combined with sulfur, phosphorus, carbon, and other nonmetals to give compounds then considered to be acids. Accordingly, he named the new element *oxygen*, meaning *acid-former*.

Occurrence. Oxygen is the most abundant element. Free oxygen occurs in ordinary air, which is approximately one-fifth oxygen by volume. Chemically combined oxygen is present in water, in many metallic oxides, and in many salts. Water is eighth-ninths oxygen by weight, and many oxides and salts contain high percentages of oxygen. The tissues of plants and animals contain from 50 to 70 per cent oxygen by weight; the human body is 65 per cent oxygen by weight.

Laboratory Preparation. Oxygen can be prepared in the laboratory by the following methods:

HEATING CERTAIN METALLIC OXIDES.

$$2 \, Ag_2O + heat \longrightarrow 4 \, Ag + O_2$$
$$\text{Silver Oxide} \longrightarrow \text{Silver} + \text{Oxygen}$$
$$2 \, BaO_2 + heat \longrightarrow 2 \, BaO + O_2$$
$$\text{Barium Dioxide} \longrightarrow \text{Barium Oxide} + \text{Oxygen}$$

HEATING CERTAIN OXYGEN-CONTAINING SALTS.

$$2 \, KNO_3 + heat \longrightarrow 2 \, KNO_2 + O_2$$
$$\text{Potassium Nitrate} \longrightarrow \text{Potassium Nitrite} + \text{Oxygen}$$
$$2 \, KClO_3 + heat \longrightarrow 2 \, KCl + 3 \, O_2$$
$$\text{Potassium Chlorate} \longrightarrow \text{Potassium Chloride} + \text{Oxygen}$$

A catalytic agent (p. 97) enables potassium chlorate to decompose more smoothly and at a lower temperature. Manganese dioxide (MnO_2) is a good catalyst for this reaction. See Figure 9-1.

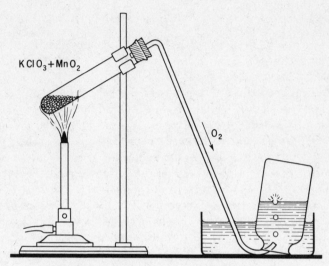

Fig. 9-1. Laboratory preparation of oxygen.

$$2 KClO_3 + (MnO_2) + heat \longrightarrow 2 KCl + O_2 + (MnO_2)$$

REACTION OF WATER ON SODIUM PEROXIDE.

$$2 Na_2O_2 + 2 H_2O \longrightarrow 4 NaOH + O_2$$
Sodium Peroxide + Water $\longrightarrow$ Sodium Hydroxide + Oxygen

CATALYTIC DECOMPOSITION OF SODIUM HYPOCHLORITE SOLUTION.

$$2 NaClO + (catalyst) \longrightarrow 2 NaCl + O_2$$
Sodium Hypochlorite $\longrightarrow$ Sodium Chloride + Oxygen

Oxygen may be prepared as a lecture demonstration by electrolyzing water containing dissolved acid or salt (see Chapter 15).

Commercial Preparation.   Oxygen can be prepared by the fractional distillation of liquid air and by the electrolysis of water. The first method is more important and is used when oxygen is required on a tonnage scale.

The old Brin Process is interesting from a chemical point of view, but is not commercially important. In this process barium oxide, heated in air to 500°C, changes to barium dioxide. Heating the dioxide to 800°C liberates one-half of the oxygen in the molecule of barium dioxide.

$$2\,BaO + O_2 \text{ (in the air)} \underset{800°C}{\overset{500°C}{\rightleftharpoons}} 2\,BaO_2$$

The arrow pointing to the left (⟵) indicates that BaO_2 will decompose (at 800°C) to give $BaO + O_2$. See Chapter 12 for examples where reverse arrows (⇌) are used.

Physical Properties.   Oxygen is a colorless, odorless, and tasteless gas. It is slightly soluble in water. One hundred cubic centimeters of water dissolves 4.9 ml of oxygen at 0°C and 760 mm pressure. Oxygen is 1.105 times heavier than air. It can be liquified by cooling it to −118°C and subjecting the cooled gas to a pressure of 50 atmospheres. Upon further cooling liquid oxygen, a white snowlike solid, is formed which melts at −218.4°C.

Chemical Properties.   Oxygen combines with the more active metals and with nonmetals. It is not very active at ordinary temperatures but is very active at elevated temperatures. For example, iron combines slowly with oxygen at room temperature but burns rapidly at high temperatures.

Uses.   Of the many uses of oxygen the following are important: It supports life. In the respiration of animals, oxygen is taken into the lungs where it is absorbed by the blood and then taken to all parts of the body. Hospitals keep tanks of compressed oxygen on hand to relieve suffocation in cases of pneumonia and other disorders. Aviators and miners carry tanks of compressed oxygen for use where there is insufficient atmospheric oxygen. Astronauts carry oxygen on their space explorations. Oxygen is essential in the burning of fuels such as gasoline and coal. It is used in oxyacetylene and oxyhydrogen welding torches. In the past few years many metallurgical processes have been improved by the use of commercial oxygen. For example, the time required to make a batch of steel in an open-hearth furnace has been shortened by 20 per cent or more.

Ozone.   Ozone is a special form of oxygen made up of three atoms of oxygen per molecule; ordinary oxygen has two atoms per molecule. It is an extremely active form of oxygen, prepared by passing oxygen between two plates which are charged at a high potential. At ordinary temperatures the yield of ozone is approximately 1 to 2 per cent. At liquid-air temperature the yield is increased to nearly 90 per cent. Under these conditions liquid ozone separates out. In nature small quantities are made by lightning and also by the action of the ultraviolet rays present in sunlight. The reaction is:

$$3\,O_2 \rightleftharpoons 2\,O_3$$

The structure is written: 117°

Pauling writes the structural formula of ozone as a hybrid structure and points out that the double bond resonates between the two forms as indicated by the brackets enclosing them.

Ozone is a pale-blue gas with a characteristic odor. It is much more active, chemically, than oxygen. It readily oxidizes silver. This is not true for oxygen.

$$2\,Ag + O_3 \longrightarrow Ag_2O + O_2$$

Ozone can be used to deodorize and sterilize water. It may also be used to bleach oils, flour, and certain cloth.

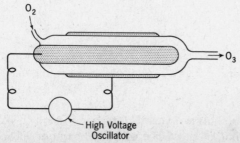

Fig. 9-2. Preparation of ozone by silent electrical discharge.

Oxygen and Oxidation: History. Men knew of and used fire for thousands of years before they understood its nature. Several conflicting and unsatisfactory theories of combustion had been suggested by the beginning of the eighteenth century.

PHLOGISTON THEORY. A relatively satisfactory explanation of combustion, known as the Phlogiston Theory, was proposed by G. E. Stahl in 1717. Briefly stated, the Phlogiston Theory is as follows. All combustible materials contain an invisible and weightless substance called phlogiston. When a substance burns, phlogiston escapes, causing the flame and smoke. The residue (ash) is the original substance minus the phlogiston. Oils, waxes, carbon, and other substances which leave very little ash upon burning were supposed to be nearly all phlogiston.

LAVOISIER'S EXPERIMENT. Lavoisier (1743–1794) overthrew the Phlogiston Theory by showing the part that oxygen takes in oxidation. By heating some mercury and air in a closed retort he found that the mercury combined with the oxygen in the air, forming a red powder. The red powder weighed as much as the mercury it contained plus the weight of the oxygen taken from the air. Upon heating the red powder to a higher temperature the oxygen was again liberated, restoring the original condition.

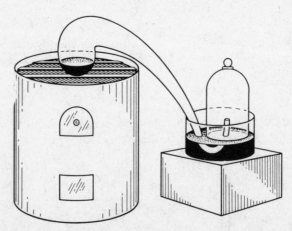

Fig. 9-3. Apparatus used by Lavoisier to show the role of oxygen in oxidation.

Oxidation. This is a chemical reaction in which oxygen combines with another substance with the liberation of heat. The reaction may be slow or rapid. If rapid the process is called *combustion*. Here, light as well as heat is produced. Slow oxidation takes place when iron rusts. Rapid oxidation or combustion takes place when a splinter of wood is burned, producing heat and light.

Note. The definition of oxidation given here is restricted. See Chapter 13 for a general discussion.

OXIDATION IN NATURE. Oxidations are the most common reactions in nature. They occur with both inorganic and organic substances. Metals are converted to oxides upon standing in the air, especially if moisture is present. Thus, if a piece of iron is left in a damp place a scale of red iron oxide forms on its surface. The decay of organic matter is a reaction in which the oxygen of the air, in the presence of bacteria and moisture, combines with the organic matter. Running water, containing organic matter, is purified through oxidation by *aeration*.

SPONTANEOUS COMBUSTION. An easily oxidizable substance surrounded by insulating materials may take fire. A pile of oily cotton waste will begin to burn if left in a poorly ventilated place. The explanation is as follows. The oil is slowly oxidized, liberating heat. Due to the insulation of the cotton waste the heat does not readily escape, so that the temperature is raised. At the higher temperature, oxygen combines with the combustible material more rapidly, liberating more heat. Owing to the continuous accumulation of heat under these conditions the kindling temperature is reached, and the substance takes fire.

Isotopes. In nature there are three isotopes of oxygen. About 99.76 per cent of oxygen atoms have a mass number of 16. Of the remaining .24 per cent, .04 per cent have a mass number of 17 and .20 per cent have a mass number of 18. (See Chapter 2 for a discussion of isotopes.)

Important Definitions

Catalytic Agent. A substance which changes the rate of a reaction. The catalyst is found to be unchanged chemically when the reaction is completed.

Kindling Temperature. The lowest temperature at which a substance ignites (takes fire).

Temperature of Combustion. The highest temperature reached during combustion.

Heat of Combustion. The total quantity of heat (measured in calories) given off when a known quantity of a substance burns.

REVIEW QUESTIONS

1. Oxygen occurs in nature as free oxygen (chemically uncombined) and in chemical combination. How can one prove this statement?
2. The atmosphere contains oxygen. How can it be removed?
3. What is a convenient method of preparing oxygen in the laboratory? Explain what happens.
4. How does oxygen take part in the burning of a fuel?
5. What is ozone? How does it differ from ordinary oxygen structurally?
6. How did Lavoisier overthrow the old Phlogiston Theory?
7. How can oxygen of good quality be prepared on a large scale?
8. What is the relationship between combustion and oxidation?
9. What is spontaneous combustion? What conditions are favorable to it?

HYDROGEN

Although the alchemists had prepared hydrogen, it was not recognized as a separate substance (properly, of course, a separate element) until 1766. In that year Cavendish, an Englishman, prepared the gas in experiments involving the action of certain metals on nonoxidizing acids. Because it burned to give water, Lavoisier, the French chemist, called the new substance *hydrogen*, meaning *water-former*.

Occurrence. Hydrogen is widely distributed in nature. It constitutes about 1 per cent by weight of the earth's crust, including the water in the oceans. Although this percentage seems small, there is a greater number of hydrogen atoms than of any other element except oxygen. Solar spectra indicate that the sun's atmosphere contains a rather high percentage of hydrogen. The amount of free hydrogen in the earth's atmosphere is negligible, but we do find small amounts in gases escaping from volcanoes and from certain oil wells. Chemically combined hydrogen occurs in water (one-ninth by weight) and in acids. It is also chemically combined in petroleum oils and in organic compounds such as alcohols, carbohydrates, and proteins.

Laboratory Preparation. Hydrogen is prepared in the laboratory in the following ways:

ELECTROLYSIS OF WATER. If a direct electric current is passed through water containing a little acid, hydrogen is liberated at the cathode.

ACTION OF METALS ON WATER. Chemically active metals such as potassium and sodium react vigorously with water to liberate hydrogen.

$$2\,Na + 2\,H_2O \longrightarrow 2\,NaOH + H_2$$
Sodium + Water $\longrightarrow$ Sodium Hydroxide + Hydrogen
$$2\,K + 2\,H_2O \longrightarrow 2\,KOH + H_2$$

ACTION OF STEAM ON IRON. Less active metals such as iron will react with water in the form of steam to liberate hydrogen.

$$3\,Fe + 4\,H_2O\ (steam) \longrightarrow Fe_3O_4 + 4\,H_2$$
Iron + Steam $\longrightarrow$ Iron Oxide + Hydrogen

ACTION OF CERTAIN METALS ON NONOXIDIZING ACIDS. Active metals such as magnesium, zinc, iron, and aluminum react with such acids as dilute sulfuric, hydrochloric, and acetic acids to liberate hydrogen. See Figure 9-4.

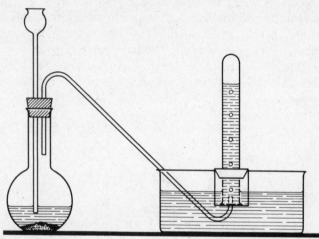

Fig. 9-4. Laboratory preparation of hydrogen by action of hydrochloric acid on zinc.

$$Zn + 2\,HCl \longrightarrow ZnCl_2 + H_2$$
$$Fe + H_2SO_4 \longrightarrow FeSO_4 + H_2$$

ACTION OF CERTAIN METALS ON ALKALIES. Aluminum and zinc react with soluble bases (alkalies) to liberate hydrogen.

$$2\,Al + 6\,NaOH \longrightarrow 2\,Na_3AlO_3 + 3\,H_2$$
Aluminum + Sodium Hydroxide $\longrightarrow$ Sodium Aluminate + Hydrogen

Commercial Preparation. Probably the least expensive method of preparing hydrogen on a large scale is by the action of steam on hot carbon. Unfortunately, carbon monoxide is also formed and must be removed.

$$C + H_2O \text{ (steam)} \longrightarrow CO + H_2$$

Passing steam and methane over a hot nickel catalyst gives hydrogen and carbon dioxide. The carbon dioxide is removed by passing the products through water under pressure.

$$CH_4 + 2\,H_2O \xrightarrow[\text{Ni}]{} CO_2 + 4\,H_2$$

The best-quality hydrogen is prepared by the electrolysis of alkali solutions. The cells are arranged so that the hydrogen being liberated at the cathode does not have an opportunity to mix with the oxygen liberated at the anode.

Although not a commercial method of preparing hydrogen and oxygen, the apparatus illustrated in Figure 9-5 is often used to prepare small amounts of the two gases. Notice that the design of the apparatus prevents the mixing of the two gases.

$$2\,H_2O \xrightarrow{\text{electrolysis}} 2\,H_2 + O_2$$

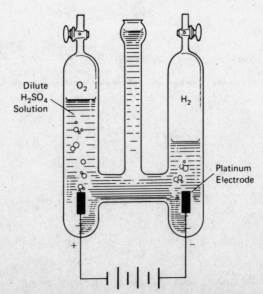

Fig. 9-5. Electrolysis of water producing hydrogen and oxygen.

Physical Properties. Hydrogen is a colorless, odorless, taste-less gas, slightly soluble in water and certain metals such as palladium. It is the lightest known substance (14.38 times less dense than air) and therefore diffuses more rapidly than any other gas. If cooled to −240°C and placed under 12.8 atmospheres pressure, hydrogen liquefies. It boils at −252.7°C under a pressure of 1 atmosphere.

It has been shown that ordinary hydrogen is made up of two forms, *ortho* and *para*, in the ratio of 3 to 1. At liquid-hydrogen temperature (in the presence of charcoal) all the ortho hydrogen is converted to the para form. The difference in the two forms of hydrogen is caused by the direction of spin of the nuclei of the two hydrogen atoms in a hydrogen molecule. This information is of interest in physical and chemical research.

Chemical Properties. Hydrogen is inert at ordinary temperatures but is chemically active at elevated temperatures. It burns in oxygen and certain other elements, and combines directly with many elements if suitable conditions prevail. Hydrogen is a reducing agent; that is, it removes oxygen from compounds. The following equations illustrate these properties. See Figure 9-6.

$$\left.\begin{array}{l} 2\ H_2 + O_2 \longrightarrow 2\ H_2O \\ H_2 + Cl_2 \longrightarrow 2\ HCl \end{array}\right\} \text{ Direct combination}$$

$$CuO + H_2 \longrightarrow Cu + H_2O \quad \text{(Hydrogen as reducing agent)}$$

Uses. Hydrogen is used to reduce certain metallic oxides, hydrogenate certain oils, and burn in the atomic-hydrogen welding torch. In the atomic-hydrogen torch, hydrogen is passed

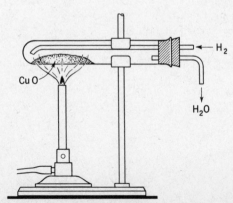

Fig. 9-6. Reduction of copper oxide with hydrogen.

through a tungsten arc, the gas dissociating into atoms. When these atoms combine to form molecules, considerable heat is liberated. All substances except carbon can be melted with such a welding torch.

Hydrogen is also used to synthesize ammonia.

$$3\,H_2 + N_2 \longrightarrow 2\,NH_3$$

This reaction is carried out commercially in the presence of a suitable catalyst, under controlled temperature and pressure.

Isotopes. Natural hydrogen consists of three isotopes: protium ($_1H^1$), heavy hydrogen or deuterium ($_1H^2$ or D), and tritium ($_1H^3$ or T). Ordinary hydrogen contains these three isotopes in the ratio:

$$H : D : T = 6{,}400 : 1 : 6 \times 10^{-15}$$

Tritium is radioactive and is used in certain types of research. Deuterium oxide (heavy water) is discussed in Chapter 10. See Chapter 2 for a discussion of isotopes.

REVIEW QUESTIONS

1. Where is one likely to find free hydrogen on the earth? on the sun?
2. What is the most important, and most readily available, source of hydrogen on earth?
3. Explain this statement: Although the earth's crust contains only 1 per cent of hydrogen, there are more atoms of hydrogen in it than of any other element except oxygen.
4. How is hydrogen prepared (*a*) in the laboratory, and (*b*) for commercial use?
5. Write equations for the preparation of hydrogen (*a*) from an acid, and (*b*) from an alkali.
6. What is the principle of the atomic-hydrogen torch?
7. How can one prove that water contains hydrogen?
8. What is the meaning of this statement: Ordinary hydrogen contains isotopes?

Chapter 10
Water and Solutions

WATER

Water. In 1781, Cavendish, an English physicist, prepared water by exploding a mixture of oxygen and hydrogen. A year or so later, Lavoisier decomposed steam by passing it over hot iron. He also showed that water is a compound of hydrogen and oxygen.

OCCURRENCE. Water is one of the most widely distributed and abundant substances found in nature. It occurs free and chemically combined. Free water makes up the oceans, lakes, and streams. It is found as a solid (ice and snow) in the colder regions. Water vapor is present in the atmosphere. A large percentage of animal and plant tissue is water, the human body containing about 70 per cent water. Chemically combined water is found in certain compounds, such as salt hydrates (e.g., $CuSO_4 \cdot 5\ H_2O$), and in certain rocks, such as hydrated iron oxide.

LABORATORY PREPARATION. Water can be prepared by the following methods:

Direct Union of Hydrogen and Oxygen. Water is formed when hydrogen burns in the air.

$$2\ H_2 + O_2 \longrightarrow 2\ H_2O$$

Decomposition of Certain Substances. If sugar is heated until decomposition takes place, water is one product.

$$C_{12}H_{22}O_{11} + heat \longrightarrow 12\ C + 11\ H_2O$$

Neutralization. Water is formed when an acid neutralizes a base.

$$H \cdot C_2H_3O_2 + NaOH \longrightarrow Na \cdot C_2H_3O_2 + H_2O$$

Acetic Acid + Sodium Hydroxide $\longrightarrow$ Sodium Acetate + Water

Reduction. Certain metallic oxides are reduced when hydrogen is passed over them at elevated temperatures. Water is one product.

$$CuO + H_2 \xrightarrow{\text{(heat)}} Cu + H_2O$$

PHYSICAL PROPERTIES.   Water is a colorless, odorless, tasteless liquid. It is the best known solvent. Its freezing point is 0°C and its boiling point is 100°C at 760 mm pressure.   At 4°C (the temperature of maximum density) 1 ml = 1 g. Water expands upon freezing. The specific heat is high. One calorie raises the temperature of 1 g of water 1°C. Its vapor pressure varies from 4.6 mm at 0°C to 760 mm at 100°C. Its heat of fusion (79.7 cal/g) and its heat of vaporization (539.6 cal/g) are high compared with those of other substances. This is probably due to hydrogen bonding.

CHEMICAL PROPERTIES.   Water is very stable through a wide range of temperatures. It combines with certain salts to form hydrates, reacts with metal oxides to form bases, reacts with oxides of nonmetals to form acids, and reacts with certain metals to form bases with the liberation of hydrogen. Water has the power to dissociate (into ions) acids, bases, and salts. Finally, it acts as a catalyst in many reactions.

The above-mentioned properties are illustrated in the following equations:

$$5 H_2O + CuSO_4 \longrightarrow CuSO_4 \cdot 5 H_2O \quad \text{(salt hydrate)}$$
$$H_2O + Na_2O \longrightarrow 2 NaOH \quad \text{(base)}$$
$$H_2O + SO_3 \longrightarrow H_2SO_4 \quad \text{(acid)}$$
$$3 H_2O + P_2O_5 \longrightarrow 2 H_3PO_4 \quad \text{(acid)}$$
$$2 H_2O + Ca \longrightarrow Ca(OH)_2 + H_2 \quad \text{(hydrogen replacement)}$$
$$HCl + \text{(water)} \longrightarrow H^+ + Cl^- \quad \text{(ionization; reaction reversible}$$
$$\text{only if water is removed)}$$

or more accurately

$$HCl + H_2O \longrightarrow H_3O^+ + Cl^- \text{ (ionization)}$$
$$H_2 + Cl_2 \xrightarrow{\text{(moisture)}} 2 HCl \text{ (catalytic reaction)}$$

MOLECULAR STRUCTURE.   Experimental evidence indicates that water molecules are not symmetrical. If they were, a straight line could be drawn through the nuclei of the three atoms, which would be arranged H:Ö:H. Actually, the two hydrogen atoms and the oxygen atom describe an angle of 105°. This means that the electrical charges are not evenly distributed in the water molecule; it is, therefore a *dipole*. One portion of the water molecule is positive with respect to the other. It can be represented as shown in Figure 10-1. It is this asymmetry, together with the

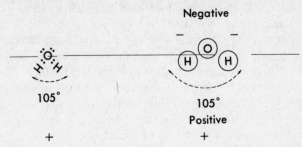

Fig. 10-1. Two methods of showing that a water molecule is not symmetrical.

phenomenon known as hydrogen bonding, that accounts for the association of water molecules into clusters.

HYDROGEN BONDS.   We usually think of hydrogen as having a valence (oxidation number) of $+1$, but there is evidence which indicates that the hydrogen atom can serve as a bridge between two atoms, especially strongly electronegative atoms such as fluorine, oxygen, and nitrogen. For example, in KHF_2 the structure is $K \cdot FHF$. See Figure 10-2a. The structure of a cluster of water molecules is represented (in two dimensions) in Figure 10-2b.

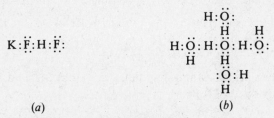

Fig. 10-2.   Hydrogen bonding in molecules.

In the $K \cdot FHF$ molecule the hydrogen atom is a bridge or bond between the two fluorine atoms. In the water cluster, the four hydrogen atoms surrounding the central oxygen atom are serving as bonds to hold the five molecules of water together as a cluster. See Chapter 4 for more on hydrogen bonding.

PURIFICATION.   The purpose for which water is used determines the method of purification. Potable water (water fit to drink) is not necessarily chemically pure. Water can be purified by the following methods:

Distillation. Distillation is a process in which water is vaporized and the vapor is condensed in a clean separate container. If properly carried out, the process will give chemically pure water.

Filtration. When water is passed through beds of sand and gravel, the suspended matter and most of the bacteria are removed. Sometimes the water is treated with lime, sodium aluminate, alum, or a mixture of these substances, and then passed through a sand filter. Gases, such as hydrogen sulfide or ammonia, can be removed from water by passing it through charcoal filters. The charcoal absorbs the gases.

Aeration. Aeration is a process in which water is sprayed into the air to oxidize the organic matter present.

Chlorination or Ozonation. By passing chlorine or ozone through water, the bacteria present are destroyed.

Boiling. When water is boiled, it is freed from its temporary hardness and from bacteria.

Ion-Exchange Resins. If water contains a dissolved salt (MA), the ions M^+ and A^- can be removed by passing the water through a vessel containing a mixture of *anion* and *cation resins*. The cation resin takes up the M^+ and liberates H^+. The anion resin takes up the A^- and liberates OH^-. The H^+ and OH^- then react to form water. No doubt the H^+ combines with H_2O to give the hydronium ion (H_3O^+), as will be considered in other chapters.

$$\text{Cation Resin} \cdot H^+ + M^+ \longrightarrow \text{Cation Resin} \cdot M^+ + H^+$$
$$\text{Anion Resin} \cdot OH^- + A^- \longrightarrow \text{Anion Resin} \cdot A^- + OH^-$$
$$H^+ + OH^- \longrightarrow H_2O$$

This water is *deionized* and may be as free from M^+ and A^- ions as carefully distilled water.

THE WATER CYCLE IN NATURE. Water evaporates from oceans, lakes, and streams, and condenses in the upper regions of the atmosphere into a fine mist. This mist is further condensed into larger drops which fall as rain. Rain water percolates through the soil back to the streams, lakes, and oceans.

USES. Water is the most widely used compound. Nearly every common substance is somewhat soluble in water. Water is sometimes spoken of as the "universal solvent." Dry crystals of table salt and silver nitrate do not react. If each salt is dissolved in water and then mixed, a reaction occurs. The reaction is between Ag^+ and Cl^-.

$$\text{NaCl (dry)} + \text{AgNO}_3 \text{ (dry)} \longrightarrow \text{No reaction}$$
$$\text{NaCl (in water)} + \text{AgNO}_3 \text{ (in water)} \longrightarrow \text{AgCl} + \text{NaNO}_3$$
$$\text{Ag}^+ + \text{Cl}^- \longrightarrow \text{AgCl}$$

Water is necessary in the formation of many compounds.

$$\text{H}_2\text{O} + \text{SO}_2 \longrightarrow \text{H}_2\text{SO}_3$$
$$\text{H}_2\text{O} + \text{Na}_2\text{O} \longrightarrow 2 \text{NaOH}$$

Water in the form of steam is used as a source of power in steam engines. Steam is also used in the preparation of hydrogen, fuel gas, and producer gas. Running streams and falls provide energy for the generation of electricity. In many parts of the world, water is used for irrigation.

COMPOSITION. Water has been studied more than any other compound. Its precise compostion has been of great interest to chemists. The composition of water is expressed in volume units and also in weight units.

Volume Units. If water is decomposed by an electric current, two volumes of hydrogen are obtained for each volume of oxygen. If a mixture of hydrogen and oxygen is made to combine by passing a spark through it, the combustion always takes place in the ratio of two volumes of hydrogen to one volume of oxygen. An excess of either gas remains uncombined.

Weight Units. Repeated analyses show that water contains 88.81 per cent oxygen and 11.19 per cent hydrogen by weight. To show this the following experiment can be performed. Pass dry hydrogen over a weighed amount of pure heated copper oxide and collect the water formed in a weighed tube of anhydrous calcium chloride. The increase in weight of the calcium chloride tube represents the weight of the water formed. Weigh the reduced copper oxide. The loss in weight represents the weight of oxygen combined with the hydrogen. The weight of the water formed, minus the weight of the oxygen, equals the weight of hydrogen in the water. An apparatus for this experiment is shown in Figure 10-3.

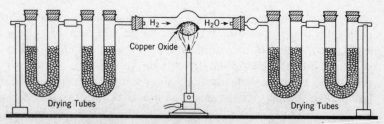

Fig. 10-3. Composition of water by weight.

Heavy Water.   The discovery of the isotope of hydrogen of mass 2 (deuterium) by H. C. Urey has led to some interesting experimental work. E. W. Washburn of the United States Bureau of Standards showed that the "heavy water" in ordinary water can be concentrated by an electrolyic method. H. S. Taylor of Princeton and H. Eyring of Utah prepared practically pure deuterium oxide by this method and studied its properties.

OCCURRENCE.   Ordinary rain water contains about one molecule of deuterium oxide (D_2O) for every 5000 molecules of ordinary water. Water from the Dead Sea, the Great Salt Lake, and the oceans is slightly richer in heavy water than is rain water. The water of crystallization of certain salts (e.g., borax tetrahydrate) is slightly richer in D_2O than is ordinary water. Finally, the concentration of D_2O in the sap from certain trees (e.g., willows) is slightly higher than it is in ordinary water.

LABORATORY PREPARATION.   Heavy water is prepared by the electrolysis of water. The "light water" is decomposed first, leaving the residue to become richer and richer in deuterium oxide as the electrolysis continues.

PHYSICAL PROPERTIES.   The properties of deuterium oxide are appreciably different from those of ordinary water, as the table below indicates.

Table 10-1

Physical Properties of Water and of Deuterium Oxide

Property	H_2O	D_2O
Freezing point	0°C	3.82°C
Boiling point	100°C	101.42°C
Temperature of maximum density	4°C	11.6°C
Maximum density	1.0000 g/ml	1.1073 g/ml
Heat of fusion (mole)	1437 cal	1501 cal
Vapor pressure (20°C)	17.5 mm	15.1 mm

CHEMICAL PROPERTIES.   Experiments have indicated that pure deuterium oxide prevents or retards the germination of certain seeds, such as the tobacco seed. Dilute solutions of D_2O have a slight inhibiting effect on the germination of seedlings. Some experiments, however, indicate that dilute solutions have a stimulating effect on certain organisms. Pure D_2O absorbs moisture from the air. Heavy water reacts with ammonia to give compounds in which the lighter hydrogen atoms are replaced by the heavier ones. Alcoholic fermentation of *d*-glucose is nine times slower in pure D_2O. Typical reactions include the following:

$$D_2O + MgCl_2 \longrightarrow MgO + 2\,DCl$$
$$2\,D_2O + 2\,Na \longrightarrow 2\,NaOD + D_2$$
$$D_2 + H_2 \longrightarrow 2\,HD$$
$$D_2 + CH_4 \longrightarrow CH_3D + HD$$

Uses. Interesting research is being done with deuterium. For example, valuable information concerning the mechanisms of surface (catalytic) reactions has been obtained from experiments involving deuterium. Also, heavy water is used as a *moderator* in nuclear reactions.

Hydrates. Many substances, especially salts, combine with water to form compounds known as *hydrates*. Hydrates are crystalline. Anhydrous copper sulfate (copper sulfate free from water) combines with water to form a blue crystalline substance called *copper sulfate hydrate*, more commonly known as *blue vitriol*. The formula for blue vitriol is $CuSO_4 \cdot 5\,H_2O$. The centered dot used in this formula indicates the water can be removed rather easily.

The stability of hydrates depends upon the temperature and the relative humidity of the atmosphere. If a salt absorbs enough moisture from the air to liquefy, it is said to be *deliquescent*. On the other hand, if crystals of a salt such as sodium carbonate are allowed to stand in a dry place, they give up their water of hydration and are said to be *efflorescent*. Anhydrous calcium chloride, in moist air, will form first a hydrate, then a liquid solution; it is deliquescent. This property of anhydrous calcium chloride makes it useful as a drying agent.

Hydrogen Peroxide. Hydrogen peroxide (H_2O_2) is a compound of hydrogen and oxygen containing 94.11 per cent oxygen. H_2O_2 contains a larger proportion of oxygen than any other compound. It is prepared as a dilute solution by the action of a dilute acid on a metallic peroxide:

$$BaO_2 + H_2SO_4 \longrightarrow BaSO_4 + H_2O_2$$

The insoluble barium sulfate ($BaSO_4$) is filtered off and, if a concentrated solution is desired, the water is evaporated under reduced pressure. This process gives about a 30 per cent solution.

Hydrogen peroxide is an oxidizing agent and is used to bleach silk, feathers, hair, and similar materials. It is used also as a mild disinfectant. The structural formula given in Figure 10-4*b* is preferred.

$$H : \overset{\cdot\cdot}{\underset{\cdot\cdot}{O}} : \overset{\cdot\cdot}{\underset{\cdot\cdot}{O}} : H$$

(a)

$$\overset{\displaystyle H}{\underset{\displaystyle H}{: \overset{\cdot\cdot}{O} : \overset{\cdot\cdot}{O} :}}$$

(b)

Fig. 10-4. Two ways of representing the structure of hydrogen peroxide.

REVIEW QUESTIONS

1. (*a*) How can one prove that water is composed of oxygen and hydrogen? (*b*) Describe an experiment that will prove that upon decomposition, water yields two volumes of hydrogen for each volume of oxygen.
2. How can water be purified on a large scale?
3. How can $CuSO_4$ be converted into $CuSO_4 \cdot 5\ H_2O$?
4. The water molecule is said to be a dipole. What does this mean?
5. Ordinarily we think of hydrogen as having a valence of $+1$. But what is the role of hydrogen in a cluster of water molecules?
6. List commercial methods for the purification of water.
7. How can a dilute table-salt solution be converted into deionized water?
8. What is heavy water? In terms of composition, how does it differ from ordinary water?
9. Distinguish between a deliquescent and an efflorescent substance.
10. What is hydrogen peroxide? How is it prepared?

SOLUTIONS

A solution is a homogeneous body, the composition of which can be varied within limits. Solutions are like compounds in that they are homogeneous. They are like mixtures in that their compositions are variable. Every solution consists of two parts, the *solvent* and the *solute*. The solvent is the substance that dissolves another substance; the solute is the substance that dissolves in the solvent. If the constituents of a solution mix in all proportions, either one may be the solvent, but that constituent in excess is usually so designated. For example, slowly add table salt to water. The salt dissolves until the limit of solubility is reached. The water is the solvent; the salt is the solute. Pour ethyl alcohol into water.

No limit of solubility is reached. The solution may contain more alcohol than water; if so, alcohol is the solvent; otherwise water is the solvent.

Types of Solutions.   There are three types of solutions: *gaseous*, *liquid*, and *solid*. Gases mix with each other in all proportions. These mixtures are homogeneous and, therefore, may be called *gaseous solutions*. For example, air is a solution of oxygen and certain other gases in nitrogen.

Gases, liquids, or solids dissolved in liquid solvents are *liquid solutions*. Air, alcohol, or sugar dissolved in water forms a liquid solution. *Solid solutions* are solutions in which a gas, a liquid, or a solid is dissolved in a solid. Hydrogen will dissolve in palladium to give a solid solution. Carbon dissolves in iron to give a solid solution.

Stability of Solutions.   The stability of solutions under various treatments is shown by the following.

FILTRATION.   Since a solution is homogeneous, the solute cannot be removed by filtration, nor will it ever settle out upon standing. Suspended material, on the other hand, will settle out or may be removed by filtration.

TEMPERATURE.   The composition of solutions is greatly af-

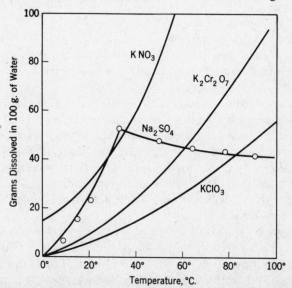

Fig. 10-5. Curves showing effect of temperature on solubility of four different salts in water.

fected by changes in temperature. The following generalizations are useful: Liquids and solids are more soluble in a given solvent with an increase in temperature. (There are many exceptions to this rule.) For example, potassium nitrate is much more soluble in hot than in cold water. Gases are less soluble in hot than in cold water. For example, ordinary drinking water will release most of the dissolved oxygen if the water is warmed. Above 32°C sodium sulfate decreases in solubility with an increase in temperature; below 32°C the hydrated salt ($Na_2SO_4 \cdot 10\ H_2O$) is stable. Between 0°C and 65°C calcium acetate decreases in solubility with an increase in temperature. See Figure 10-5.

PRESSURE. At constant temperature, the weight of a gas dissolved by a given volume of a liquid is proportional to the pressure. This is known as *Henry's law*. Suppose 1 liter of water will dissolve 2 grams of gas at 20°C and 1 atmosphere pressure. If the temperature remains constant, and the pressure is increased to 2 atmospheres, 4 grams of gas will dissolve. Gases such as hydrogen chloride and ammonia do not obey Henry's law. These two gases have a strong chemical attraction for water.

Effect of Solutes on Solvents. When a solute is added to a solvent, the properties of the solvent are changed. When a solute dissolves in a solvent, heat is evolved or absorbed. This is called the *heat of solution*. If concentrated sulfuric acid is poured into water, the solution gets warm. This shows that heat is evolved in the formation of the solution. If ammonium chloride is dissolved in water, the solution becomes cooler, showing that heat is being absorbed.

Solids and certain liquids dissolved in liquids give solutions that *boil at temperatures above that of the pure solvent*. The boiling point of a salt solution is higher than that of pure water. Solids and liquids dissolved in liquids give solutions that *freeze at temperatures below that of the pure solvent*. In winter, enough ethylene glycol is added to the water in the radiator of an automobile to give a solution that does not freeze under winter weather conditions.

The *vapor pressure* on a liquid surface (the pressure due to its own vapor) is lowered by the addition of a nonvolatile solute. At 20°C the vapor pressure of water is 17.54 mm. If table salt is dissolved in the water, the vapor pressure is less than 17.54 mm. The amount it is lowered depends on the quantity of salt in the solution.

The *viscosity*, which is the resistance to flow of a liquid, may be increased or decreased by the addition of solutes. The viscosity of water is increased by the addition of a quantity of sugar.

Note. If the viscosity of water at 25°C is .00893, then the *fluidity* of water at 25°C is 1/.00893 = 111.9+.

Kinds of Solutions. Solutions are distinguished from each other by the following terms.

DILUTE. A dilute solution is one which contains a small proportion of solute. A gram of table salt dissolved in a liter of water is a dilute solution.

CONCENTRATED. A concentrated solution is one which contains (relatively) a large proportion of solute. Twenty grams of table salt dissolved in 100 ml of water is a concentrated solution.

SATURATED. A saturated solution is one in which the molecules of the solute in solution are in equilibrium with undissolved molecules. For example, with the temperature remaining constant, add so much salt to water that part of it remains undissolved. This is a saturated solution. If some of the excess salt should dissolve, a corresponding amount of salt will come out of solution, thus maintaining the equilibrium.

SUPERSATURATED. A supersaturated solution is one in which more solute is in solution than is present in a saturated solution of the same substances at the same temperature and pressure. Saturate a few mililiters of water with sodium thiosulfate at 20°C. Heat the solution to 40°C and add more of the salt. Carefully cool the solution to 20°C. The salt that went into the solution at the higher temperature does not come out of solution. This solution is supersaturated. It is possible to determine whether a solution is unsaturated, saturated, or supersaturated by adding a small crystal of the solute to the solution. If the crystal dissolves, the solution is unsaturated. On the other hand, if the crystal causes the formation of more crystals of solute, the solution is supersaturated.

Methods of Expressing Concentrations. The concentration of a solution may be expressed in a number of ways. The following are in common use.

WEIGHT OF SOLUTE PER VOLUME OF SOLUTION. This is usually expressed in grams per liter. The brine contained 10 grams of salt per liter of solution.

PERCENTAGE COMPOSITION. Percentage composition is the

weight of solute in 100 grams of solution. The solution contained 10 per cent sugar.

MOLARITY.   A molar solution contains one mole (the molecular or formula weight in grams) of solute in a liter of solution. Thus, the formula weight of table salt (sodium chloride) is 58.442. A molar table salt solution, therefore, contains 58.442 g in a liter of solution.

NORMALITY.   A normal solution contains one equivalent of the solute in a liter of solution. An equivalent of an acid in a liter is that weight of the acid necessary to furnish one gram of replaceable hydrogen (H^+). Thus, 36.461 g of hydrochloric acid [which furnishes one gram of replaceable hydrogen or one gram of hydrogen ion (H^+)] in a liter of solution is a normal solution. Likewise the equivalent weight of a base contains 17 g of hydroxyl group (OH^{-1}) in a liter of solution.

Table 10-2 shows the relation between molar and normal solutions. The figures are in grams per liter of solution. The atomic weights were rounded off in arriving at molecular weights used here. Note that for sulfuric (H_2SO_4) and phosphoric (H_3PO_4) acids the molar weight is divided by 2 and by 3 respectively to give the number of grams of these acids (in a liter of solution) necessary to make a normal solution. This is because a mole of H_2SO_4 contains 2 equivalents of H^{+1} whereas a mole of H_3PO_4 contains 3 equivalents of H^{+1}. Likewise the weight of a mole of $Ba(OH)_2$ is divided by 2 because a mole of $Ba(OH)_2$ contains 2 equivalents of OH^{-1}.

Table 10-2
Relation between Molar and Normal Solutions

Substance	Formula	Molar	Normal
Nitric acid	HNO_3	63	63
Sulfuric acid	H_2SO_4	98	49
Phosphoric acid	H_3PO_4	98	32.66
Sodium hydroxide	NaOH	40	40
Barium hydroxide	$Ba(OH)_2$	171.36	85.68

In the preparation of solutions of known concentrations the use of quantitative apparatus is required. These include pipets, volumetric flasks, burets, and analytical balances. See Figures 10-6*a,b,c*.

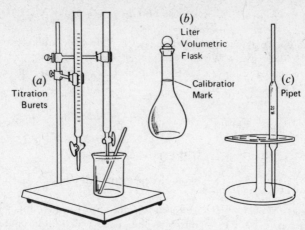

Fig. 10-6. Apparatus used in preparing a solution of precise concentration.

MOLAL SOLUTIONS. A molal solution contains one mole of a solute in 1000 g of solvent (usually water). A mole of cane sugar, sucrose, contains 342 g so this quantity of sugar in 1000 g of water is a molal solution. Molal solutions are considered in discussing the colligative properties of solutions. See Chapter 11.

STANDARD SOLUTIONS. Any solution whose composition is known is a standard solution. For example, a molar, a normal, or a molal solution of an acid or a base is a standard solution.

Problems Based on Molarity, Normality, and Molality

Molarity Problem. A liter of a water-ethyl alcohol solution contains 9.2 grams of alcohol. What is its molarity?

Solution: The molecular weight of alcohol (C_2H_5OH) is 46 so a molar solution contains 46 g/liter.

Let x = molarity of the solution.

$$\frac{x}{9.2\,g} = \frac{1M}{46\,g} \qquad x = \frac{1M \times 9.2}{46} = 1M \times .2 = .2M \quad Ans.$$

Normality Problem. Exactly one liter of hydrochloric acid solution contains 9.125 g of HCl. What is its normality? Since 36.5 g HCl in a liter of solution is a normal solution (see definition above) then our solution is considerably less than normal.

Solution: Let x = the normality.

$$\frac{x}{9.125\,g} = \frac{1N}{36.5\,g} \qquad x = \frac{1N \times 9.125}{36.5} = 0.25N \quad Ans.$$

Molality Problem. A solution contains 9 g glucose ($C_6H_{12}O_6$) in 250 g water. What is its molality?

Solution: 9 g of glucose in 250 g water has the same molality as 36 g glucose in 1000 g water. We know from the formula of glucose that 180 g glucose in 1000 g water is a molal solution. Let x = the molality of the solution. Then

$$\frac{x}{36\,g} = \frac{1M}{180\,g} \qquad x = \frac{1M \times 36}{180} = 0.2\,\text{Molal} \quad Ans.$$

REVIEW QUESTIONS

1. Describe a method of making a supersaturated solution of sodium acetate.
2. If a glass of cold water is allowed to stand in a warm room, a large number of gas bubbles collect on the inside wall of the container. Explain.
3. How can one distinguish between an unsaturated, a saturated, and a supersaturated solution?
4. In what respect is a solution like a compound? Like a mixture?
5. For Fig. 10-5 explain the unusual curve for sodium sulfate. *Hint:* Note the change in composition that occurs at 32°C.
6. State Henry's law.
7. Distinguish between a molar and a normal solution of (*a*) hydrochloric acid, and (*b*) sulfuric acid.
8. Distinguish between a standard solution and a molal solution.

Chapter 11
Solutions of Nonelectrolytes and of Electrolytes

Nonelectrolytes. A nonelectrolyte is a substance that, when dissolved in water, gives a solution that is a nonconductor of electric current. In general, if nonvolatile solutes are used, the properties of their solutions depend on the *number* of solute particles present, say in 1000 g of water, and not on their chemical composition or their size. These properties are referred to as *colligative properties.* Such properties include (1) the lowering of the freezing point, (2) the raising of the boiling point, (3) the lowering of the vapor pressure, and (4) increasing the osmotic pressure of such solutions. One mole (expressed in grams) of any solute contains 6.02×10^{23} particles (molecules). This number is known as *Avogadro's number.* See Table 11-1.

Table 11-1
The Colligative Properties of Certain Solutions

Substance	Formula	Weight of 1 Mole	Number of Particles	Weight of Solvent	Freezing Point	Boiling Point
Urea	$Co(NH_2)_2$	60 g	6.02×10^{23}	1000 g water	$-1.86°C$	$+100.518°C$
Glucose	$C_6H_{12}O_6$	180 g	6.02×10^{23}	1000 g water	$-1.86°C$	$+100.518°C$
Sucrose	$C_{12}H_{22}O_{11}$	342 g	6.02×10^{23}	1000 g water	$-1.86°C$	$+100.518°C$

Note: $-1.86°C$ and $+0.518°C$ are sometimes referred to as the *freezing point constant* and the *boiling point constant* respectively.

Problem. This problem is based on data obtained from Table 11-1. What is the freezing point of a solution that contains 6 g urea in 1000 g of water?

Solution: The freezing point constant is $-1.86°C$. Six grams of urea is 0.1 mole so the solution should freeze at $-0.186°C$. Using an equation and letting x = the freezing point of the solution:

$$\frac{x}{6\,g} = \frac{-1.86°C}{60\,g} \qquad x = \frac{-1.86°C \times 6}{60} = -0.186°C \quad Ans.$$

VAPOR PRESSURE.   One mole of any solute dissolved in 1000 grams of water lowers its vapor pressure a definite amount. *Raoult's law* states that the depression of the vapor pressure is proportional to the number of molecules of solute dissolved in a given weight of solvent.

OSMOTIC PRESSURE.   If a concentrated solution is separated from some of the pure solvent by a semipermeable membrane, the solvent molecules pass through the membrane and dilute the solution. The force which brings about this dilution is called *osmotic pressure*. Solutions of equal molecular concentration have the same osmotic pressure. For example, the osmotic pressure of a molal solution of urea, $CO(NH_2)_2$, will be the same as the osmotic pressure of a molal solution of sucrose, $C_{12}H_{22}O_{11}$.

Properties of Solutions of Electrolytes.   Electrolytic solutions consist chiefly of acids, bases, and salts dissolved in water. The boiling points, freezing points, vapor pressures, and osmotic pressures of these solutions are abnormal when compared with solutions of nonelectrolytes. A molal sugar solution (a non-electrolyte) freezes at $-1.86°C$, but a molal solution of sodium chloride (an electrolyte) freezes at $-3.42°C$. As pointed out above, the lowering of the freezing point depends on the number of particles (molecules or units) present in the solution. Since the freezing point of molal sodium chloride is considerably below that of the molal sugar solution, it appears that the sodium chloride solution contains more units (particles) than the sugar solution. This is true, as will be explained below.

CONDUCTIVITY.   Substances (acids, bases, and salts) which in solution show abnormal physical properties are conductors of an electric current. They are called *electrolytes*. If two metallic electrodes are placed in a solution of electrolytes and a storage battery or some other source of electricity is connected to them, a current passes through the solution, causing its decomposition. This process is called *electrolysis* and may be shown by the presence of decomposition products at the electrodes. For example, if an electric current is passed through a hydrochloric acid solution, chlorine is set free at the anode and hydrogen escapes at the

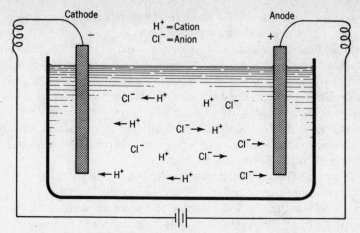

Fig. 11-1.   Electrical conduction of solutions of electrolytes.

cathode. Dry acids, bases, or salts do not conduct the current unless they are fused (melted). Also, if they are dissolved in solvents other than water, they are nonconductors. (Exceptions are known.)

RATE OF REACTIONS.   Reactions between electrolytes are very rapid. If a solution of silver nitrate is mixed with a solution of table salt, the white precipitate of silver chloride appears immediately.

Classical Theory of Electrolytes.   In 1887, the Swedish chemist Arrhenius announced his theory of electrolytic dissociation. It may be stated as follows:

Electrolytes (acids, bases, and salts) when dissolved in water break up into smaller, electrically charged units called *ions*. These are charged atoms or groups of atoms.

Dissociation (or ionization) takes place as the substance dissolves. It is evident that the solvent has something to do with dissociation, since acids, bases, and salts dissolved in solvents other than water do not ionize or are ionized very slightly.

Hydrogen and metallic ions are positively charged and are called *cations*. The nonmetals and acid radicals are negatively charged and are called *anions*.

The degree of dissociation of a substance depends upon the concentration of the solution. As the solution becomes more dilute, the substance (solute) becomes more completely ionized.

There is an equilibrium between the undissociated molecules

and the charged ions. The positive and negative ions attract each other, but water tends to keep them apart.

This theory is quite satisfactory for weak electrolytes, but unsatisfactory for strong electrolytes. Weak electrolytes are only slightly ionized in solution, whereas strong electrolytes are probably 100 per cent ionized in solution. We can calculate the percentage of ionization of a weak electrolyte in solution by the lowering of the freezing point method, but such calculations are meaningless for solutions of strong electrolytes. The next paragraph or two will explain why this is so.

Modern Theory of Strong Electrolytes.   Electrovalent substances (ionic compounds) exist as ions in the solid (crystalline) state. When these substances are dissolved in water, the ions separate from the crystal and diffuse through the solution. The force of attraction between the ions is small because of the high *dielectric constant* of the water. The salts and some of the alkalies belong to this class.

Covalent compounds, such as HCl, HNO_3, and other acids, give up protons to solvent molecules. Thus, a water molecule attached to a proton becomes a positive hydronium ion (H_3O^+), and a proton attached to NH_3 becomes an ammonium ion (NH_4^+). Examples follow.

$$HCl + H_2O \rightleftharpoons H_3O^+ + Cl^-$$
$$HNO_3 + H_2O \rightleftharpoons H_3O^+ + NO_3^-$$
$$H_2SO_4 + H_2O \rightleftharpoons H_3O^+ + HSO_4^-$$
$$HCl + NH_3 \rightleftharpoons NH_4^+ + Cl^- \text{ (in water)}$$

DEBYE-HÜCKEL THEORY.   The electrolyte is 100 per cent dissociated (ionized) in solution. The electrolyte appears to be only partially ionized (as determined by conductivity measurements or by freezing point experiments) because the ions are not able to move freely through the solution. The reason for this is that each ion is surrounded by more ions of opposite charge than of the same charge. This means that the ions are subjected to electrical forces which restrict their freedom.

The decrease in equivalent conductivity of strong electrolytes with increasing concentration is attributed to a decreased migration velocity of the ions rather than to a decrease in the number of ions. The migration velocity of ions with higher valences will be less than that of ions with univalence, because the attracting forces between ions will be greater.

ACTIVITY. In order to correct for the electrical effects of the ions upon each other, it is necessary to multiply the concentration of the ions by a factor which is called the *activity coefficient*. The concentration times the activity coefficient equals the activity of the particular ion.

Explanation of Behavior of Electrolytes. The boiling points, freezing points, vapor pressures, and osmotic pressures of solutions of nonelectrolytes depend on the number of solute molecules per unit volume. One mole of any solute in 1000 grams of water produces the same effect because there are the same number of molecules present. If a mole of HCl is added to 1000 g of water, 6.02×10^{23} H_3O^+ ions and 6.02×10^{23} Cl^- ions are present. Each ion is about as effective as a dissolved molecule in changing the properties of the solution. Since electrovalent compounds (salts) are not composed of molecules, it is necessary to assume that a mole of a salt is that weight in grams which corresponds to the simplest formula. Thus, for sodium chloride 58.44 g is a mole, based on the formula NaCl.

INSTANTANEOUS REACTIONS. These are reactions between ions. For example, there is a white precipitate formed the instant a silver nitrate solution mixes with a table salt solution. In this case, silver ions react with chloride ions to give insoluble silver chloride. An elementary course in qualitative analysis is largely a study of the methods of separating and identifying the common cations and anions. See Appendixes VIII and IX.

ELECTROLYSIS. When an electric current passes through a solution of an electrolyte, the current is carried by the ions. The positive ions (cations) are attracted to the negative electrode called the *cathode*, and the negative ions (anions) are attracted to the positive electrode called the *anode*.

Ions and Ionic Reactions. Ions are represented by adding superscript plus $(+)$ or minus $(-)$ signs to the symbol of the element or the formula of the radical in question. Thus, sodium ion is Na^+; calcium ion is Ca^{++}; phosphate ion is PO_4^{---}. (Some textbooks show sodium ion as Na^{+1}, calcium ion as Ca^{+2}, phosphate ion as PO_4^{-3}. Both methods are satisfactory.) The number of charges on an ion is equal to the electrovalence (oxidation number) of the atom or radical. From the standpoint of atomic structure, an ion is an atom (or group of atoms) that has lost or gained one or more electrons. Thus, the sodium ion is the sodium atom minus one electron from its outer energy level. The chloride ion is the chlorine atom with an extra electron.

ACTION OF WATER ON ACIDS.   The proton, dissociated from the acid molecule, combines with a molecule of water, forming the hydronium ion, H_3O^+.

ACTION OF WATER ON SALTS.   When salts are added to water, the positive and negative ions diffuse from the crystal lattice and become surrounded with water molecules.

PROTON-BASE REACTION.   This is a reaction in which a proton (or, more likely, H_3O^+) combines with a base. If the base is the hydroxyl ion, 13,780 calories of heat are liberated for each 18 g of water formed.

HYDROLYSIS.   Hydrolysis is a reaction between a substance (salt) and water. Covalent compounds such as starch are broken down into simpler substances.

$$(C_6H_{10}O_5)x + x\,H_2O \rightleftharpoons x \cdot C_6H_{12}O_6$$
$$\text{starch} \qquad\qquad\qquad \text{glucose}$$

Electrovalent compounds such as salts may react with water to give solutions that are acidic or basic to indicators.

Aluminum Chloride, $Al^{+++}(Cl^-)_3$.

$$Al^{+++} + 3\,Cl^- + 7\,H_2O \rightleftharpoons Al(OH)_3 + 3\,H_3O^+ + 3\,Cl^- + H_2O\ (acidic)$$
$$Al^{+++} + 6\,H_2O \rightleftharpoons Al(OH)_3 + 3\,H_3O^+\ (acidic)$$

Sodium Carbonate, $(Na^+)_2\,CO_3^{--}$.

$$2\,Na^+ + CO_3^{--} + 2\,H_2O \rightleftharpoons H_2CO_3 + 2\,OH^- + 2\,Na^+\ (basic)$$
$$CO_3^{--} + H_2O \rightleftharpoons HCO_3^- + OH^-$$

Sodium Chloride, Na^+Cl^-.

$$Na^+ + H_2O \longrightarrow \text{No reaction}$$
$$Cl^- + H_2O \longrightarrow \text{No reaction}$$

As a rule, hydrolytic reactions reach equilibrium after a small percentage of salt has been used up in the reaction.

MODERN EXPLANATION OF HYDROLYSIS.   A salt solution is neutral, or acidic, or basic for the following reasons:

Neutral.   The solution is neutral if the cation is a weak acid and the anion is a weak base. (See definition on p. 63). NaCl in water is neutral because Na^+ is weakly acidic and Cl^- is weakly basic.

Acidic. The solution is acidic if the acid ion is stronger than the basic ion. NH_4Cl solution is acidic because NH_4^+ is a stronger acid than Cl^- is a base.

Basic. The solution is basic if the basic ion is stronger than the acidic ion. $NaC_2H_3O_2$ solution is basic because the Na^+ is weakly acidic whereas $C_2H_3O_2^-$ is a rather strong base.

HYDRONIUM ION CONCENTRATION IN SOLUTIONS. The hydronium ion concentration $[H_3O^+]$ of water solutions varies from 1×10^{-14} gram-ion per liter in a strong basic solution to 1 gram-ion in a strong acid solution. A solution whose $[H_3O^+]$ and $[OH^-]$ are 1×10^{-7} gram-ion per liter is neutral. A solution whose $[H_3O^+]$ is greater than 1×10^{-7} is acidic, whereas one whose $[H_3O^+]$ is less than 1×10^{-7} is basic. We say that hydrolysis has taken place in any salt solution where the $[H_3O^+]$ is greater than or less than 1×10^{-7} gram-ion per liter.

Although the H_3O^+ (hydronium ion) is found in all water solutions of acids, there is still some tendency on the part of chemists to think of water solutions of acids as containing hydrogen ions (H^+).

The pH of a Water Solution. Pure water is slightly ionized into H_3O^{+1} and OH^{-1}. The reaction is:

$$H_2O \rightleftharpoons H^+ + OH^{-1}$$

Adding these together

$$\frac{H_2O + H^+ \longrightarrow H_3O^{+1}}{2 H_2O \rightleftharpoons H_3O^{+1} + OH^{-1}}$$

The reverse arrows ($\rightleftharpoons$) mean that the forward and reverse reactions take place simultaneously. At equilibrium the concentration of H_3O^{+1} is 1×10^{-7} mole/liter and likewise the concentration of the OH^{-1} is 1×10^{-7} mole/liter. The equation is

$$\frac{[H_3O^{+1}][OH^{-1}]}{[H_2O][H_2O]} = K$$

Since the concentration of the water is essentially constant, the above equation can be written $[H_3O^+][OH^-] = K_w$ where K_w is known as the water constant or the *ion product for water*. In pure water we have $[1 \times 10^{-7}][1 \times 10^{-7}] = 1 \times 10^{-14}$. If the $[H_3O^+]$ is higher than 1×10^{-7} mole/liter the solution is acidic. If the value is less than 1×10^{-7} the solution is basic. Remember that the *product* of the $[H_3O^+]$ and $[OH^-]$ must equal 1×10^{-14}. This means that in every water solution there are hydronium and hydroxyl ions present. Whether the solution is acidic or basic depends on their concentrations. Since calculations involving small numbers such as 1×10^{-5} or 2.5×10^{-6} are rather inconvenient, the use of pH in expressing acidity or basicity was invented. *The pH of a solution is defined as the logarithim (base 10) of the reciprocal of the hydronium ion concentration expressed in moles per liter.* Mathematically we write:

$$pH = -\log_{10} [H_3O^+] \quad \text{or} \quad pH = \log_{10} \frac{1}{[H_3O^+]}$$

HYDRONIUM ION
CONCENTRATION

$[H_3O^+]$	pH
1×10^{-7}	7
1×10^{-6}	6
1×10^{-5}	5
1×10^{-9}	9

Problem. What is the pH of a solution if its hydronium ion concentration is .005M?

Solution. The $[H_3O^{+1}] = 5 \times 10^{-3}$

$$pH = \log \frac{1}{5 \times 10^{-3}} = \log \frac{10^3}{5}$$

$$\log 10^3 - \log 5 = 3 - .7 = 2.3 = pH$$

Problem. What is the $[H_3O^{+1}]$ of a solution that has a pH of 7.5?

Solution: $7.5 = \log \frac{1}{[H_3O^+]}$

The antilogarithm of 7.5 is 3.16×10^7.

$$\frac{1}{[H_3O^+]} = 3.16 \times 10^7$$

$$[H_3O^+] = \frac{1}{3.16 \times 10^7}$$

$$= .316 \times 10^{-7}$$
$$= 3.16 \times 10^{-8}$$

The hydroxyl ion concentration $[OH^{-1}]$ of a solution can be expressed as the pOH of the solution thus:

$$pOH = -\log_{10}[OH^-] \quad \text{or} \quad pOH = \log_{10}\frac{1}{[OH^-]}$$

We have seen that $[H_3O^+][OH^-] = 1 \times 10^{-14}$. It follows that

$$pH + pOH = 14$$

Problem. What is the pOH of a solution whose $[OH^-] = 2 \times 10^{-8}$?

Solution: $pOH = \log_{10}\dfrac{1}{[OH^-]} \quad \text{or} \quad pOH = \log_{10}\dfrac{1}{2 \times 10^{-8}}$

$$= \log\frac{10^8}{2}$$

$$\log 10^8 - \log 2 = 8 - .3 = 7.7 = pOH$$

Problem. What is the $[OH^{-1}]$ of a solution whose pOH is 8.2?

Solution: $pOH = 8.2 = \log\dfrac{1}{[OH^-]}$

The antilogarithm of 8.2 is 1.58×10^8.

$$\frac{1}{[OH^-]} = 1.58 \times 10^8$$

$$[OH^-] = \frac{1}{1.58 \times 10^8}$$

$$= .633 \times 10^{-8}$$

$$= 6.33 \times 10^{-9}$$

Table 11-2

The pH and pOH of a Number of Substances

Substance	$[H_3O^+]$	$[OH^-]$	pH	pOH
Pure water	10^{-7}	10^{-7}	7	7
Milk of magnesia	10^{-9}	10^{-5}	9	5
Orange juice			4.7	9.3
Baking soda			8.5	5.5
Lemon juice			2.5	11.5
1 molar HCl solution	10^0	10^{-14}	0	14
1 molar NaOH solution	10^{-14}	10^0	14	0

Problem. If the pH of orange juice is 4.7 what is its $[H_3O^+]$?

Solution: $4.7 = \log\dfrac{1}{[H_3O^+]}$

The antilogarithm of 4.7 is 5.0×10^4.

$$\frac{1}{[H_3O^{+1}]} = 5.0 \times 10^4$$

$$[H_3O^+] = \frac{1}{5.0 \times 10^4}$$

$$= .2 \times 10^{-4}$$
$$= 2.0 \times 10^{-5}$$

Indicators. An indicator is a substance that changes color at a definite pH (hydronium ion—or hydroxyl ion—concentration). The more common indicators change color in solutions in which the hydronium ion concentration is near that of pure water (1×10^{-7} gram-ion per liter, or pH = 7).

In titrating acids with bases, use an indicator which will change color when the solution is of the same $[H_3O^+]$ as that of a solution made by dissolving some of the (formed) salt in water. (Review hydrolysis.)

Titration	Reaction	Indicator
NaOH vs. HCl	none	Litmus, methyl orange, or phenolphthalein
NaOH vs. H·C₂H₃O₂	basic	Phenolphthalein
NH₄OH vs. HCl	acidic	Methyl orange

Table 11-3

Colors of Indicators

	Acidic	Neutral	Basic
pH⟶	1 2 3 4 5 6	7 8 9 10 11	12 13 14 ⟵pH
Litmus	Red⟶	T⟵Blue	
Phenolphthalein	Colorless⟶	T⟵Red	
Methyl orange	Pink-red⟶T⟵	Orange	
Methyl violet	Y,G,B⟶T⟵	Violet	
Alizarin yellow	Colorless⟶	T⟵Yellow	

T = Transition

Y = Yellow   G = Green   B = Blue

Buffer Solutions. A buffer solution is obtained when anions of a weak acid are added to a solution of that acid and when cations of a weak base are added to a solution of that base. These buffer solutions, or buffers, resist appreciable changes in pH when hydronium ion (H_3O^+) or hydroxyl ion (OH^-) is added.

Example. To make a buffered acetic acid solution, a suitable quantity of sodium acetate ($NaC_2H_3O_2$) is added to the solution. This buffered solution has a definite pH. If a few drops of tenth normal hydrochloric acid are added to it, the pH remains very nearly the same. The following equation explains what happens:

$$H_3O^+ + C_2H_3O_2^- \rightleftharpoons HC_2H_3O_2 + H_2O$$

The added H_3O^+ ions react with the excess acetate ions ($C_2H_3O_2^-$) to give un-ionized acetic acid. Consequently the hydronium ion (H_3O^+) concentration does not change appreciably and therefore the pH remains nearly the same.

Note. There are many examples of buffer action in nature. Buffers in human blood, for example, keep its pH very nearly at 7.4.

REVIEW QUESTIONS

1. For nonelectrolytic solutions, what is meant by a boiling point constant? A freezing point constant?
2. Water solutions of acids, bases, and salts are called electrolytic solutions. Why?
3. What is the fundamental reaction that takes place when an acid neutralizes a base?
4. Distinguish between a strong acid and a weak acid.
5. What are the important postulates of the Arrhenius theory of electrolytic dissociation?
6. What are ions? anions? cations?
7. What is the modern theory of electrolytic dissociation for strong electrolytes?
8. What happens when a salt dissolves in water? What happens when a salt reacts with water?
9. What is meant by a proton-base reaction?
10. What is meant by the pH of a solution?
11. What is the pH of a solution if the hydronium ion concentration is 7×10^{-5}? *Ans.* pH = 4.16.
12. In titrating a strong base with a strong acid, what two indicators would give satisfactory results? Check with Table 11-3.

Chapter 12
Chemical Equilibrium

Irreversible and Reversible Reactions. Chemical reactions may be classified under the following headings.

IRREVERSIBLE. An irreversible chemical reaction is one which goes to completion. That is, the products of the reaction do not react directly to give the original substances. Burn magnesium in oxygen:

$$2\,Mg + O_2 \longrightarrow 2\,MgO$$

Decompose sugar by heating:

$$C_{12}H_{22}O_{11} + heat \longrightarrow 12\,C + 11\,H_2O$$

Some chemists maintain that all chemical reactions are reversible. The reason why a reaction such as the oxidation of magnesium appears to be complete is that no experimental method is sensitive enough to detect the reverse reaction. The equilibrium constant for such reactions can be readily calculated.

REVERSIBLE. The products formed in reactions of this type may react directly with each other to give the original substances. Reversible reactions may be divided into two groups.

Complete. The reaction goes to completion, or nearly so, because an insoluble precipitate is formed, or a gas is liberated.

In the reaction between silver ion and chloride ion, the silver chloride formed is very slightly soluble; consequently, the reverse reaction is very slight. This is indicated by a heavy arrow pointing to the right and a light arrow pointing left.

$$Ag^{+1} + Cl^{-1}\,(water) \rightleftharpoons AgCl\,(water)$$

In the reaction between sodium chloride and sulfuric acid, the hydrogen chloride formed escapes as a gas. The reaction is incomplete if the gas is not allowed to escape.

$$NaCl + H_2SO_4 \rightleftharpoons NaHSO_4 + HCl$$

Incomplete. In incomplete reactions, the products react to give the original substances. At first, the forward reaction is rapid and the reverse reaction slow. Then, as more molecules of the products are formed, and fewer molecules of the original substances remain, the reverse reaction increases in speed and the speed of the forward reaction decreases. Finally, both reactions are taking place at the same speed. Examples:

$$Cl_2 + 2 H_2O \rightleftharpoons H_3O^{+1} + Cl^{-1} + HClO$$
$$SO_2 + H_2O \rightleftharpoons H_2SO_3$$

Factors Affecting Equilibrium. Chemical equilibrium is characterized by the sameness in the rates of the two opposing reactions. In order to displace the equilibrium, it is necessary to change one of the reaction rates with respect to the other.

TEMPERATURE. The rate of a chemical reaction is usually doubled for each 10 Centigrade degrees rise in temperature. Where this is true, the chemist says that the *temperature coefficient* is 2.

Assume that when solutions of sodium thiosulfate ($Na_2S_2O_3$) and hydrochloric acid (HCl) are mixed, one gram of sulfur is precipitated in 10 minutes at 20°C and in 5 minutes at 30°C. That is:

$$\frac{\text{Time for precipitating 1 g sulfur at 20°C}}{\text{Time for precipitating 1 g sulfur at 30°C}} = \frac{10}{5} = 2$$

Increasing the temperature of a system in chemical equilibrium increases the rates of both reactions, but not equally. This is due to the fact that one of the reactions will always be *endothermic* and the other *exothermic*. (See p. 167.) The law which describes what happens is a special case of Le Chatelier's Principle and is known as *van't Hoff's law*. It states that when the temperature of a system in equilibrium is raised, the equilibrium point is displaced in the direction that absorbs heat.

LIGHT. Many reactions are caused, or accelerated, by light. These are called *photochemical reactions*. It is difficult to determine the effect of light on chemical equilibrium since the opposing reactions may not be sensitive to the same wave length of light. Hydrogen peroxide is preserved in dark-brown bottles to prevent its decomposition by light. Silver chloride is reduced to metallic silver by light.

PRESSURE. An increase in pressure has very little effect on the rate or completeness of reactions taking place in the solid or liquid states. Many gaseous reactions, on the other hand, are appreciably affected by changes in pressure. If equilibrium reactions involve substantial volume changes, then a change of pressure can shift the equilibrium point considerably. This is explained by the Principle of Le Chatelier. See p. 131.

$$N_2 + 3 H_2 \rightleftharpoons 2 NH_3$$
$$\text{4 vols.} \qquad \text{2 vols.}$$

For the above reaction, an increase of pressure favors the formation of ammonia.

$$N_2 + O_2 \rightleftharpoons 2 NO$$
$$\text{2 vols.} \qquad \text{2 vols.}$$

In the second reaction, an increase of pressure has no effect because there is no volume change.

CATALYSTS.   A catalyst is a substance which aids in a chemical reaction without itself being permanently changed. At the end of the reaction, the catalyst may be recovered and used again. The catalyst does not change the final equilibrium, since it alters the rates of the two opposing reactions by equal amounts; it simply changes the rate at which equilibrium is approached.

Many commercial processes for the manufacture of various substances depend on the use of catalysts. For example, wood alcohol is made by the direct combination of hydrogen and carbon monoxide in the presence of a catalyst, zinc oxide.

Positive Catalysts.   A positive catalyst increases the rate of a chemical reaction. Manganese dioxide is a positive catalyst for the decomposition of potassium chlorate.

$$2 KClO_3 \, (400°C) \longrightarrow 2 KCl + 3 O_2 \, (\text{slowly})$$
$$2 KClO_3 + [MnO_2] \, (400°C) \longrightarrow$$
$$2 KCl + 3 O_2 + [MnO_2] \, (\text{rapidly})$$

Negative Catalysts.   These are substances which cut down the rate of chemical reactions. The rate of decomposition of hydrogen peroxide is reduced greatly by the addition of sulfuric or phosphoric acids. Negative catalysts are often called *inhibitors*.

CONCENTRATION.   The *law of mass action* states that the rate of a chemical reaction is proportional to the molecular concentration of the reacting substances.

Case 1:   Increase in concentration of one of the reacting substances. Assume that a state of equilibrium has been reached in the following:

$$H_2 + I_2 \rightleftharpoons 2\,HI$$

The forward and reverse reactions are of the same velocity. If, now, an excess of hydrogen is added, the rate of the forward reaction is increased for the time being because the number of molecules of hydrogen in unit volume is greater. The equilibrium is said to have shifted to the right, i.e., in the direction of the HI.

Case 2: Decrease in concentration of one of the reacting substances. One substance may be removed from the field of action by formation of a precipitate. In the following reaction, the copper hydroxide, $Cu(OH)_2$, is insoluble.

$$CuSO_4 + 2\,NaOH \longrightarrow Cu(OH)_2 + Na_2SO_4$$

One substance may be removed by the formation of gas. In the following reaction, the HCl is a gas and escapes.

$$NaCl + H_2SO_4 \longrightarrow NaHSO_4 + HCl\uparrow$$

By regulating the amount of water used in a reaction mixture, the equilibrium point can be shifted over a considerable range.

Ions may be removed through the formation of slightly ionized substances. Thus, H_3O^{+1} will react with OH^{-1} and un-ionized water will form.

$$H_3O^{+1} + Cl^{-1} + Na^{+1} + OH^{-1} \longrightarrow Na^{+1} + Cl^{-1} + 2\,H_2O$$

PHYSICAL STATE. The physical state of a substance has a direct bearing on the rate at which it will react with another substance. For example, a ball of iron weighing one gram reacts very slowly with oxygen. It may take days for the iron to be completely oxidized. On the other hand, if the gram of iron is in the form of a fine powder, it reacts with oxygen so rapidly that the mass may be heated to incandescence.

Le Chatelier's Principle (Law). This is a qualitative rule which enables one to predict the effect of temperature, pressure, and concentration changes on chemical reactions. It may be stated as follows: *A system in equilibrium, if disturbed by external factors, such as temperature or pressure, will adjust itself in such a way that the effect of the disturbing factors will be reduced to a minimum.* Thus, when a system in equilibrium is subjected to an increase in pressure, it adjusts itself so that it will occupy less volume. This will offset the pressure increase. If ice is placed under an increased pressure, it melts because the water obtained from a given mass of it occupies less volume. In the formation of ammonia from hydrogen and nitrogen, the product of the reaction (NH_3) occupies less volume than the two uncombined gases. An increase in pressure, therefore, favors the production of ammonia.

The Equilibrium Constant. If, when equilibrium has been reached, the product of the concentrations of the substances produced (in the reaction) is divided by the product of the concentrations of the unreacted (starting) substances, a number is obtained called the *equilibrium constant* (K).

A specific example is used in the following derivation of the equilibrium constant. Ethyl alcohol and acetic acid react to give ethyl acetate and water. The reaction is incomplete, so we may write it as follows:

$$C_2H_5OH + H \cdot C_2H_3O_2 \rightleftharpoons C_2H_5 \cdot C_2H_3O_2 + H_2O$$

At equilibrium, the forward and reverse reaction velocities are equal. According to the mass law, the velocity of a reaction at any moment is proportional to the molecular concentrations of the reacting molecules. In the above reaction, by introducing a proportionality constant (k) the velocity of the forward reaction (V_1) is:

$$V_1 = k_1 \times [C_2H_5OH] \times [H \cdot C_2H_3O_2]$$

The brackets enclosing the formulas are used as a symbol of molecular concentration. Thus $[C_2H_5OH]$ is read, "the molecular concentration of ethyl alcohol."

The velocity of the reverse reaction (V_2) is:

$$V_2 = k_2 \times [C_2H_5 \cdot C_2H_3O_2] \times [H_2O]$$

where k_2 is a proportionality constant. Since at equilibrium the velocities of the forward and reverse reactions are equal, we may write:

$$k_1 \times [C_2H_5OH] \times [H \cdot C_2H_3O_2] =$$
$$k_2 \times [C_2H_5 \cdot C_2H_3O_2] \times [H_2O]$$

This may also be written:

$$\frac{k_1}{k_2} = K = \frac{[C_2H_5 \cdot C_2H_3O_2] \times [H_2O]}{[C_2H_5OH] \times [H \cdot C_2H_3O_2]}$$

CALCULATION OF K.   When 1 mole of ethyl alcohol and 1 mole of acetic acid are mixed and the reaction is allowed to reach equilibrium, we find upon analysis the following:

$$[C_2H_5OH] = 1/3 \text{ mole} \qquad [H \cdot C_2H_3O_2] = 1/3 \text{ mole}$$
$$[C_2H_5 \cdot C_2H_3O_2] = 2/3 \text{ mole} \qquad [H_2O] = 2/3 \text{ mole}$$

Substituting these values in the equation we have:

$$K = \frac{2/3 \times 2/3}{1/3 \times 1/3} = \frac{\frac{4}{9}}{\frac{1}{9}} = 4$$

See Figure 12-1.

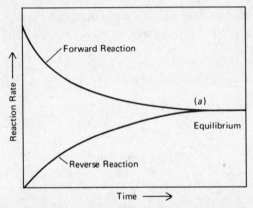

Fig. 12-1. A schematic representation of an equilibrium reaction: $A + B \rightleftharpoons C + D$. At the beginning the forward reaction rate is high but the reverse reaction rate is very slow. As time passes the forward reaction goes slower and the reverse reaction goes faster. At point (a) the two reactions are at the same rate and a state of equilibrium exists.

APPLICATIONS. Suppose, in the above reaction, the concentration of the acetic acid is greatly increased. If the value of K remains constant, there must be a change in the concentrations of the other substances. The concentration of the ethyl alcohol is decreased, and the concentrations of the ethyl acetate and the water are increased. In other words, the forward reaction is made more nearly complete by increasing the concentration of the acetic acid.

Ionization Constant (K_i). An ionization constant is the equilibrium constant for an ionic equilibrium. The ionization constant is the product of the concentration of the ions divided by the concentration of the un-ionized molecules of solute.

The ionization constant (K_i) for acetic acid is obtained as follows:

$$H \cdot C_2H_3O_2 + H_2O \rightleftharpoons H_3O^+ + C_2H_3O_2^-$$

Therefore, the equation for the equilibrium constant (K) is:

$$K = \frac{[H_3O^+] \times [C_2H_3O_2^-]}{[H \cdot C_2H_3O_2] \times [H_2O]}$$

In another form we have:

$$K[H_2O] = K_i = \frac{[H_3O^+] \times [C_2H_3O_2^-]}{[H \cdot C_2H_3O_2]}$$

Since the concentration of the water does not change appreciably, the product $K[H_2O]$ is a constant which we shall call the ionization constant (K_i).

Problem 1. A molar solution of acetic acid is .42 per cent ionized. What is the ionization constant?
Solution:

$$\frac{.0042 \times .0042}{1 - .0042} = .0000177 = K_i$$

Problem 2. A 0.1 molar solution of acetic acid is 1.33 per cent ionized. What is the ionization constant K_i?
Solution:

$$\frac{[.1 \times .0133][.1 \times .0133]}{.1 - .00133} = \frac{.00133 \times .00133}{.09867} = .0000179 = K_i$$

Compare the answers to Problems 1 and 2. It is apparent that the values for K_i are essentially the same. That is, K_i is a constant. It should be noted that the mass law does not hold for strong electrolytes. For example, there is no ionization constant for HCl or other strong acids.

Table 12-1

Determination of Ionization Constant for Acetic Acid

Exp. No.	$[H \cdot C_2H_3O_2]$ Start	$[H \cdot C_2H_3O_2]$ Eq.	$[H_3O^+]$ Eq.	$[C_2H_3O_2^-]$ Eq.	Constant
1	00980	00938	000415	000415	1.83×10^{-5}
2	00344	00320	000240	000240	1.81×10^{-5}
3	001364	001216	000148	000148	1.80×10^{-5}
4	0000280	0000130	0000150	0000150	1.73×10^{-5}

Solubility Product Constant ($K_{s.p.}$). The product of the concentrations of the ions of a slightly soluble salt is equal to a number called the *solubility product constant*. In a saturated solution of silver chloride, the dissolved salt is 100 per cent ionized, and so the concentration of the chloride ion multiplied by the concentration of the silver ion is equal to a constant $K_{s.p.}$.

Problem 1. The solubility of AgCl in water is 1×10^{-5} mole per liter. What is the solubility product constant?

Solution: A liter of a saturated solution contains 1×10^{-5} gram-ion of silver and 1×10^{-5} gram-ion of chloride ion. The $K_{s.p.}$ is then:

$$[Ag^+][Cl^-] = K_{s.p.}$$
$$(1 \times 10^{-5})(1 \times 10^{-5}) = K_{s.p.} = 1 \times 10^{-10}$$

The concentration of the two ions need not be the same. It is the *product* of the concentrations that gives the value of the constant. Thus, if $[Cl^-]$ is high, the $[Ag^+]$ will be small enough so that the product does not exceed 1×10^{-10}.

Problem 2. The solubility of silver chromate is 7.5×10^{-5} mole per liter. Calculate the solubility product constant ($K_{s.p.}$).

Solution: This salt ionizes as follows:

$$Ag_2CrO_4 \rightleftharpoons 2\,Ag^+ + CrO_4^{--}$$

Since there are two silver ions produced from each molecule of silver chromate, the equation is written:

$$[Ag^+]^2[CrO_4^{--}] = K_{s.p.}$$

To get the silver ion concentration, we must take $(2 \times 7.5 \times 10^{-5})$ and substitute in the above formula thus:

$$[2 \times 7.5 \times 10^{-5}]^2[7.5 \times 10^{-5}] = K_{s.p.} = 1.7 \times 10^{-12}$$

Problem 3. The solubility product constant ($K_{s.p.}$) for barium sulfate is 1.21×10^{-10}. What is the solubility of the salt in grams per liter?

Solution:

$$[Ba^{++}] \times [SO_4^{--}] = [BaSO_4]$$
$$\text{Let } x = [Ba^{++}] \text{ or } [SO_4^{--}] \text{ or } [BaSO_4]$$
$$\text{Then } x \text{ times } x = x^2 = 1.21 \times 10^{-10}$$
$$x = \sqrt{1.21 \times 10^{-10}} = 1.1 \times 10^{-5} \text{ mole per liter}$$

To convert moles per liter to grams per liter, multiply moles by the molecular weight. The molecular (formula) weight of barium sulfate = 233.41. Consequently:

$$233.41 \times 1.1 \times 10^{-5} = 2.567 \times 10^{-3} \text{ gram per liter}$$

REVIEW QUESTIONS

1. Distinguish between an irreversible chemical reaction and a reversible reaction.

2. What factors favor a reaction going to completion?

3. What is meant when one says that the temperature coefficient of a certain reaction is 2?

4. Give examples of reactions that are affected by:
   (a) an increase in temperature, (b) an increase in pressure, (c) the action of light, (d) the action of a catalyst.

5. State Le Chatelier's Principle.

6. Derive the equilibrium constant equation for the reversible reaction:

$$ROH + HA \rightleftharpoons RA + H_2O$$

7. Distinguish between weak and strong electrolytes in terms of ionization constant.

8. Explain what is meant by the statement: "A salt (CA) is very slightly soluble in water. Therefore, its $K_{s.p.} = [C^+][A^-]$ is a very small number."

9. A saturated solution of $PbSO_4$ contains 1.35×10^{-4} mole per liter. Calculate its solubility product.

Chapter 13
Oxidation and Reduction

Oxidation. Oxidation is a process in which an atom *loses* one or more electrons. Thus, if a sodium atom loses an electron, the atom changes to an ion. Examples of oxidation are:

$$Na \longrightarrow Na^+ + electron$$
$$Ba \longrightarrow Ba^{++} + 2\ electrons$$
$$Al \longrightarrow Al^{+++} + 3\ electrons$$

In each of the above equations, the loss of electrons from neutral atoms is an oxidation process. If an ion loses an electron, this also is an oxidation process. Thus:

$$Fe^{++} \longrightarrow Fe^{+++} + electron$$

Reduction. Reduction is a process in which an atom or ion *gains* one or more electrons. It is apparent that this is just the reverse of oxidation. The following examples will help to make this clear:

$$Na^+ + e \longrightarrow Na$$
$$Fe^{+++} + e \longrightarrow Fe^{++}$$
$$S + 2\ e \longrightarrow S^{--}$$

Oxidation and Reduction Mutually Dependent. In chemical reactions where oxidation occurs, reduction occurs at the same time. If some atoms lose electrons, other atoms gain them. For example, in the reaction:

$$2\ Na + Cl_2 \longrightarrow 2\ Na^+ + 2\ Cl^-$$

each sodium atom loses an electron and each chlorine atom gains an electron. The number of electrons gained must be equal to the number lost.

$$AlCl_3 + 3\ Na \longrightarrow 3\ NaCl + Al$$

The aluminum ion gained three electrons and, therefore, it required three sodium atoms to furnish three electrons for the reaction. The aluminum ion was *reduced* to metallic aluminum, and the sodium atoms were *oxidized* to sodium ions.

Agents. An *oxidizing agent* is a substance containing an atom that *gains* electrons. A *reducing agent* is a substance that contains an atom that *loses* electrons. In the previous example, the aluminum ion is the *oxidizing* agent and the sodium atom is the *reducing* agent. One should keep in mind that an oxidizing agent or a reducing agent can be a molecule or an ion. Thus, MnO_2 is an oxidizing agent for the preparation of chlorine from hydrogen chloride. Also, the MnO_4^- ion is an oxidizing agent in an acid solution.

Electronegativity. The halogens, oxygen, and nonmetals are electronegative. Linus Pauling arbitrarily assigned a value of 4 to fluorine, the most electronegative element. On Pauling's scale, all elements (except the chemically inert ones, helium, argon, etc.) have assigned values greater than zero. The following table illustrates how Pauling's values vary from nonmetals to metals.

Table 13-1

Electronegativity Table

Most Electronegative

Fluorine	4.0
Oxygen	3.5
Nitrogen	3.0
Chlorine	3.0
Bromine	2.8
Carbon	2.5
Sulfur	2.5
Hydrogen	2.1
Copper	1.9
Manganese	1.6
Aluminum	1.5
Magnesium	1.2
Sodium	0.9
Potassium	0.8
Rubidium	0.8
Cesium	0.7

Least Electronegative (Electropositive)

Note. This table is the same as Table 4-2 on p. 45. It is given here as a matter of convenience to the student.

In the table we may think of an element such as sulfur as electronegative with respect to those below it and, relatively speaking, electropositive to those above it. Likewise an element that is electropositive in one compound may be electronegative in another. The following illustrates this.

SELECTED COMPOUND	MOST ELECTRONEGATIVE ELEMENT
$AlCl_3$	Cl
MgO	O
CH_4	C
CCl_4	Cl
OF_2	F

It is customary to think of the least electronegative element as being positive with respect to another element or elements present. Thus in CH_4 the hydrogen atom may be thought of as being electropositive, and in CCl_4 the carbon atom is considered to be electropositive.

Oxidation Number. (See Chapter 5 for more on oxidation number.) The oxidation number of an atom is defined as the charge which the atom *appears* to have when its valence electron or electrons are assigned to the most electronegative atom in the molecule. Thus in MnO (where oxygen is the electronegative atom) the oxidation number of manganese is +2; in MnO_2 it is +4; and in Mn_2O_7 it is +7. In each of these three oxides the oxidation number of the oxygen atom is −2. In the compound Mg_3N_2 the oxidation number of the magnesium atom is +2 and the nitrogen atom is a −3.

The following rules are useful in determining oxidation numbers.

1. A free element (not chemically combined) has an oxidation number of zero (0). Thus the oxidation number of the sodium atom (Na) is zero.

2. For simple (one atom) ions the oxidation number is equal to the ionic charge. The bromide ion is Br^{-1}, the lithium ion is Li^{+1}. For polyatomic ions such as nitrate (NO_3^{-1}) and ammonium (NH_4^{+1}), the oxidation number is the same as its electrovalence.

3. For most oxygen-containing compounds, the oxidation number of each oxygen atom is −2. Peroxides are exceptions.

4. The oxidation number of hydrogen is $+1$. Certain hydrides are exceptions.

5. In molecules the sum of the positive oxidation numbers and the negative oxidation numbers must equal zero.

Example: In sodium sulfate (Na_2SO_4) oxygen is most electronegative and sulfur and sodium are considered to be electropositive. Each of the four oxygen atoms has acquired two extra electrons, or a total of eight electrons (eight minus charges). The sulfur atom furnished six of these electrons and the two sodium atoms furnished the other two. Hence the eight positive charges are balanced by the eight negative charges.

Oxidation and Reduction Equations. Many of these equations are easily balanced (see examples given above). There are, however, a number of equations in this class which are difficult to balance. Suggestions for balancing such equations are given here. Specific examples are used for this purpose. It is understood that the final products are known in every case.

STEPS IN BALANCING. (1) Write the skeleton (unbalanced) equation. (2) Determine the oxidizing and reducing agents. (3) Take the proper number of molecules of the reducing and oxidizing substances to equalize the electrons taken up and given up. (4) See that the same number of atoms of each element are on each side of the equation sign.

Example 1.

$$KMnO_4 + H_2S + HCl \longrightarrow KCl + MnCl_2 + H_2O + S$$

$KMnO_4$ is the *oxidizing* agent and H_2S is the *reducing* agent. The manganese atom changes in valence from $+7$ to $+2$ and the sulfur atom changes from -2 to 0. This means that each manganese atom gains 5 electrons and each sulfur atom loses 2 electrons. It is evident that one sulfur atom cannot supply the electrons for one manganese atom. Two sulfur atoms furnish but 4 electrons, and 3 atoms furnish one extra electron. The way out, of course, is to find the least common multiple of 5 and 2 (10). This tells us that 5 sulfur atoms furnish enough electrons for 2 manganese atoms. All that remains is to add enough HCl (in this reaction, 6 molecules) to take care of 2 potassium and 2 manganese atoms.

$$2\,KMnO_4 + 5\,H_2S + 6\,HCl \longrightarrow$$
$$2\,KCl + 2\,MnCl_2 + 8\,H_2O + 5\,S$$

Example 2.

$$P + HNO_3 + H_2O \longrightarrow NO + H_3PO_4$$

Elementary phosphorus is the reducing agent and nitric acid is the oxidizer. Phosphorus loses 5 electrons and nitrogen gains 3 electrons. This means that it will take 3 phosphorus molecules (atoms) to furnish enough electrons to reduce 5 molecules of nitric acid. Balanced, the equation is:

$$3\,P + 5\,HNO_3 + 2\,H_2O \longrightarrow 5\,NO + 3\,H_3PO_4$$

OXIDATION NUMBER METHOD. The same equation can be balanced using the oxidation number method. Thus:

The oxidation number of elemental phosphorus is zero.

The oxidation number of phosphorus in H_3PO_4 is $+5$.

The oxidation number of nitrogen in HNO_3 is $+5$.

The oxidation number of nitrogen in NO is $+2$.

This means that phosphorus in changing from 0 to $+5$ loses 5 electrons and nitrogen in changing from $+5$ to $+2$ gains 3 electrons. The lowest common multiple of 5 and 3 is 15. Therefore 3 P atoms (molecules) will lose 15 electrons, and 5 HNO_3 molecules, in changing to NO molecules, will gain these 15 electrons. So as before we have:

$$3\,P + 5\,HNO_3 + 2\,H_2O \longrightarrow 3\,H_3PO_4 + 5\,NO$$

Example 3.

$$Cu + HNO_3 \longrightarrow Cu(NO_3)_2 + H_2O + NO$$

Elementary copper loses 2 electrons and nitrogen gains 3 electrons. This means that 3 copper atoms will furnish the 6 electrons necessary to reduce 2 nitric acid molecules. Notice that copper nitrate is one of the products. In other words, part of the nitric acid used is not acting as an oxidizing agent. Balanced, we have:

$$3\,Cu + 8\,HNO_3 \longrightarrow 3\,Cu(NO_3)_2 + 4\,H_2O + 2\,NO$$

Since many prefer equations written in the ionic form, the above examples are rewritten ionically. Here, as before, the electrons given up must equal those taken up.

(1) $2\,MnO_4^- + 5\,H_2S + 6\,H_3O^+ \longrightarrow 2\,Mn^{++} + 14\,H_2O + 5\,S$

(2) $3\,P + 5\,NO_3^- + 5\,H_3O^+ \longrightarrow 5\,NO + 3\,H_3PO_4 + 3\,H_2O$

(3) $3\,Cu + 2\,NO_3^- + 8\,H_3O^+ \longrightarrow 3\,Cu^{++} + 2\,NO + 12\,H_2O$

MODIFIED METHOD (USING IONS). The steps are: (1) Write the equation for the change that takes place when the oxidizing ion acquires electrons. (2) Write the equation for the change that occurs when the reducing ion loses electrons. (3) Multiply each equation by that number which will equalize the electrons gained and lost. (4) Add the equations and cancel the electrons. (5) See that the same number of atoms of each element are on each side of the equation. (6) Count the charges on each side of the equation. The positives should equal the positives (or negatives should equal negatives).

Example 4. Sulfide ion (S^{--}) in an acid solution is oxidized by permanganate ion (MnO_4^-).

(a) $$MnO_4^- + 5\,e + 8\,H_3O^+ \longrightarrow Mn^{++} + 12\,H_2O$$

(b) $$S^{--} \longrightarrow S + 2\,e$$

Multiply (a) by 2 and (b) by 5.

(c) $$2\,MnO_4^- + 10\,e + 16\,H_3O^+ \longrightarrow 2\,Mn^{++} + 24\,H_2O$$

(d) $$5\,S^{--} \longrightarrow 5\,S + 10\,e$$

Add (c) and (d) and cancel electrons.

$$2\,MnO_4^- + 5\,S^{--} + 16\,H_3O^+ + \cancel{10\,e} \longrightarrow$$
$$2\,Mn^{++} + 5\,S + 24\,H_2O + \cancel{10\,e}$$

In this equation, there are 2 atoms of manganese, 24 of oxygen, 5 of sulfur, 48 of hydrogen, and 10 electrons on each side of the equation sign. The total charge on the left is $+4$; on the right $+4$. As a rule, the oxygen atoms present in the oxidizing ion combine with protons to form water. In reaction (a) above, the proton is furnished by the hydronium ion (H_3O^+). The balancing of this equation is essentially the same when H^+ is used in place of H_3O^+.

Equivalent Weights of Oxidizing and Reducing Agents. For an oxidizing agent, the equivalent weight is its mole weight divided by the total number of electrons taken up by one atom (or one molecule) when it is reduced. In an acid solution, the MnO_4^- ion is reduced to Mn^{++}. Thus:

$$MnO_4^- + 5\text{ electrons} + 8\,H_3O^+ \longrightarrow Mn^{++} + 12\,H_2O$$

The equivalent weight of, say, $KMnO_4$ is, therefore, 1/5 mole and a normal solution contains 1/5 mole per liter.

The equivalent weight of a reducing agent is its mole weight divided by the total number of electrons lost per atom (or molecule) when it is oxidized. Iron(II) ion (Fe^{++}) is oxidized to iron(III) ion (Fe^{+++}) by a suitable oxidizing agent. Thus:

$$Fe^{++} - 1 \text{ electron} \longrightarrow Fe^{+++}$$

or

$$Fe^{++} \longrightarrow Fe^{+++} + e$$

The equivalent weight of $FeSO_4$ is, therefore, its mole weight divided by 1, and so a normal solution contains 1 mole per liter of solution.

Unusual Equations.   Two somewhat unusual equations will be discussed.

Example 1.   What happens when potassium chlorate oxidizes cane sugar ($C_{12}H_{22}O_{11}$) at an elevated temperature? We know that the sugar molecules and chlorate molecules are destroyed and in their place we obtain CO_2, H_2O, and KCl. There are two assumptions to make in balancing this oxidation-reduction equation. These have to do with the loss and gain of electrons during the reaction.

(1) It is obvious that in changing from $KClO_3$ to KCl, each chlorine atom gains 6 electrons. Where do these electrons come from? From the carbon atoms in the cane sugar? What is the *oxidation number* of C in $C_{12}H_{22}O_{11}$? The rule cited above tells us that oxygen is the most electronegative atom and, therefore, we can assign 2 extra electrons for each oxygen atom. Eleven oxygen atoms will control 22 electrons. Let us assume these come from the 22 hydrogen atoms. This will leave carbon with an oxidation number of zero. This is a reasonable assumption to make, and it gives the correct answer.

$$12 \times 4\,e = 48\,e$$

$$8\,KClO_3 + C_{12}H_{22}O_{11} + \text{heat} \longrightarrow 8\,KCl + 12\,CO_2 + 11\,H_2O$$

$$8 \times 6\,e = 48\,e$$

Checking: Total electrons gained = total electrons lost.

12 carbon atoms *lose* $4 \times 12 = 48$ electrons.
8 chlorine atoms *gain* $6 \times 8 = 48$ electrons.

(2) If we assume in this reaction that the sugar is first de-

composed by heat into water and carbon, then we know that the carbon formed must be oxidized to give us CO_2.

(a) $\qquad$ $C_{12}H_{22}O_{11}$ + heat $\longrightarrow$ 12 C + 11 H_2O

(b) $\qquad$ 12 C + 8 $KClO_3$ $\longrightarrow$ 8 KCl + 12 CO_2

Adding (a) + (b):

$$8\ KClO_3 +\ C_{12}H_{22}O_{11} +\ \text{heat} \longrightarrow 8\ KCl + 12\ CO_2 + 11\ H_2O$$

Example 2.   When sodium sulfite is heated under favorable conditions, two products are formed: sodium sulfate and sodium sulfide. This means that *some* molecules are oxidized while *others* are reduced. The equation is:

$$4\ Na_2SO_3 +\ \text{heat} \longrightarrow 3\ Na_2SO_4 +\ Na_2S$$

The equation is balanced as written, but we can rewrite it showing the details of the oxidation and reduction.

$$\begin{array}{ccc} +4 & 2\ e\ \text{lost} & +6 \end{array}$$

$$Na_2SO_3 + 3\ Na_2\overset{}{SO_3} +\ \text{heat} \longrightarrow Na_2\overset{}{SO_4} +\ Na_2S$$

$$\begin{array}{ccc} +4 & 6\ e\ \text{gained} & -2 \end{array}$$

Since 6 electrons are required to convert +4 S to −2 S, we need 3 molecules of Na_2SO_3 to furnish them.

REVIEW QUESTIONS

1. In chemical reactions, why does reduction always accompany oxidation?

2. In the change, $FeCl_2$ to $FeCl_3$, we say that the iron(II) ion has been oxidized to the iron(III) ion. Why is this oxidation, even though oxygen does not take part in the change?

3. Distinguish between an oxidizing and a reducing agent.

4. Balance the following oxidation-reduction chemical equations:
   (a) $Mg + O_2 \longrightarrow MgO$
   (b) $Fe^{++} + H^+ + MnO_4^- \longrightarrow Fe^{+++} + Mn^{++} + H_2O$
   (c) $Cu + H^+ + NO_3^- \longrightarrow Cu^{++} + H_2O + NO$
   (d) $H_2SO_4 + H_2S \longrightarrow H_2O + SO_2 + S$

5. What is the equivalent weight of potassium dichromate ($K_2Cr_2O_7$) to be used to oxidize HCl to free chlorine?

6. How many grams of metallic sodium are required to liberate 9 g of pure aluminum from aluminum chloride?

7. What is the oxidation number of each metal atom in the following compounds? Cr_2O_3, $TiCl_4$, Ag_2O, $HMnO_4$.

Chapter 14
Nuclear Chemistry

The nuclei of atoms are not involved in ordinary chemical reactions. Here, the planetary (valence) electrons of atoms determine, through interaction, what chemical combinations take place and whether or not the molecules thus formed are stable. In any ordinary chemical reaction, a modest amount of energy is absorbed or liberated. In nuclear reactions, on the other hand, the compositions (and structures) of atomic nuclei are changed to give atoms of other elements. In these reactions, the quantity of energy liberated or absorbed is of a much higher order of magnitude. Indeed, it is the large quantities of energy involved which make certain nuclear reactions so important for military and industrial use.

Atomic transmutations occur in two ways. First, the nuclei of certain atoms are changed by natural radioactive processes. Second, the nuclei of certain atoms are *transmuted* by bombardment with high-speed projectiles such as protons, neutrons, deuterons, or alpha particles. Using projectiles such as these, transmutation gives two types of atoms: those that are artificially radioactive, and those that are not radioactive.

Natural Radioactivity. Radioactivity is a process in which the nuclei of certain atoms disintegrate. This disintegration cannot be hastened or retarded by ordinary physical or chemical means. In 1896, Becquerel found that uranium, and minerals containing uranium, have the following properties: they will "expose" a photographic plate even though the plate is covered to exclude light, and they will discharge an electroscope. In 1898, Pierre and Marie Curie found that pitchblende contained a substance many times more radioactive than pure uranium. After a long series of fractional crystallizations, radium chloride and radium bromide were isolated. In 1910, Madame Curie isolated metallic radium. This element is about a million times more radioactive than uranium. Since radium is in the second group of the periodic table, its ordinary physical and chemical properties are very similar to those of barium, strontium, and calcium.

Radioactive Disintegration. The atoms of radioactive substances are unstable. Three types of "rays" are shot out from the nucleus: alpha (α) rays, beta (β) rays, and gamma (γ) rays.

The alpha rays are helium atoms minus two electrons, or helium nuclei. They have an initial velocity of 10,000 to 20,000 miles per second, but have very slight penetrating power.

The beta rays (or electrons) have an initial velocity of 100,000 miles per second. They penetrate thin sheets of metal.

The gamma rays are electromagnetic waves, like X rays, but of higher frequency. They penetrate relatively thick layers of metal and have a velocity equal to that of light. See Figure 14-1.

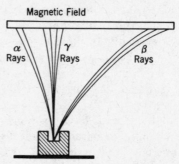

Fig. 14-1. Effect of a magnetic field on alpha, beta, and gamma rays.

Energy is liberated when a radioactive element disintegrates. It has been estimated that one gram of radium liberates 120 calories of heat per hour.

The Uranium Radioactive Series. Table 14-1 shows the relation of radium to other radioactive substances. It also gives information concerning: the time of half-life (the time for one-half of the atoms of a given quantity of an element to decompose), the atomic weight and atomic number of each element, and the type of radiation causing the change in each case. There are other radioactive series, such as the thorium series and the actinium series. Nature's way of transmuting certain elements of higher atomic weights is through natural radioactivity.

Sources of Radioactive Substances. Uranium, the element with which radium and other radioactive substances are associated, is found in various parts of the world. Low-grade deposits are found in Colorado, Utah, Austria, and Australia.

Table 14-1

The Uranium Disintegration Series

Uranium $^{238}_{92}$U	Lead $^{214}_{82}$Pb
$\alpha \downarrow 4.5 \times 10^9$ years	$\beta \downarrow 26.8$ minutes
Thorium $^{234}_{90}$Th	Bismuth $^{214}_{83}$Bi
$\beta \downarrow 24.5$ days	$\beta \downarrow 19.7$ minutes
Protoactinium $^{234}_{91}$Pa	Polonium $^{214}_{84}$Po
$\beta \downarrow 1.14$ minutes	$\alpha \downarrow 1.5 \times 10^{-4}$ seconds
Uranium $^{234}_{92}$U	Lead $^{210}_{82}$Pb
$\alpha \downarrow 2.7 \times 10^5$ years	$\beta \downarrow 22.3$ years
Thorium $^{230}_{90}$Th	Bismuth $^{210}_{83}$Bi
$\alpha \downarrow 8.3 \times 10^4$ years	$\beta \downarrow 5$ days
Radium $^{226}_{88}$Ra	Polonium $^{210}_{84}$Po
$\alpha \downarrow 1.6 \times 10^3$ years	$\alpha \downarrow 140$ days
Radon $^{222}_{86}$Rn	Lead $^{206}_{82}$Pb
$\alpha \downarrow 3.825$ days	(Stable)
Polonium $^{218}_{84}$Po	
$\alpha \downarrow 3.05$ minutes	

Richer deposits are found in the Belgian Congo and in the northern part of western Canada. Carnotite ($K_2O \cdot 2\ UO_3 \cdot V_2O_5 \cdot 3\ H_2O$) and pitchblende ($U_3O_8$) associated with other oxides and sulfides are the two most important minerals. A few years ago, uranium ores were mined for their radium content. Now uranium is one of the most important metals. This is true because of its use in atomic energy projects. Radium is still important for use in medicine, as in the treatment of cancer and certain other disorders (it can also be used to sterilize seeds and to kill germs but this use is not important); in paints and varnishes, to render them visible in the dark; and in scientific research, to obtain information regarding atomic structures and to explain the nature of electricity.

Artificial Radioactivity. Since 1934, many ordinary elements have been made radioactive by shooting high-speed projectiles such as alpha particles, deuterons, protons, and neutrons into their nuclei. More than 500 radioactive substances have been produced. Their period of half-life ranges from a fraction of a second for some to several years for others.

USING PROJECTILES FROM NATURAL RADIOACTIVE SUBSTANCES. Alpha particles (from radium) will change ordinary aluminum into radioactive phosphorus. The equation is:

$$^{27}_{13}Al + {}^{4}_{2}He \longrightarrow {}^{30}_{15}P + {}^{1}_{0}n$$

This equation is read: aluminum atoms of atomic number 13 and mass 27 react with helium ions (alpha particles) of atomic number 2 and mass 4 to give phosphorus atoms of atomic number 15 and mass 30, plus a neutron of atomic number 0 and mass 1.

The phosphorus is radioactive and changes into silicon with the liberation of a positron.

$$^{30}_{15}P \longrightarrow {}^{30}_{14}Si + {}^{0}_{+1}e$$

$$^{0}_{+1}e = \text{positive electron (positron)}$$

USING PROJECTILES PRODUCED IN THE LABORATORY. Huge machines such as cyclotrons, synchrotrons, and linear accelerators have been built which impart high velocities to particles such as alpha particles, deuterons, protons, and electrons. Using these high-velocity particles, we can obtain such interesting results as the following:

1. $^{12}_{6}C + {}^{2}_{1}H \longrightarrow {}^{13}_{7}N + {}^{1}_{0}n$

      $^{13}_{7}N \longrightarrow {}^{13}_{6}C + {}^{0}_{+1}e$ (half life = 9.9 minutes)

2. $^{34}_{16}S + {}^{2}_{1}D \longrightarrow {}^{32}_{15}P + {}^{4}_{2}He$ (h.l. = 14.22 days)

3. $^{23}_{11}Na + {}^{2}_{1}D \longrightarrow {}^{24}_{11}Na + {}^{1}_{1}H$ (h.l. = 15.0 hours)

4. $^{29}_{14}Si + {}^{4}_{2}He \longrightarrow {}^{32}_{15}P + {}^{1}_{1}H$ (h.l. = 14.22 days)

The Cyclotron.   This is the "atom-smashing" device invented by E. O. Lawrence at the University of California. It is a huge electromagnet between the poles of which is a flat circular metallic box which can be evacuated. Within the box are two hollow electrodes, each shaped like a capital letter *D*. The electrodes are separated by a gap and charged to a high potential (50,000–100,000 volts) by a high-frequency oscillator. When hydrogen (or deuterium) is introduced into the box, the gas is ionized by a hot filament located in the space between the two electrodes. The positive ions produced will be attracted by the negative electrode, and since they are moving in a magnetic field, they will travel in a curved path. As they reach the space between electrodes, the charge on each electrode is reversed and the ion is attracted by the other electrode. This process is repeated many thousands of times, the ions describing a spiral path with speeds finally reaching values from 30,000 to 85,000 miles per second. Finally, these high-velocity ions are drawn out of the apparatus and allowed to strike suitable targets, such as the nuclei of atoms. See Figure 14-2.

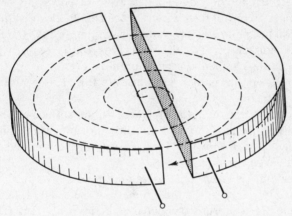

Fig. 14-2. The D's of a Cyclotron.

Nonradioactive Transmutation. When high-speed projectiles, such as alpha particles and protons, strike the nuclei of atoms, transmutation may take place wherein nonradioactive atoms are formed. Three examples are given here:

When beryllium atoms are bombarded with high-speed alpha particles, ordinary carbon ($^{12}_{6}C$) is formed and a neutron is obtained. It was this reaction that led to the discovery of neutrons by Chadwick in 1932.

$$^{9}_{4}Be + ^{4}_{2}He \longrightarrow ^{12}_{6}C + ^{1}_{0}n$$

The bombardment of nitrogen with alpha particles gives an isotope of oxygen and hydrogen. This experiment was first performed by Rutherford in 1919.

$$^{14}_{7}N + ^{4}_{2}He \longrightarrow ^{17}_{8}O + ^{1}_{1}H$$

When lithium is bombarded with high-speed protons, helium is produced.

$$^7_3\text{Li} + {}^1_1\text{H} \longrightarrow 2\,{}^4_2\text{He}$$

Radioactive Tracers.   Since radioactive substances also have ordinary chemical properties, they can be used to study the mechanisms of physical and chemical processes. To detect the presence of radioactive material at various stages in a process, an instrument known as a *Geiger counter* is used. It consists of two parallel plates (electrodes) carrying opposite charges. Particles (shot out from a radioactive substance) upon passing between the plates will be attracted to one of the plates, or will produce ions that will be attracted to one of the plates. This causes an electrical impulse which can be amplified as an audible sound, or be made to move a sensitive pointer on a dial. Most prospectors carry portable Geiger counters in their search for uranium ores.

The following examples will illustrate the uses of radioactive substances as tracers:

Radioactive lead atoms when placed on the surface of ordinary lead will diffuse into the lead and become somewhat uniformly distributed. This is shown by cutting slices off the lead and testing it with a Geiger counter.

If an animal is given food containing radioactive phosphorus, a radioactive analysis of the animal's bones will disclose the amount of phosphorus assimilated, the time this requires, and other valuable information.

In the author's research laboratory, radioactive cobalt was plated on mild steel. This in turn was heated in a furnace to allow the cobalt to diffuse into the steel. The depth of penetration of the cobalt was then determined by analysis of suitable samples in the Geiger counter.

Synthesis of New Elements.   Several new elements have been made in the laboratory by suitable nuclear reactions. The equations in Table 14-2 illustrate how some of these elements were made. They are part of the actinide series, of which actinium is the prototype. Other members of the series include thorium, protactinium, and uranium. (See the periodic table for complete list.) Those with atomic numbers greater than 92 are called *transuranium* elements.

Elements 97 and 98 were made with the help of the 60-inch cyclotron by Seaborg and his associates at the University of California. Element 97 is named *berkelium* in honor of Berkeley, California, and element 98 is called *californium* in honor of the

Table 14-2

Reactions Producing Transuranium Elements

Element Produced	Atomic Number	Reaction
Neptunium (Np)	93	$^{238}_{92}U + ^{1}_{0}n \longrightarrow ^{239}_{93}Np + ^{0}_{1}e*$
Plutonium (Pu)	94	$^{238}_{92}U + ^{2}_{1}H \longrightarrow ^{238}_{93}Np + 2^{1}_{0}n$
		$^{238}_{93}Np \longrightarrow ^{238}_{94}Pu + ^{0}_{-1}e$
Americium (Am)	95	$^{239}_{94}Pu + ^{1}_{0}n \longrightarrow ^{240}_{95}Am + ^{0}_{-1}e$
Curium (Cm)	96	$^{239}_{94}Pu + ^{4}_{2}He \longrightarrow ^{242}_{96}Cm + ^{1}_{0}n$
Berkelium (Bk)	97	$^{241}_{95}Am + ^{4}_{2}He \longrightarrow ^{243}_{97}Bk + 2^{1}_{0}n$
Californium (Cf)	98	$^{242}_{96}Cm + ^{4}_{2}He \longrightarrow ^{245}_{98}Cf + ^{1}_{0}n$

$*^{0}_{-1}e$ = negative electron $^{0}_{1}e$ = positive electron (positron)

state of California. Berkelium was made by bombarding americium with alpha particles. Californium was produced by bombarding an isotope of curium with alpha particles. Both elements are radioactive. Elements 99 and 100 were discovered in the "fallout" from a nuclear explosion and were named *einsteinium* and *fermium* in honor of two great scientists, Einstein and Fermi. Elements 101 and 102 have been produced and are named *mendelevium* and *nobelium* in honor of Mendeleev, famous for his work on the periodic table of the elements, and Alfred Nobel, who developed dynamite and whose great fortune goes to support the coveted Nobel Prizes in the sciences, arts, and literature. Element 103 has been detected and has been named *lawrencium* in honor of Dr. E. O. Lawrence, inventor of the cyclotron. The Russian scientist Kurchatov claimed he prepared element 104 (named *kurchatovium*) by bombarding plutonium with neon.

$$^{242}_{94}Pu + ^{22}_{10}Ne \longrightarrow ^{260}_{104}Ku + 4^{1}_{0}n$$

Nuclear Fission. This is a process in which an unstable nucleus of a heavy atom, for example, an isotope of uranium ($^{235}_{92}U$), splits into two parts of approximately equal mass with the simultaneous liberation of several neutrons and a large quantity of energy. Two examples are given:

Stable $^{235}_{92}U$ is subjected to bombardment with slow-moving neutrons. If a neutron is captured, $^{236}_{92}U$ is formed. This is unstable and undergoes fission.

$$^{235}_{92}U + ^{1}_{0}n \longrightarrow ^{236}_{92}U$$

$$^{236}_{92}U \longrightarrow ^{90}_{36}Kr + ^{144}_{56}Ba + 2\,^{1}_{0}n + energy$$

In the following example, the common isotope of uranium is converted into plutonium ($^{239}_{94}Pu$). When a neutron strikes this isotope, it is converted into $^{240}_{94}Pu$, which undergoes fission.

$$^{238}_{92}U + ^{1}_{0}n \longrightarrow ^{239}_{92}U$$

$$^{239}_{92}U \longrightarrow ^{239}_{92}Np + ^{0}_{-1}e$$

$$^{239}_{93}Np \longrightarrow ^{239}_{94}Pu + ^{0}_{-1}e$$

$$^{239}_{94}Pu + ^{1}_{0}n \longrightarrow ^{240}_{94}Pu$$

$$^{240}_{94}Pu \longrightarrow ^{138}_{56}Ba + ^{88}_{38}Sr + 14\,^{1}_{0}n + energy$$

NUCLEAR CHAIN REACTION. A neutron strikes $^{235}_{92}U$ nucleus, causing fission to take place with the liberation of several neu-

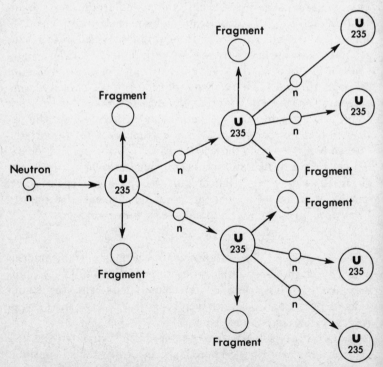

Fig. 14-3. A nuclear chain reaction.

trons and considerable energy. If some of the liberated neutrons strike other $^{235}_{92}$U nuclei, more fission occurs and more energy and neutrons are set free. These neutrons in turn will strike more nuclei with similar results. This is a chain reaction. If a small piece of uranium is used, most of the neutrons will escape and no chain reaction will occur. If a larger piece of uranium is used (optimum size), a chain reaction will take place. The release of energy when fission occurs in a nuclear chain reaction is the basis of the atomic bomb. See Figure 14-3 and Figure 14-4.

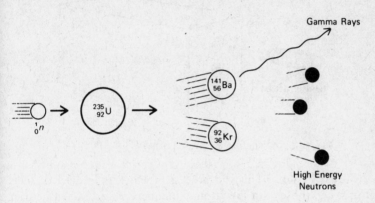

Fig. 14-4. Schematic fission of uranium nucleus.

The fragments produced in this chain reaction may be the nuclei of such atoms as barium, krypton, molybdenum, tin and others. Thus

$$^{235}_{92}\text{U} + ^{1}_{0}\text{n} \longrightarrow ^{141}_{56}\text{Ba} + ^{92}_{36}\text{Kr} + 3\,^{1}_{0}\text{n} + \text{energy}$$

$$^{235}_{92}\text{U} + ^{1}_{0}\text{n} \longrightarrow ^{103}_{42}\text{Mo} + ^{131}_{50}\text{Sn} + 2\,^{1}_{0}\text{n} + \text{energy}$$

THE ATOMIC PILE.   In one form this is a rectangular block of graphite containing openings or slots into which metallic uranium rods (sealed in aluminum cans) are placed. The graphite (called the moderator) serves to slow down the neutrons set free during fission so that they may be captured by other uranium nuclei to keep the process going. Sometimes, cadmium rods are inserted into slots in the graphite block. Neutrons are absorbed by cadmium. The use of such rods makes it possible to control the rate of reaction in the pile. Large quantities of heat are set free in an atomic pile. This heat can be used to operate machines in an industrial plant. See Figure 14-5.

In its more practical form, an atomic pile is called a *nuclear reactor* and is used for obtaining controlled nuclear energy. Reactors are used as a source of power. The submarine U.S.S. "Nautilus" is one of many nuclear-powered vessels that has traveled thousands of miles without refueling. Large aircraft may be nuclear-powered in the future.

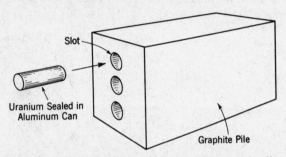

Fig. 14-5. A schematic representation of an atomic pile.

Nuclear Fusion. This is a process in which the nuclei of light atoms, i.e., those with low atomic numbers, fuse to form heavier nuclei. The fusion takes place with the liberation of considerable energy. However, a fusion reaction requires a very high temperature and this can (at present) be obtained only with the help of an atomic fission reaction.

The high temperatures which prevail on the sun can bring about nuclear fusion reactions such as:

$$^1_1H + ^1_1H \longrightarrow ^2_1H + ^{\ 0}_{-1}e + energy$$

$$^1_1H + ^2_1H \longrightarrow ^3_2He + energy$$

$$^3_2He + ^3_2He \longrightarrow ^4_2He + 2\,^1_1H + energy$$

The high heat energy liberated in these reactions is measured in electron volts. An electron volt is equal to the kinetic energy of an electron accelerated by a potential difference of 1 volt. Or, one electron volt is equal to 23,053 calories per mole.

Mass Defect. The helium nucleus is made up of 2 protons and 2 neutrons. The mass of the proton is 1.00758, and that of the neutron is 1.00893. Therefore, the mass of the helium nucleus should be $2 \times 1.00758 + 2 \times 1.00893$, which is 4.03302. Actually, the helium nucleus weighs only 4.00277. The difference (.03025) between these values is called the *mass defect*. According to modern theory, it represents the quantity of mass that was converted into energy when a helium nucleus was made. The release of energy when fusion takes place is the basis of the hydrogen bomb. Although details are not available, tritium (the scarcest isotope of hydrogen) and deuterium are probably used in the fusion reaction. The following equations represent possible fusion reactions:

Let $_1^1H$ = protium, $_1^2H$ = $_1^2D$ = dueterium, $_1^3H$ = $_1^3T$ = tritium

$$_1^2D + _1^3T \longrightarrow _2^4He + _0^1n + energy$$

$$_3^6Li + _0^1n \longrightarrow _2^4He + _1^3T + energy$$

Energy and Mass. According to Einstein's equation

$$E = MC^2$$

(where E is the energy, M the mass, and C the velocity of light) mass and energy are convertible one into the other. The laws of conservation of mass and of energy can be restated to include this fact. In a system undergoing change, the mass plus the energy of the substances undergoing change will be equal to the mass plus the energy of the products of the reaction.

$$Matter_1 + Energy_1 = Matter_2 + Energy_2$$
$$\text{(before reaction)} \qquad \text{(after reaction)}$$

If a pound of mass were completely converted into energy, about 10 billion kilowatt hours would be obtained. If one mass unit (i.e., one proton) were converted into energy, 931 million electron volts would be obtained.

Ordinary Chemical Changes. Here the classical mass law appears to be obeyed because experimental methods are not sensitive enough to detect a loss or gain in mass during chemical change. Thus, when carbon dioxide is formed by burning carbon in oxygen, heat energy is liberated.

$$C + O_2 \longrightarrow CO_2 + 94,200 \text{ calories}$$
$$(12 \text{ g}) + (32 \text{ g}) \longrightarrow (44 - a) \text{ g}$$

where a represents the mass converted into energy. This a is so small that it cannot be detected by ordinary means.

Note. In nuclear reactions, so far studied, much of the energy comes from the rearrangement of nucleons (protons and neutrons). Only small fractions of the protons or neutrons are converted into energy.

REVIEW QUESTIONS

1. Distinguish between ordinary chemical reactions and nuclear reactions.
2. (a) Name three elements that possess natural radioactivity. (b) Name ten elements that do not possess natural radioactivity. (c) What is the most highly radioactive element found in nature?
3. Distinguish between alpha, beta, and gamma rays.
4. What is meant by the statement: "The half-life of radium is 1,590 years"?
5. By what radioactive changes (steps) can uranium atoms change into lead atoms?
6. (a) In what minerals is uranium found? (b) Why is radium always found associated with uranium in nature?
7. Describe one method of changing an ordinary element into a radioactive element.
8. Describe the properties of neutrons.
9. What are radioactive tracers? Suggest a method whereby carbon 14 ($^{14}_{6}C$) can be used as a tracer.
10. What is a nuclear chain reaction?
11. Distinguish between the terms *fission* and *fusion* as they are used in nuclear chemistry.
12. Write an equation for the action of alpha particles on metallic aluminum.
13. What are transuranium elements?
14. Plutonium (Pu) is bombarded with helium to give curium (Cm). Give the equation.
15. What is (a) $^{238}_{94}Pu$, (b) $^{1}_{0}n$, (c) $^{0}_{-1}e$?

Chapter 15
Electrochemistry

That division of chemistry concerned with the conversion of electrical energy into chemical energy or the conversion of chemical energy into electrical energy is called *electrochemistry*.

Electrical Terms. An electrical *conductor* is a substance which allows electrons to flow through it. If there is no chemical action, it is a conductor of the *first class*. If a chemical reaction occurs during the passage of the current, it is a conductor of the *second class*. A current passing through a copper wire is first-class conduction; a current passing through a copper sulfate solution with a chemical reaction taking place at the electrodes is second-class conduction.

A *coulomb* is a current of one ampere flowing for one second. It takes 96,500 seconds to deposit one equivalent of silver from a silver nitrate solution using a current of one ampere (96,500 coulombs).

An *ampere* is that current which, when passing through a solution of silver nitrate, deposits .001118 g of silver per second.

The *electromotive force* (EMF) is the driving force which sends a current through a resistance. The unit is the volt.

A *volt* is that difference in potential (EMF) necessary to force a current of one ampere through a resistance of one ohm.

An *ohm* is the resistance to an electric current offered by a column of mercury (at 0°C) 14.4521 g in mass of a uniform cross-sectional area and 106.3 cm in length.

Ohm's law states that the current strength in a circuit is directly proportional to the electromotive force and inversely proportional to the resistance.

$$\text{Amperes} = \frac{\text{volts}}{\text{ohms}}$$

A *watt* is an electric power unit which equals 1 ampere times 1 volt.

$$\text{Watts} = \text{amperes} \times \text{volts}$$

Kilowatt-hours are obtained by the following:

$$\text{Kilowatt-hour} = \frac{\text{volts} \times \text{amperes} \times \text{hours}}{1000}$$

Faraday's Laws. These laws were discovered by Michael Faraday in 1833.

FIRST LAW. In electrolysis, the quantities of substances set free at the electrodes are directly proportional to the quantity of electricity passing through the solution.

SECOND LAW. A given quantity of electricity sets free the same number of equivalents of substances at the electrodes.

FIRST AND SECOND LAWS COMBINED. One equivalent of a substance is deposited at each electrode by the passage of 96,500 coulombs of electricity through a solution. This quantity is called one *faraday* (F). For example, 96,500 coulombs of electricity will deposit 107.87 g of silver or 31.77 g of copper.

Electrode Potentials. When a strip of metal (an electrode) is placed in water, there is a tendency for the metal to go into solution as ions, with the simultaneous accumulation of electrons on the metal strip. This process produces an electrical potential difference between metal and solution called an *electrode potential.* When equilibrium has been reached between the metal and a 1 molar solution of its ions, the value obtained is termed the *standard electrode potential.* Unfortunately, this potential cannot be measured directly. But if the metal strip, immersed in a solution of its ions, is made a *half cell* and this half cell is properly combined with another half cell, a potential difference between the two half cells can readily be determined. As a matter of convenience, and also because the actual potential difference between hydrogen gas and a 1 molar solution of hydrogen ions is small, the hydrogen half cell is used as one half of each cell and the other half is the metal whose electrode potential is being determined. Likewise, the potential of the hydrogen half cell is assumed to be zero, and so the actual voltage across the cell is the electrode potential of that metal.

Hydrogen Electrode. A hydrogen electrode consists of an electrode (strip) of metallic platinum coated with a layer of finely divided spongy platinum, called *platinum black*, immersed in a 1 molar solution of H^+ or H_3O^+ (furnished by hydrochloric acid). Hydrogen gas, at a pressure of one atmosphere, is allowed to bubble over the platinized platinum electrode. The reaction is:

$$H_2 + 2\,H_2O \rightleftharpoons 2\,H_3^+O + 2\,e$$

More simply, this is:

$$H_2 \rightleftharpoons 2\,H^+ + 2\,e$$

The potential (E) is assumed to be zero.

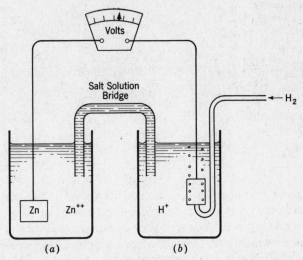

Fig. 15-1.   Apparatus for measuring electrode potentials.

Figure 15-1 represents the method of determining electrode potentials. In vessel (*a*) is a zinc electrode immersed in a 1 molar zinc ion solution. This is a half cell. In (*b*) is a hydrogen half cell. Hydrogen gas at 1 atmosphere pressure bubbles over a platinum electrode covered with platinum black. Since the voltage of the hydrogen half cell is assumed to be zero, the potential across the cell (.762 volt) is the electrode potential for zinc. The half cell reaction is:

$$Zn \rightleftharpoons Zn^{+2} + 2\,e$$

Standard Electrode Potentials.   The method as described here was used to determine the standard electrode potentials given in the following incomplete table:

Table 15-1

Some Standard Oxidation Potentials

Half Cell Reaction	E^0 Volts
$K \rightleftharpoons K^{+1} + e$	$+2.92$
$Ca \rightleftharpoons Ca^{+2} + 2e$	$+2.87$
$Na \rightleftharpoons Na^+ + e$	$+2.71$
$Mg \rightleftharpoons Mg^{+2} + 2e$	$+2.34$
$Al \rightleftharpoons Al^{+3} + 3e$	$+1.67$
$Zn \rightleftharpoons Zn^{+2} + 2e$	$+0.762$
$Fe \rightleftharpoons Fe^{+2} + 2e$	$+0.44$
$Ni \rightleftharpoons Ni^{+2} + 2e$	$+0.236$
$Pb \rightleftharpoons Pb^{+2} + 2e$	$+0.13$
$H_2 \rightleftharpoons 2H^{+1} + 2e$	Zero
$Cu \rightleftharpoons Cu^{+2} + 2e$	-0.34
$2I^{-1} \rightleftharpoons I_2 + 2e$	-0.54
$Fe^{+2} \rightleftharpoons Fe^{+3} + e$	-0.77
$Ag \rightleftharpoons Ag^{+1} + e$	-0.80
$2Br^{-1} \rightleftharpoons Br_2 + 2e$	-1.06
$2Cl^{-1} \rightleftharpoons Cl_2 + 2e$	-1.36
$Au \rightleftharpoons Au^{+++} + 3e$	-1.50

As given in Appendix XI, the activity series of the metals is essentially the same as the electrochemical series, or the electromotive force series. Since the loss of one or more electrons by an atom is *oxidation* and the gain of one or more electrons by an atom is *reduction*, the half cell reactions in the above table represent *oxidation* in the direction of the upper arrows and *reduction* in the direction of the lower arrows. In the table, the potassium atom is most easily oxidized and the gold atom is the most difficult to oxidize. Likewise, the gold ion (Au^{+3}) is most easily reduced to a gold atom and a potassium ion (K^{+1}) is more difficult to reduce to the atom. Let us give an example or two to illustrate how the table can be used. We shall make use of the values of E^0 to make our examples quantitative.

(1) Under favorable conditions, will an oxidation-reduction reaction take place if potassium atoms are placed in contact with aluminum ions (Al^{+++})? In other words, does the following reaction go as written?

$$3K + Al^{+3} \longrightarrow 3K^{+1} + Al$$

Writing the half cell reactions and including the values for E^0, we have:

$$3 K \longrightarrow 3 K^{+1} + 3 e \qquad +2.92 \text{ volts (from table)}$$
$$Al^{+3} + 3 e \longrightarrow Al \qquad -1.67 \text{ volts}$$

Adding the two:

$$3 K + Al^{+3} \longrightarrow 3 K^{+1} + Al \qquad +1.25 \text{ volts}$$

It should be kept in mind that the potential is the same whether one K atom changes to $K^+ + e$ or 3 K atoms change to $3 K^+ + 3 e$. Also, since the Al^{+++} is accepting electrons, the sign of the potential will change from plus to minus. Since the answer is $+1.25$ volts, we can say definitely the reaction as written does take place.

(2) Does the following reaction take place?

$$Cu + Pb^{+2} \longrightarrow Cu^{+2} + Pb$$

Writing the half cell reaction, we have:

$$Cu \longrightarrow Cu^{+2} + 2 e \qquad -.34 \text{ volt}$$
$$Pb^{+2} + 2 e \longrightarrow Pb \qquad -.13 \text{ volt}$$

$$Cu + Pb^{+2} \longrightarrow Cu^{+2} + Pb \qquad -.47 \text{ volt}$$

Since the voltage is negative, the reaction will not take place.

Cells and Batteries. These are devices that convert chemical energy into electrical energy. From Table 15-1, we see there should be a number of cells (and batteries) possible. However, the ones discussed here are among those that have proved to be practical and, therefore, of commercial value.

THE DANIELL CELL. The Daniell cell is made as follows: A strip of zinc is placed into a solution of dilute zinc sulfate. A strip of copper is placed into a saturated solution of copper sulfate. The two solutions are separated by a porous partition which allows SO_4^{--} to diffuse through it. If a wire connects the two electrodes, i.e., the zinc and copper strips, a current flows. If a voltmeter is included in the circuit, a potential difference is observed; if an ammeter is included, the quantity (amperes) passing through the solution may be observed.

The reactions at the electrodes are as follows. At the zinc electrode, zinc atoms change to zinc ions (Zn^{++}) by the loss of electrons. These electrons collect on the zinc electrode and then flow through the connecting conductor to the copper electrode. Copper ions, from the solution of copper sulfate through a process of diffusion, come in contact with the copper electrode, where they receive these electrons to make metallic copper. The reactions are:

At zinc electrode: $\qquad\qquad Zn \longrightarrow Zn^{+2} + 2\,e$

At copper electrode: $\qquad Cu^{+2} + 2\,e \longrightarrow Cu$

One form of the Daniell cell is called the *gravity cell*. In this form, the saturated copper sulfate solution is in the bottom of a glass jar and the dilute zinc sulfate solution is floated on the top of the copper sulfate. With care the two solutions do not mix because of the differences in their specific gravities. See Figure 15-2b.

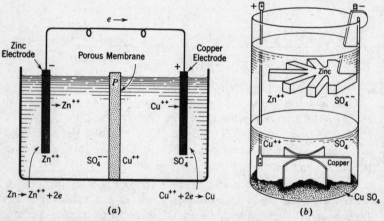

Fig. 15-2. Two different arrangements of a Daniell cell. The one on the right is the gravity (Daniell) cell.

THE LEAD STORAGE BATTERY. This is a battery used in automobiles to convert chemical energy into electrical energy. Its power starts the engine, furnishes light for night driving, operates the radio and the air conditioner, and opens and closes windows. Fortunately a discharged battery can be recharged by forcing electrical energy (at suitable EMF) through the battery in the opposite direction. Consequently each automobile is equipped

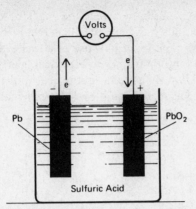

Fig. 15-3. Schematic lead storage battery.

with a generator or an alternator to keep the battery charged. This battery contains a number of sponge lead plates properly spaced between a like number of lead(IV) oxide plates immersed in a sulfuric acid solution of 1.25–1.30 specific gravity. See Figure 15-3. On discharge of the battery we have:

At the anode (– pole)

$$Pb \longrightarrow Pb^{+2} + 2e$$

The lead ions thus produced react with sulfate ion of the sulfuric acid.

$$Pb^{+2} + SO_4^{-2} \longrightarrow PbSO_4$$

At the cathode (+ pole)

$$PbO_2 + 4H^+ + 2e \longrightarrow Pb^{+2} + 2H_2O$$

The lead ions produced here also react with SO_4^{-2} to give $PbSO_4$. The complete *discharge* reaction is:

$$Pb + PbO_2 + 4H^+ + 2SO_4^{-2} \longrightarrow 2PbSO_4 + 2H_2O$$

When the battery is being charged we have:

$$2PbSO_4 + 2H_2O \xrightarrow{\text{charging}} Pb + PbO_2 + 4H^+ + 2SO_4^{-2}$$

Or using two arrows, one for charging and the other for discharging we have:

$$Pb + PbO_2 + 4H^+ + 2SO_4^{-2} \underset{\text{charge}}{\overset{\text{discharge}}{\rightleftarrows}} 2PbSO_4 + 2H_2O$$

THE FUEL CELL. Cells in which the reactant chemicals can be supplied continuously and the products removed continuously are called *fuel cells*. The main purpose of a fuel cell is to convert chemical energy (continuously) into electrical energy. A fuel cell will operate continuously as long as the reactants, including the catalysts, are not poisoned. Figure 15-4 is a schematic representa-

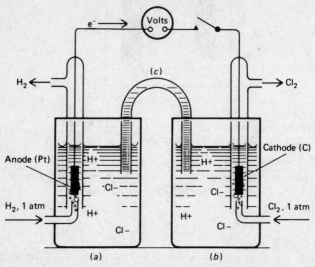

Fig. 15-4. Schematic representation of a fuel cell. Vessels (*a*), (*b*) and inverted U tube (*c*) contain molar hydrochloric acid solution.

tion of a hydrogen-chlorine fuel cell. Vessels (*a*) and (*b*) and U tube (*c*) contain molar hydrochloric acid. The platinum anode contains a coating of finely divided platinum (sometimes called platinum black). The carbon cathode contains a layer of absorbed chlorine gas. The half cell reactions are:

$$\text{Anode:} \quad H_2 \longrightarrow 2\,H^{+1} + 2e$$
$$\text{Cathode:} \quad Cl_2 + 2e \longrightarrow 2\,Cl^{-1}$$
$$\overline{\text{Total cell reaction} \quad H_2 + Cl_2 \longrightarrow 2\,H^{+1} + 2\,Cl^{-1}}$$

The cell voltage is

$$H_2 \longrightarrow 2\,H^{+1} + 2\,e^- = 0 \text{ volts}$$
$$Cl_2 + 2\,e^- \longrightarrow 2\,Cl^{-1} = \underline{+1.36}$$
$$\text{Total} \quad +1.36 \text{ volts}$$

DRY CELL (DRY BATTERY). A zinc cylinder is coated on the inside with a thick layer of moist ammonium chloride. This serves as the negative electrode. In the center is a graphite rod surrounded with manganese dioxide. This is the positive electrode. This cell will give a voltage of 1.5 to 1.6 volts. The cell reactions are:

Negative pole: $Zn \longrightarrow Zn^{+2} + 2\,e$

Positive pole: $2\,NH_4^+ + 2\,MnO_2 + 2\,e \longrightarrow 2\,NH_3 + H_2O + Mn_2O_3$

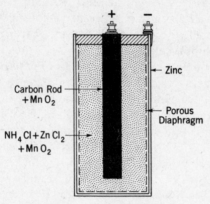

Fig. 15-5. A dry cell (battery).

REVIEW QUESTIONS

1. Define: ampere, volt, ohm, faraday.
2. Distinguish between first- and second-class conduction.
3. State Faraday's laws.
4. Explain what is meant by (a) a half cell, (b) a hydrogen electrode, (c) a half cell reaction; write one.
5. What is meant by the term "electrochemical series"? Compare this with the activity series.
6. Describe the Daniell cell.
7. Make a simple drawing of a dry cell.
8. Write equations for (a) the charging of a lead storage battery, (b) the discharging of a lead storage battery.
9. What is a fuel cell? How does it differ from an ordinary Daniell cell?

Chapter 16
Thermal Energy and
Chemical Reactions

Thermochemistry. When chemical reactions take place, there is a quantitative relationship between the chemical reaction and the heat liberated or absorbed during the reaction. A study of these relationships is called *thermochemistry*.

Before going on we shall define some of the terms used in this branch of chemistry. A *calorie* (cal) is that quantity of heat necessary to raise the temperature of 1 g of water 1°C. In reactions and processes where many calories are involved, we use a larger unit called the *kilogram calorie* (k or kcal) which is equivalent to 1000 small calories. In the British Empire, the thermal energy unit used is the *British Thermal Unit* (BTU), which is that quantity of heat necessary to raise the temperature of one pound of water 1°F.

Heat of Physical Changes. When substances change from one physical state to another, heat is liberated or absorbed. The following definitions are important in expressing these changes.

SPECIFIC HEAT. *Heat capacity*, or as it is sometimes called the *specific heat*, is that quantity of heat (in calories) required to raise the temperature of 1 g of a substance 1°C, or that quantity of heat necessary to raise the temperature of 1 mole of a substance 1°C. It is understood that no phase change takes place.

Table 16-1

The Heat Capacity (Specific Heat) of a Few Metals at 20°C

Metal	Per gram
Aluminum	0.214 cal
Bismuth	0.294 cal
Copper	0.092 cal
Iron	0.107 cal
Zinc	0.0925 cal

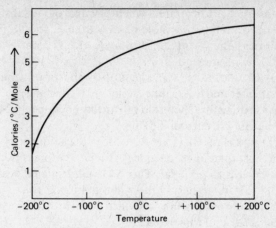

Fig. 16-1. Heat capacity of aluminum at various temperatures (cal/°C/mole).

Figure 16-1 is a graph showing the relationship between temperature and heat capacity for aluminum.

LAW OF DULONG AND PETIT.   This law states that the specific heat of a solid element times its atomic weight is equal to a constant (approximately 6.2). This relationship is useful in fixing the approximate atomic weight of an element if its specific heat is known. What is the approximate atomic weight of lead if its specific heat is .03?

$$.03 \times \text{at. wt.} = 6.2$$
$$\text{at. wt.} = \frac{6.2}{.03} = 206.7$$

HEAT OF VAPORIZATION AND FUSION.   It takes heat to change a unit weight of a substance (say 1 g of water at its boiling point) into vapor. This quantity of heat, in calories, is its *heat of vaporization*. In the case of water, the heat of vaporization is 540 calories. Likewise, the heat (in calories) required to change a gram of a solid into a gram of liquid is its *heat of fusion*. Thus, the heat of fusion of water is 79 calories.

Heat of Chemical Reaction.   The heat of reaction is the heat absorbed or liberated during chemical change. If heat is liberated, the change is *exothermic*. If heat is absorbed, the change is *endothermic*.

HEAT OF FORMATION.   The heat of formation is that quantity

of heat absorbed or liberated when a mole of a substance is formed. The formation of a mole of CO_2 from C and O_2 liberates 94,240 calories. The formation of a mole of NO from N_2 and O_2 absorbs 21,600 calories.

HEAT OF COMBUSTION.　The heat of combustion is that quantity of heat liberated when one mole of a substance is burned (completely oxidized). One mole of carbon liberates 94,240 calories upon burning to carbon dioxide.

MEASUREMENTS OF HEAT OF REACTION.　The quantity of heat set free or absorbed in a chemical reaction is measured in an apparatus called a *calorimeter*. This is a well-insulated vessel provided with a sensitive thermometer and a stirring rod. A Dewar flask (a thermos bottle) when provided with a thermometer and stirring rod is a useful calorimeter.

When the tube containing A and B (Fig. 16-2) is inverted a reaction takes place. If heat is liberated, it will warm the water.

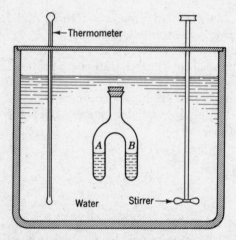

Fig. 16-2. A simple calorimeter.

The rise in temperature is measured by the thermometer. Knowing the amount of water present and the temperature rise, it is relatively simple to determine the heat liberated. The stirrer insures uniform temperature in all parts of the calorimeter.

Thermochemical Equations and Calculations.　If heat is liberated when a reaction takes place (exothermic reaction) the products must contain less heat than the starting substances. Consequently the sign is *negative*. Likewise the sign is *positive* for an endothermic reaction because heat must be added to bring about the reaction. For reactions carried out in an open vessel, i.e., at constant pressure, we say that the heat of reaction is equal to the change in heat content or the enthalpy (H) of the system. By system we mean a portion of material (nature) that has been isolated or set apart for study. If H_1 is the heat content (enthalpy) of the system before reaction and H_2 is the content (enthalpy) of the system after the reaction, then:

$$H_2 - H_1 = \Delta H$$

where ΔH equals the change in heat content of the system. If the reaction is exothermic, ΔH is negative; if the reaction is endothermic the sign of ΔH is positive.

Example 1.   When carbon dioxide is formed from the elements the reaction is exothermic, so we have:

$$C_{(s)} + O_{2(g)} = CO_{2(g)} \ldots \ldots \Delta H = -94,240 \text{ cal}$$

This means 1 mole (12 g) of carbon plus 1 mole of oxygen gives one mole of carbon dioxide and 94,240 calories of heat liberated.

Example 2.   Hot carbon will react with water (steam) to give hydrogen gas and carbon monoxide gas

$$C_{(s)} + H_2O_{(v)} = H_2 + CO \ldots \ldots \Delta H = +31,400 \text{ cal}$$

The Law of Hess (Law of Constant Heat Summation) states that the heat involved in a chemical reaction, if no external work is done, is the same whether the reaction takes place in one or in several steps. This makes it possible to calculate the heat of reaction for those reactions where it is experimentally difficult or impossible to make the measurements.

Preparation of ammonium chloride by two methods gives the same heat of formation.

First Method
  (*a*) 1 mole NH_3 in 200 moles water $\Delta H = -8,400$ cal
  (*b*) 1 mole HCl in 200 moles water $\Delta H = -17,300$ cal
  (*c*) Mix (*a*) + (*b*) $\Delta H = -12,300$ cal

                                        Total $= -38,000$ cal

Second Method

(a) 1 mole HCl gas + 1 mole NH_3
gas = 1 mole NH_4Cl $\Delta H = -42,100$ cal

(b) 1 mole of NH_4Cl in 400 moles water $\Delta H = +4,100$ cal

Total $= -38,000$ cal

Thermochemical equations can be added and subtracted as if they were ordinary algebraic equations. What is the heat of formation of carbon monoxide (CO) from the elements? The heat of formation of CO_2 from $C + O_2$ is known. Also the heat of formation of CO_2 from CO and O_2 is known. Therefore if we subtract one equation from the other, the desired information can be obtained.

(a) $C + O_2 \longrightarrow CO_2$ ....$\Delta H = -94,240$ cal
(b) $CO + \frac{1}{2}O_2 \longrightarrow CO_2$ ....$\Delta H = -67,700$ cal

Subtract (b) from (a). Change sign of each term in (b).

$$C - CO + \tfrac{1}{2}O_2 \longrightarrow \ ....\Delta H = -26,540 \text{ cal}$$

Transposing:

$$C + \tfrac{1}{2}O_2 \longrightarrow CO \ .....\Delta H = -26,540 \text{ cal}$$

Table 16-2

Heats of Formation of Several Compounds*

Compound	ΔH (kcal per mole)
$CO_{2(g)}$	-94.24
$CO_{(g)}$	-26.54
$KCl_{(s)}$	-97.55
$CH_{4(g)}$	-17.9
$C_2H_{2(g)}$	$+54.2$
$HI_{(g)}$	$+6.2$
$CCl_{4(l)}$	-33.2
$C_2H_{4(g)}$	$+12.5$
$NO_{(g)}$	$+21.6$
$SO_{2(g)}$	-71.0
$H_2O_{(l)}$	-68.3
$CH_3OH_{(l)}$	-57.0

*1 kcal = 1000 calories

Problem. What is the value for the ΔH for the reaction

$$C_2H_{4(g)} + H_2O_{(l)} \longrightarrow C_2H_5OH_{(l)}$$

Given:

(a) $C_2H_5OH_{(l)} + 3 O_{2(g)} \longrightarrow 2 CO_{2(g)} + 3 H_2O_{(l)} \ldots \Delta H = -327$ kcal

(b) $C_2H_{4(g)} + 3 O_{2(g)} \longrightarrow 2 CO_{2(g)} + 2 H_2O_{(l)} \ldots \Delta H = -337.3$ kcal

Solution: Subtract (a) from (b). By doing this we eliminate oxygen, carbon dioxide, and 2/3 of the water. To do this change the sign of each substance in equation (a). Thus

$-(a)$ $-C_2H_5OH_{(l)} - 3O_{2(g)} \longrightarrow -2CO_{2(g)} - 3H_2O_{(l)} \ldots \Delta H = +327$ kcal

(b) $\underline{C_2H_{4(g)} + 3O_{2(g)} \longrightarrow 2CO_{2(g)} + 2H_2O_{(l)} \ldots \Delta H = -337.3 \text{ kcal}}$

 $C_2H_{4(g)} + H_2O_{(l)} \longrightarrow C_2H_5OH_{(l)} \ldots \Delta H = -10.3$ kcal

REVIEW QUESTIONS

1. Define each of the following terms: (a) thermochemistry, (b) calorie, (c) British Thermal Unit, (d) specific heat, (e) heat capacity.
2. Distinguish between (a) exothermic and endothermic reactions, (b) heat of formation and heat of combustion.
3. How are heats of reactions measured in the laboratory?
4. State the Law of Hess, or, as it is sometimes called, the Law of Constant Heat Summation.
5. State in some detail how to determine the heat of formation of carbon dioxide from the elements.
6. Suggest a method of determining the heat of neutralization of a strong acid by a strong base.
7. How would you determine the specific heat of metallic zinc?
8. The heat of formation of CO_2 is given as $\Delta H = -94.24$ kcal per mole. Why is the sign negative?
9. What is the meaning of enthalpy?

Chapter 17
The Colloidal State

In 1862, Thomas Graham, an English chemist, pointed out that certain substances will diffuse through an animal or vegetable membrane, whereas others will not. Those that passed through were the compounds that were obtainable in crystalline form. They were called *crystalloids*. The substances which did not crystallize were the ones which did not pass through the membrane. These were called *colloids*, meaning "gluelike." When a substance is divided into very small particles which are larger than individual molecules, and are scattered (dispersed) through certain media, it is said to be in the *colloidal state*. Colloidal particles vary in diameter from a few millionths of a millimeter to over a hundred millionths of a millimeter.

Colloidal Systems. Following are the important possible colloidal systems with an example of each.

DISPERSED SYSTEM	EXAMPLE
Solid in a solid	Ruby glass (gold dispersed in glass)
Solid in a liquid	Ferric oxide dispersed in water
Solid in a gas	Ordinary smoke
Liquid in a solid.................	Water dispersed in opal
	Water in butter?
Liquid in a liquid	Emulsion of kerosene in water
Liquid in a gas	Mist or fog
Gas in a solid.....................	Air dispersed in meerschaum
Gas in a liquid...................	Froths and foams

Of the possible colloidal systems, two are considered here: solids dispersed in liquids, and liquids dispersed in liquids.

Solids Dispersed in Liquids. There are two general types of solids dispersed in liquids: *lyophobic* and *lyophilic*. A lyophobic colloidal system is one in which there is very little attraction between the colloidal particles and the dispersion medium. Sometimes, these colloids are called "solvent-haters." If water is the dispersion medium, they are called *hydrophobic* colloidal systems.

Two examples are iron(III) oxide dispersed in water and gold particles dispersed in water. A lyophilic colloidal system is one in which there is considerable attraction between the colloidal particles and the dispersion medium. These are the "solvent-lovers." If water is the dispersion medium, they are called *hydrophilic* colloidal solutions. Soap dispersed in water and gelatin dispersed in water are examples. A colloidal system that is liquid (fluid) is called a *sol*. The concentration of the dispersed phase is usually low. Many sols are of the lyophobic type. A colloidal system, usually of the lyophilic type, that has "set" is called a *gel*. The more concentrated solutions set upon standing for a time. Jellies made from fruit juices are gels.

Preparation of Colloidal Systems. There are two general methods of preparing colloidal systems: by *condensation* and by *dispersion*. In the condensation method, many molecules of a substance come together to form an aggregate of colloidal size. These particles may be obtained by chemical action under suitable conditions. If a solution of arsenic trioxide is treated with hydrogen sulfide water, the arsenic sulfide molecules formed combine to give colloidal arsenic sulfide particles. Also, colloidal iron(III) oxide is obtained by the addition of $FeCl_3$ to hot water. The reaction is one of hydrolysis.

By the dispersion method, large particles are subdivided until colloidal particles are formed. Dispersion is obtained by mechanical disintegration, peptization by liquids and ions, and emulsification. Thus, if sand is ground until the particles are very small, they will remain permanently suspended in a dispersion medium.

Properties of Colloidal Systems. In the following paragraphs, the properties of colloidal systems common to both lyophobic and lyophilic types are considered first. Then the properties peculiar to each type are considered separately. The following properties are common to both types: The boiling and freezing points of colloidal systems are essentially the same as those of the dispersion media. The path of a beam of light passed through a colloidal solution is visible. This is called the *Tyndall effect*. The *Brownian motion* (the disordered movement of the particles in a colloidal solution) is observable with an ultramicroscope. Colloidal particles will not pass through an animal or vegetable membrane but substances in true solution will pass through. This process is called *dialysis*. Colloidal particles have a high adsorptive capacity, especially for ions. See Figures 17-1 and 17-2, p. 174.

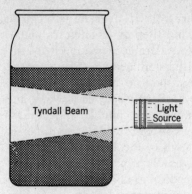

Fig. 17-1. The Tyndall effect.

Properties of Lyophobic Colloids. (1) The concentration of the dispersed phase is rather low. (2) The particles possess an electric charge. Iron(III) oxide particles are positively charged; arsenic trisulfide particles are negatively charged. Such sols are easily coagulated by electrolytes or by colloids of opposite charge. (3) There is very little attraction between the particles and the solvent. (4) The sol is not easily reversible. If the solvent is added to the coagulated substance, the sol usually cannot be restored. (5) Lyophobic colloids are made from substances whose molecular weights are low. (6) There are three stabilizing factors: (*a*) The particles are all charged the same; therefore, they do not collide. (*b*) The particles are in rapid motion. (*c*) Due to their

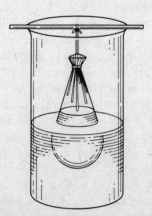

Fig. 17-2. A collodion sack used in dialysis.

small size, the particles are not appreciably affected by gravity. (7) Coagulation by electrolytes takes place in the order of the number of charges on the coagulating ion. Aluminum ions are more effective than barium ions, and barium ions more effective than sodium ions in coagulating arsenic trisulfide particles. (8) The viscosity of these solutions is practically the same as that of the dispersion medium.

Properties of Lyophilic Colloids. (1) The concentration of the dispersed phase can be high, much higher than is the case with lyophobic colloids. (2) The particles are charged, but the sign of the charge depends on the acidic or basic nature of the solvent. Such sols are not easily coagulated. They may be "salted out" by comparatively large concentrations of electrolytes. (3) The attraction between the colloidal substance and the dispersion medium is pronounced. (4) The sol is reversible. The solvent may be removed by evaporation; then upon adding more solvent, the sol is reformed. (5) Lyophilic colloids are usually substances of high molecular weight. Gelatin is about 20,000 and albumin 35,000. (6) Although lyophilic colloids show the Brownian motion, and are charged electrically, the most important stabilizing factor is the marked affinity between the particles and the dispersion medium. (7) Lyophilic colloids are not coagulated by electrolytes unless comparatively large amounts are added. This type of coagulation is called *salting out*. (8) The viscosity of lyophilic colloidal solutions is comparatively high. A 1 per cent soap solution is many times more viscous than water.

Protective Colloids. Lyophobic colloids can be stabilized by the addition of protective colloids. These are generally lyophilic colloids. The stabilizing effect is produced by the formation of a protective film (by adsorption) around each particle. The protected colloid acquires the properties of the colloid added. If gelatin solution is added to colloidal iron(III) oxide, the sol is protected and is not easily coagulated by electrolytes.

Liquids Dispersed in Liquids. When two immiscible liquids are shaken, droplets of one liquid are dispersed in the other. Such a colloidal system is called an *emulsion*. There are two general types of emulsions: temporary and permanent.

A temporary emulsion is one that breaks down quickly. This is due to the high surface tension of at least one of the liquids. If a small quantity of kerosene is shaken with water, a temporary emulsion is formed. Upon standing, it soon breaks down.

A permanent emulsion is one in which the droplets of the dispersed phase do not coalesce (grow into larger drops). The addition of a third substance, called an *emulsifying agent*, is necessary to make the emulsion permanent. The purposes of the emulsifying agent are: to lower the surface tension of the dispersing medium, and to form a protective film around each droplet of the dispersed phase. If a small quantity of soap solution is added to a kerosene and water mixture and shaken vigorously, a permanent emulsion is formed. The cleansing action of soap depends upon the fact that it emulsifies grease on the skin or clothing and adsorbs some of the particles of dirt, which can then be washed away with water. Many articles of food, such as milk and mayonnaise dressing, are emulsions.

Practical Colloidal Problems.   A knowledge of colloidal chemistry is important in the manufacture of photographic film, paints, varnishes, inks, and rubber goods. In the past few years water-based paints have been greatly improved by making use of the principles of colloid chemistry. These paints are now produced in amounts measured in millions of gallons per year. *Latex paints* are probably the most widely used of the water-based paints and are dispersions of resinous particles in water. These paints have several attractive properties. They are quick-drying, have low odor, are durable, and retain their color, and the paint brush can be easily cleaned by washing in water.

Colloidal chemistry also is important in the concentration of low-grade ores by the "flotation process," in the clarification and purification of water, and in the precipitation of obnoxious smokes and fumes (*Cottrell Process*). Since the human body is made up of many colloidal fluids and materials, many of the advances in medical science, no doubt, will be made through a knowledge of colloidal chemistry.

Recently our country has become very much concerned with the pollution of the air we breathe and the filthy condition of the water in our lakes and streams. Chemists believe that the application of colloid chemistry to pollution problems will help solve many of them.   Certainly the use of Cottrell precipitators has helped clean up the atmosphere in industrial centers where smokes and fumes are produced. See Figure 17-3.

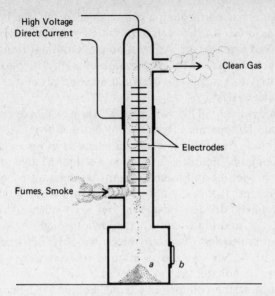

Fig. 17-3. Cottrell precipitator (schematic). (*a*) Precipitated smoke and fume particles. (*b*) Clean-out door.

Important Definitions

Cataphoresis. When a difference in potential is set up between two electrodes dipping into a colloidal solution, the charged particles will slowly migrate to the electrode of opposite charge. This process is called *cataphoresis* and is used to determine the sign of charge on colloidal particles.

Peptization. If a colloidal solution, partially or completely coagulated, is more highly dispersed by the addition of a certain reagent, it is said to have been *peptized*. Sometimes a dilute solution of an electrolyte will do this.

Electro-endosmosis. If a potential difference is applied between the water in a container and water in a porous cup that dips into the container, the water passes through the cup in the direction of the negative electrode. This movement of water through the porous cup under the influence of the electric current is called *electro-endosmosis*.

Adsorption. This is a process in which molecules, atoms, or ions become attached to the surface of a solid or a liquid. If ordinary air, contaminated with an obnoxious gas such as hydrogen sulfide, is passed through a charcoal gas mask, the hydrogen sulfide is *adsorbed* on the surface of the charcoal particles and is thus removed from the air. Adsorption plays an important role in certain reactions such as the preparation of methyl alcohol by the reaction of hydrogen gas and carbon monoxide gas on the surface of a suitable catalyst.

Chromatography. This is a process in which substances in solution can be separated from the solvent and from each other by adsorption. As a simple example let us suppose an aqueous solution contains complex substances (*A* and *B*) that are very similar in properties and therefore hard to separate by ordinary chemical means. If this solution is poured into a glass tube containing a finely divided adsorbent material such as calcium carbonate, the solution will percolate down through the tube. Let us assume molecules of *A* are more strongly adsorbed than molecules of *B*. Now as the solution percolates down through the calcium carbonate, the molecules of *A* will be adsorbed first, near the top of the column as a band. Somewhat below the *A* band another band containing molecules of *B* will form. If the *A* and *B* bands have different colors they can be identified easily. If not, other appropriate means must be used to identify them.

REVIEW QUESTIONS

1. A certain yellow solution is either colloidal arsenic sulfide or a dilute solution of potassium chromate. How can one determine what it is?
2. Distinguish between Brownian motion and the Tyndall effect.
3. What are the characteristics of a lyophobic colloidal solution?
4. What is a gel? A sol? A peptizing agent?
5. How can one separate substances in true solution from particles in colloidal suspension?
6. What are the stabilizing factors of a lyophilic colloidal solution?
7. Explain what is meant by: (*a*) protective colloid; (*b*) coagulation; (*c*) cataphoresis.
8. Distinguish between a temporary and a permanent emulsion.
9. What can one do to ordinary milk to make it coagulate?
10. What role does adsorption have in chromatography?
11. How does a Cottrell precipitator remove dust particles from air?

Chapter 18
The Halogen Family

Fluorine (9) $1s^2 2s^2 2p^5$
Chlorine (17) $1s^2 2s^2 2p^6 3s^2 3p^5$
Bromine (35) [*   ] $4s^2 4p^5$
Iodine (53) [   ] $5s^2 5p^5$
Astatine (85) [   ] $6s^2 6p^5$

*To save space and time the inner electron shells are not shown. The $s^2 p^5$ electrons are the valence electrons.

The halogens belong to group VII, family A of the periodic table. These elements combine with metals to form salts; hence the name *halogen*, which means *salt-former*. Their physical and chemical properties are closely related to their atomic structures. Chlorine and bromine each have two isotopes. Astatine (At), although predicted years ago, was not known until 1940. It was prepared in the laboratory by the bombardment of bismuth with alpha particles.

$$^{209}_{83}Bi + {}^4_2He \longrightarrow {}^{210}_{85}At + 3\,{}^1_0n$$

Astatine is radioactive and has a half-life of 8.3 hours. The element has never been isolated from natural sources. Consequently its physical and chemical properties are not included in this chapter.

Preparation. The general laboratory method of preparation of a halogen is to treat any metallic halide with sulfuric acid and an oxidizing agent.

$$2\,KBr + 3\,H_2SO_4 + MnO_2 \longrightarrow 2\,KHSO_4 + MnSO_4 + 2\,H_2O + Br_2$$
$$2\,NaCl + 3\,H_2SO_4 + MnO_2 \longrightarrow 2\,NaHSO_4 + MnSO_4 + 2\,H_2O + Cl_2$$

Iodine and bromine can also be prepared by replacement.

$$2\,NaI + Br_2 \longrightarrow 2\,NaBr + I_2$$
$$2\,KBr + Cl_2 \longrightarrow 2\,KCl + Br_2$$

Physical Properties. In passing from fluorine to iodine in the halogen family, there is a gradual change in physical properties. Table 18-1 summarizes this information.

Table 18-1

Physical Properties of the Halogens

	Fluorine	Chlorine	Bromine	Iodine
Color	pale yellow	greenish yellow	red	purple to black
Physical State	gas	gas	liquid	solid
Melting Point	−223°C	−102°C	−7°C	114°C
Boiling Point	−188°C	−35°C	59°C	183°C
Atomic Radius	0.72Å	0.99Å	1.14Å	1.33Å
Ionization Energy	17.34 v	12.95 v	11.8 v	10.6 v
Heat of Fusion Kcal/Mole	0.37	1.53	2.52	3.74
Electro- negativity	4.0	3.0	2.8	2.5
Density g/cm^3			3.12 (liquid)	4.94

All except fluorine are slightly soluble in water, but they are quite soluble in many organic solvents. Fluorine is so active that it decomposes water and other solvents. Iodine sublimes; that is, when heated and cooled it changes directly from solid to vapor and from vapor to solid without passing through the liquid state. However, iodine does melt and boil under suitable experimental conditions. See Figure 18-1.

$$\text{Iodine crystals} \; \underset{\text{cooling}}{\overset{\text{heating}}{\rightleftharpoons}} \; \text{Iodine vapor}$$

Chemical Properties. In passing from fluorine to iodine, there is a gradual change in the chemical activity. Fluorine is the most active and iodine the least active.

VALENCE. The common valence (oxidation number) of the

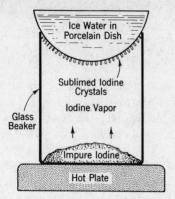

Fig. 18-1. Purification of iodine by sublimation.

group is −1. This is because their atoms have 7 electrons in their outer energy levels and thus have a tendency to acquire another electron to give the atom a stable structure. Halogen compounds are known in which the valence (oxidation number) of the halogen is +3, +5, or +7. (See Chapter 13.)

DISPLACEMENT SERIES.   Any halogen will replace a less active halogen from its compounds (halides).

$$2\,NaI + Br_2 \longrightarrow 2\,NaBr + I_2$$
$$2\,NaBr + Cl_2 \longrightarrow 2\,NaCl + Br_2$$
$$2\,NaCl + Br_2 \longrightarrow no\ reaction$$

REACTION WITH HYDROGEN.   All the halogens combine directly with hydrogen. The stability of these compounds decreases in passing from hydrogen fluoride to hydrogen iodide.

$$Cl_2 + H_2 \longrightarrow 2\,HCl$$
$$Br_2 + H_2 \longrightarrow 2\,HBr$$

METALLIC HALIDES.   All the halogens react with metals to form halides.

$$Mg + Cl_2 \longrightarrow MgCl_2$$

NONMETALLIC HALIDES.   Many nonmetals combine directly with the halogens.

$$2\,P + 3\,Br_2 \longrightarrow 2\,PBr_3$$

OTHER REACTIONS.   In the following equations let X = any halogen.

$$X_2 + CO \longrightarrow COX_2$$
$$X_2 + SO_2 \longrightarrow SO_2X_2$$
$$X_2 + H_2S \longrightarrow 2\,HX + S$$
$$X_2 + H_2O \rightleftharpoons HX + HXO$$

Fluorine. The earth's crust contains about 0.1 per cent fluorine. It is found in minerals such as fluorite (Fluorspar) (CaF_2) in Illinois, and cryolite (Na_2AlF_6) in Greenland and Iceland. The phosphate mineral called *apatite* contains fluorine. Its formula is $CaF_2 \cdot 3\,Ca_3(PO_4)_2$.

PREPARATION. Fluorine is prepared by the electrolysis of fused sodium or potassium hydrogen fluoride, using a copper vessel and graphite electrodes. Moissan, the French chemist, was the first to prepare fluorine. He electrolyzed a solution of potassium acid fluoride dissolved in liquid HF in a platinum alloy vessel. See Figure 18-2.

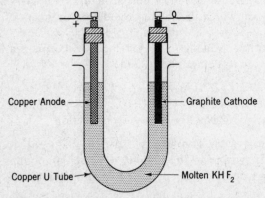

Fig. 18-2. Preparation of fluorine by electrolysis.

PROPERTIES. Fluorine is a light-yellow gas, 1.31 times denser than air. Fluorine is the most active element. It decomposes water with the liberation of oxygen and traces of ozone. It does not attack the noble metals.

Compounds of Fluorine. Hydrogen fluoride (hydrofluoric acid, HF) is the most important compound of fluorine. It is prepared in the laboratory by treating any metallic fluoride with a non-volatile acid. Hydrogen and fluorine will combine directly to give hydrogen fluoride, but the method is impracticable.

$$CaF_2 + H_2SO_4 \longrightarrow CaSO_4 + 2\,HF$$

Hydrogen fluoride is a colorless gas which is very soluble in water (100 g of saturated solution contains 35 g of the gas). It attacks silicon dioxide or substances such as glass. This property makes it necessary to keep hydrofluoric acid in wax or plastic bottles. The reaction is:

$$SiO_2 + 4\,HF \longrightarrow SiF_4 + 2\,H_2O$$

The gas forms a constant boiling solution with water. In water it is very slightly dissociated (0.15%) into ions.

Hydrofluoric acid is used for etching glass and for dissolving complex silicates. The compound CCl_2F_2, called "Freon-12" compound, is odorless and nontoxic, and is used as a refrigerant. It is prepared from CCl_4 and SbF_3 according to the following equation:

$$3\,CCl_4 + 2\,SbF_3 \longrightarrow 3\,CCl_2F_2 + 2\,SbCl_3$$
$$\text{("Freon-12")}$$

It is claimed that one part per million of fluorine in drinking water prevents the decay of teeth.

Chlorine.   Chlorine was first prepared in 1774 by the Swedish chemist Scheele, by heating muriatic acid (HCl) with an oxidizing agent such as manganese dioxide. The new gas was called *oxymuriatic acid gas* because it was thought to contain oxygen. Chlorine never occurs free in nature. Of the many compounds of chlorine, sodium chloride is the most important. It is found in the oceans and salt lakes. Ocean water contains about 3 per cent sodium chloride. The Dead Sea in Palestine and the Great Salt Lake in Utah are practically saturated solutions of sodium chloride. Salt wells and beds of rock salt are found in various places, especially in Michigan.

PREPARATION.   Hydrochloric acid is treated with an oxidizing agent. Manganese dioxide (MnO_2) and potassium permanganate ($KMnO_4$) are suitable oxidizing agents. See Figure 18-3, p. 184.

$$MnO_2 + 4\,HCl \longrightarrow MnCl_2 + 2\,H_2O + Cl_2$$
$$2\,KMnO_4 + 16\,HCl \longrightarrow 2\,KCl + 2\,MnCl_2 + 8\,H_2O + 5\,Cl_2$$

Electrolysis of Brine.   This is the commercial method of preparing chlorine. The Nelson cell is shown in Figure 18-4, p. 185.

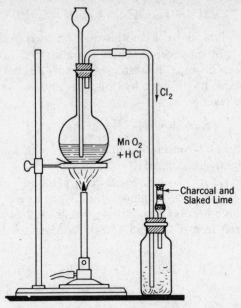

Fig. 18-3. An apparatus used in preparing chlorine in the laboratory.

Action of Acid on Bleaching Powder.   The equation is:

$$CaOCl_2 + H_2SO_4 \longrightarrow CaSO_4 + H_2O + Cl_2$$

Deacon Process.   Hydrogen chloride is oxidized by the oxygen of the air in the presence of a suitable catalyst. This method was formerly of commercial importance.

PHYSICAL PROPERTIES.   Chlorine is a greenish-yellow gas, 2.5 times denser than air. It has a strong irritating odor. One volume of cold water will dissolve two volumes of chlorine. Chlorine can be liquefied by applying a pressure of 6 atmospheres.

CHEMICAL PROPERTIES.   Active metals burn in chlorine. It combines directly with metals and nonmetals. It reacts with water or soluble bases to form hypochlorous acid and hydrochloric acid or their salts. Chlorine replaces bromine and iodine from their salt solutions. In the presence of sunlight (or arc-light) chlorine combines with hydrogen with explosive violence. The chemical properties of chlorine are closely related to the structure of its atoms.

$$Mg + Cl_2 \longrightarrow MgCl_2$$
$$2P + 5Cl_2 \longrightarrow 2PCl_5$$

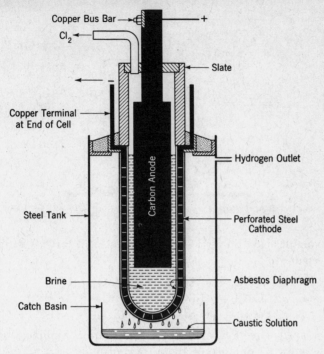

Fig. 18-4. Cross section of the Nelson cell for the preparation of chlorine by the electrolysis of brine.

$$H_2O + Cl_2 \rightleftharpoons HCl + HClO$$
$$2\,NaBr + Cl_2 \longrightarrow 2\,NaCl + Br_2$$

USES. Chlorine is used as a bleaching agent for cotton goods and in the manufacture of bleaching powder. It is used extensively in the purification of water; germs are killed by low concentrations of chlorine in water. Chlorine was used as a poisonous gas during World War I. Some of the compounds now prepared for use in insecticides contain chlorine. Dichlorodiphenyltrichloroethane $[(C_6H_4Cl)_2 \cdot CHCCl_3]$, commonly called DDT, contains 5 atoms of chlorine per molecule.

Hydrogen Chloride. Small quantities of hydrogen chloride occur in the gastric juice of the stomach. Traces have also been detected in the gases escaping from volcanoes.

PREPARATION. The methods of preparation of hydrogen chloride are:

General Method. The chloride of a metal is treated with a

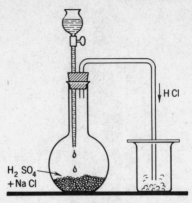

Fig. 18-5. Laboratory preparation of hydrogen chloride.

nonvolatile acid. Sulfuric acid is the best acid for this purpose. See Figure 18-5.

$$NaCl + H_2SO_4 \longrightarrow NaHSO_4 + HCl$$
$$CaCl_2 + H_2SO_4 \longrightarrow CaSO_4 + 2\,HCl$$
$$2\,FeCl_3 + 3\,H_2SO_4 \longrightarrow Fe_2(SO_4)_3 + 6\,HCl$$

Direct Combination. Hydrogen and chlorine combine directly. In the presence of sunlight and a trace of moisture, the reaction takes place with explosive violence.

$$H_2 + Cl_2 \longrightarrow 2\,HCl$$

Physical Properties. Hydrogen chloride is a colorless gas with a sharp, penetrating odor. When inhaled, it has a suffocating effect. It is very soluble in water; one volume of water dissolves more than 500 volumes of the gas. It is 1.26 times denser than air. At 0°C it may be liquefied by applying a pressure of 28 atmospheres. Pure hydrogen chloride is a nonconductor of an electric current. Dissolved in water, it is an excellent conductor because the molecules are ionized. Hydrogen chloride dissolved in water forms a constant boiling solution. This solution is used as a standard in the preparation of normal hydrochloric acid (see Chapter 30).

Hydrogen chloride is a covalent molecule. Since, however, the electron pair is closer to the chlorine atom than to the hydrogen nucleus, the molecule has an appreciable dipole moment. Dissolved in benzene or toluene, hydrogen chloride is a nonconductor, because the molecules are not dissociated into ions.

CHEMICAL PROPERTIES.   Hydrogen chloride is very stable. It does not burn or support combustion. It combines with ammonia to form ammonium chloride.

Hydrochloric Acid.   This is a solution of hydrogen chloride in water. A concentrated solution has a specific gravity of 1.2 and contains about 40 per cent HCl. Hydrochloric acid is highly ionized and, therefore, is a strong acid. It reacts readily with metals above hydrogen in the activity series. Typical reactions are:

$$Mg + 2\,HCl \longrightarrow MgCl_2 + H_2$$
$$CaCO_3 + 2\,HCl \longrightarrow CaCl_2 + H_2O + CO_2$$
$$Na_2SO_3 + 2\,HCl \longrightarrow 2\,NaCl + H_2O + SO_2$$

The acid is used in the manufacture of dyes; for extracting glue from bones; in the preparation of glucose from starch; in the preparation of soap; and for cleaning the oxidized scales on metal surfaces.

The test for hydrochloric acid and other soluble chlorides makes use of silver ion (Ag^+). The silver ion forms an insoluble silver chloride with chloride (Cl^-) ion.

$$Ag^+ + Cl^- \longrightarrow AgCl \text{ (curdy, white precipitate)}$$

Bromine.   This halogen has been known since 1826. It was first prepared from salts obtained by the evaporation of sea water. Besides being found in sea water, bromine has been found in certain salt wells in Michigan and also in the Stassfurt salt deposits in Germany.

PREPARATION.   Commercially, bromine is prepared by the replacement method. In the laboratory it is prepared by the oxidation of hydrobromic acid. Bromine is now successfully extracted from sea water on a commercial scale.

$$MgBr_2 + Cl_2 \longrightarrow MgCl_2 + Br_2 \text{ (commercial)}$$
$$4\,HBr + MnO_2 \longrightarrow MnBr_2 + 2\,H_2O + Br_2$$
$$2\,NaBr + 3\,H_2SO_4 + MnO_2 \longrightarrow 2\,NaHSO_4 + MnSO_4 + 2\,H_2O + Br_2$$

PROPERTIES.   Bromine is a heavy, red liquid with a disagreeable odor. Its vapor irritates the eyes and throat. Bromine is slightly soluble in water but readily soluble in carbon disulfide

(CS_2). It reacts directly with hydrogen, the metals, the nonmetals, and water. It also replaces iodine from iodide solutions.

$$Br_2 + H_2O \longrightarrow HBr + HBrO$$
$$Br_2 + 2\,NaI \longrightarrow 2\,NaBr + I_2$$

Hydrobromic Acid. Pure hydrogen bromide (hydrobromic acid) cannot be made by the action of H_2SO_4 on NaBr because HBr is oxidized to bromine and water by sulfuric acid. To make pure HBr, water is allowed to react with PBr_3. The following equations show what happens:

(a) $2\,NaBr + 2\,H_2SO_4 \longrightarrow 2\,NaHSO_4 + 2\,HBr$
(b) $2\,HBr + H_2SO_4 \longrightarrow 2\,H_2O + SO_2 + Br_2$

Adding (a) and (b) to give (c):

(c) $2\,NaBr + 3\,H_2SO_4 \longrightarrow 2\,NaHSO_4 + 2\,H_2O + SO_2 + Br_2$
(d) $PBr_3 + 3\,H_2O \longrightarrow 3\,HBr + H_3PO_3$ (no free bromine)

PROPERTIES. Hydrogen bromide (HBr) is a colorless gas that dissolves in water to give a strong acid. This compound is more readily oxidized than HCl.

USES. Hydrogen bromide (hydrobromic acid) is used to prepare bromides, such as silver bromide (AgBr) used in photography, and as a reducing agent in organic chemistry. Salts of hydrobromic acid (KBr and NaBr) are used as sedatives by the medical profession.

Iodine. The earth's crust contains about 0.001 per cent iodine. It is widely distributed as shown in the following: Sodium iodide, potassium iodide, and magnesium iodide are found in sea water. Iodine is found in sea weeds such as kelp. Sodium iodate $(NaIO_3)$ is found in the Chile saltpeter beds; this is the important commercial source.

COMMERCIAL PREPARATION. Iodine is liberated from the iodate solution (which remains after sodium nitrate has been extracted from the saltpeter) by the action of the sulfites of sodium.

$$2\,NaIO_3 + 3\,Na_2SO_3 + 2\,NaHSO_3 \longrightarrow 5\,Na_2SO_4 + H_2O + I_2$$

LABORATORY PREPARATION. An iodide is treated with dilute sulfuric acid and an oxidizing agent such as manganese dioxide.

$$2\,NaI + 2\,H_2SO_4 + MnO_2 \longrightarrow 2\,NaHSO_4 + MnSO_4 +$$
$$2\,H_2O + I_2$$

Iodine is purified by sublimation. The iodine is vaporized by heating. The pure vapor is condensed (as crystals) on a cold surface (see Figure 18-1, p. 181).

PHYSICAL PROPERTIES. Iodine is a shiny, dark-gray crystalline solid that gives off violet vapors upon heating. It is slightly soluble in water but very soluble in a solution of potassium iodide, probably because of the formation of a complex KI_3 (or $KI \cdot I_2$). Iodine dissolves in alcohol (tincture of iodine), carbon tetrachloride, and carbon disulfide.

CHEMICAL PROPERTIES. Iodine combines directly with metals and nonmetals.

$$2\,Na + I_2 \longrightarrow 2\,NaI$$
$$2\,P + 3\,I_2 \longrightarrow 2\,PI_3$$

Iodine turns starch blue. This is a test for iodine.

USES. Iodine is used to make iodides, to prepare iodoform, in dye manufacturing, and to make tincture of iodine. It is an excellent antiseptic.

Compounds of Iodine. Hydrogen iodide (hydriodic acid, HI), salts of HI, and iodoform, CHI_3, are the important compounds of iodine. Iodoform is an important antiseptic; potassium iodide and silver iodide are used in photography and in medicine. The human body contains a small quantity of iodine. A compound called *iodothyrin*, found in the thyroid gland, is necessary to prevent goiter and similar disorders. "Iodized" table salt is on the market; it is thought to be useful in preventing goiter.

Oxygen Compounds of the Halogens. Some interesting chemical reactions take place when oxygen-containing halogen compounds are formed.

OXIDES. There are three important oxides, oxygen difluoride (OF_2), chlorine monoxide (Cl_2O), and iodine pentoxide (I_2O_5). The first is prepared by passing fluorine gas through cold dilute sodium hydroxide.

$$F_2 + 2\,NaOH \longrightarrow 2\,NaF + OF_2 + H_2O$$

This oxide is interesting because in it oxygen is electropositive to chlorine. See p. 45 for the table on electronegativity. Chlorine monoxide (Cl_2O) is prepared by bubbling chlorine through a suspension of HgO in water.

$$2\,Cl_2 + 2\,HgO \longrightarrow HgO \cdot HgCl_2 + Cl_2O$$

It liquifies at 5°C and reacts with water to give hypochlorous acid.

$$Cl_2O + H_2O \longrightarrow 2\,HClO$$

Iodine pentoxide (I_2O_5) is used as an oxidizing agent and is prepared by heating iodic acid to 200°C.

$$2\,HIO_3 \longrightarrow H_2O + I_2O_5$$

OXYACIDS.   Hypochlorous acid (HClO) is prepared by:
(*a*) adding Cl_2 to cold water.

$$Cl_2 + H_2O \rightleftharpoons HCl + HClO$$

(*b*) adding Cl_2O to water.

$$Cl_2O + H_2O \longrightarrow 2\,HClO$$

(*c*) treating potassium hypochlorite with sulfuric acid.

$$2\,KClO + H_2SO_4 \longrightarrow K_2SO_4 + 2\,HClO$$

It is an oxidizing (and bleaching) agent.
Chloric acid is prepared by adding H_2SO_4 to barium chlorate.

$$Ba(ClO_3)_2 + H_2SO_4 \longrightarrow 2\,HClO_3 + BaSO_4$$

It is a strong but unstable acid.
It converts iodine to iodic acid.

$$2\,HClO_3 + I_2 \longrightarrow 2\,HIO_3 + Cl_2$$

OXYSALTS OF HALOGENS.   Sodium hypochlorite (NaClO) is prepared by passing chlorine gas through cold dilute NaOH.

$$2\,NaOH + Cl_2 \longrightarrow NaCl + NaClO + H_2O$$

Sodium hypochlorite is an oxidizing agent and a bleacher. Commercially, NaClO is sold as Clorox and Purex and is used to whiten clothes in the washing process.

Potassium chlorate ($KClO_3$) is prepared by passing Cl_2 into hot KOH solution.

$$3\,Cl_2 + 6\,KOH \longrightarrow KClO_3 + 5\,KCl + 3\,H_2O$$

It is used in making matches, fireworks, and explosives. It is also used as an oxidizing agent.

REVIEW QUESTIONS

1. From the standpoint of atomic structure, what is the difference between chlorine and bromine?
2. What is a general method of preparing a halogen?
3. Write an equation to show how one halogen can replace another.
4. What is meant by the statement, "Iodine sublimes"?
5. How can one prepare: (a) HCl, (b) HBr, (c) HF?
6. The compound CCl_2F_2 is called *Freon*. How is it made?
7. What is the action of HF on silica? Write an equation to illustrate this action.
8. How is chlorine manufactured commercially?
9. What happens when hot iron (turnings) is placed in a bottle of chlorine gas?
10. Why cannot pure HBr be prepared by the action of sulfuric acid on NaBr?
11. How can iodine be liquefied?
12. Why is oxygen difluoride written with the oxygen coming first in the formula?
13. How is a dilute solution of hypochlorous acid made?
14. Give the equation for the preparation of chloric acid.
15. What is the action of heat on potassium chlorate?

Chapter 19
The Nitrogen Family

Nitrogen (7) $1s^2 2s^2 2p^3$
Phosphorus (15) [   ] $3s^2 3p^3$
Arsenic (33) [   ] $4s^2 4p^3$
Antimony (51) [   ] $5s^2 5p^3$
Bismuth (83) [   ] $6s^2 6p^3$

These elements belong to group VA. Nitrogen is a gas; the others are solids. In passing from phosphorus to bismuth, there is a transition from nonmetallic to metallic properties. All except bismuth exist in more than one crystalline form. The important physical properties are summarized in Table 19-1.

Table 19-1

Physical Properties of the Nitrogen Family

	Nitrogen	Phosphorus	Arsenic	Antimony	Bismuth
Color	colorless	white red (violet) black	silvery gray	silvery gray	silvery gray
Physical State	gas	solid	solid	solid	solid
Melting Point	−210°C	44°C (white)	814°C (36 atm)	631°C	271°C
Boiling Point	−196°C	280°C (white)	610°C (sublimes)	1380°C	1560°C
Atomic Radius	0.74 Å	1.10 Å	1.21 Å	1.41 Å	1.52 Å
Ionization Energy	14.5 v	11.0 v	10 v	8.6 v	8 v
Heat of Fusion cal/g	6.2	5.0		38.3	12.5
Electro- negativity	3.0	2.1	2.0	1.8	1.9
Density g/cm³ at 20°C	1.165×10^{-3}	1.83	5.73	6.62	9.8

Nitrogen. Nitrogen, a comparatively inert gas, is the first member of Group VA of the periodic table. The 5 electrons in the highest energy level of the nitrogen atom account for its valence and other atomic properties.

OCCURRENCE. Nitrogen occurs free and in the combined state. Dry air contains 78 per cent nitrogen by volume. This amounts to more than 20,000,000 tons over each square mile of the earth's surface. Nitrogen is present in ammonium salts, nitrates, and nitrites found in fertile soils; in animal and vegetable matter such as proteins; and in certain natural deposits such as sodium nitrate (Chile saltpeter).

PREPARATION. Nitrogen is extracted from the air and also prepared from compounds.

From Air. If phosphorus is burned in a vessel of air over water, the oxygen is removed as phosphoric pentoxide. This oxide dissolves in the water and nitrogen remains.

$$4 P + 5 O_2 \longrightarrow 2 P_2O_5$$
$$P_2O_5 + 3 H_2O \longrightarrow 2 H_3PO_4$$

From Liquid Air. If liquid air is allowed to boil, the nitrogen escapes first. This is an important commercial method of preparing the gas.

From Compounds. Certain compounds, upon decomposition, yield pure nitrogen. For example, if a mixture of an ammonium salt and sodium nitrite is heated, ammonium nitrite is formed. This compound is unstable and breaks down to give water and nitrogen.

$$NH_4Cl + NaNO_2 \longrightarrow NaCl + NH_4NO_2$$
$$NH_4NO_2 \longrightarrow 2 H_2O + N_2$$

PROPERTIES AND USES. Nitrogen is a colorless, odorless, tasteless gas that is very slightly soluble in water. It is slightly lighter than air. One liter weighs 1.2596 g. Nitrogen is comparatively inactive chemically. It does not burn or support combustion. At suitable temperatures and pressures, it combines with oxygen, certain metals, and hydrogen.

$$N_2 + O_2 \longrightarrow 2 NO$$
$$3 Mg + N_2 \longrightarrow Mg_3N_2$$
$$3 H_2 + N_2 \longrightarrow 2 NH_3 \text{ (catalyst needed here)}$$

Nitrogen is used as a food by plants, to prepare ammonia by

direct combination of the elements, to make calcium cyanamide, and to synthesize nitric acid.

Ammonia.   Ammonia (NH_3) has been known for a long time. Its water solutions have been called *spirits of hartshorn.* In 1774, Priestley collected the gas over mercury and made a study of its properties. He called the gas "alkaline air." In 1785, Berthollet proved that ammonia contains hydrogen and nitrogen.

PREPARATION.   In the presence of a suitable catalyst and under proper temperature and pressure control, ammonia is prepared by the direct combination of the elements. This is the Haber Process (see p. 201).

Ammonia is a by-product of the destructive distillation of coal. It may also be prepared by the action of superheated steam on calcium cyanamide.

$$CaCN_2 + 3\,H_2O \longrightarrow CaCO_3 + 2\,NH_3$$

If any ammonium salt is mixed with a base and heated, ammonia is liberated. This is one laboratory method. Figure 19-1 illustrates another.

$$2\,NH_4Cl + Ca(OH)_2 \longrightarrow CaCl_2 + 2\,NH_4OH$$
$$NH_4OH\,(heat) \longrightarrow NH_3 + H_2O$$

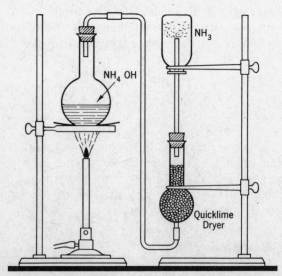

Fig. 19-1.   Apparatus for preparing ammonia.

PROPERTIES. Ammonia is a colorless gas with a strong pungent odor. It is easily liquefied and solidified, and is very soluble in water. One volume of water will dissolve more than 1,000 volumes of NH_3 under standard conditions.

Action of Heat. Ammonia is decomposed by heat.

$$2 NH_3 \text{ (heat)} \longrightarrow N_2 + 3 H_2$$

Action of Acids. Ammonia reacts directly with acids to form ammonium salts.

$$HCl + NH_3 \longrightarrow NH_4Cl$$
$$HNO_3 + NH_3 \longrightarrow NH_4NO_3$$

Action of Water. Ammonia reacts with water to form ammonium hydroxide, a weak base.

$$NH_3 + H_2O \longrightarrow NH_4OH$$

Action of Oxygen. In the presence of a catalyst, ammonia reacts with oxygen to form nitric oxide, which subsequently may be converted into nitric acid. This is the Ostwald Process.

$$4 NH_3 + 5 O_2 \text{ (catalyst)} \longrightarrow 4 NO + 6 H_2O$$

See paragraph on the commercial preparation of nitric acid (p. 199).

USES. Ammonia is used for refrigeration, for making nitric acid, and in the preparation of sodium bicarbonate by the Solvay Process. Dilute water solutions are used as a cleaning fluid. Liquid ammonia is an important solvent used in certain chemical investigations.

Manufacture of Ice. At the present time, a large percentage of the ice used in the United States is made in refrigeration machines. The physical properties of ammonia (and certain other compounds) make these machines possible. The steps in making ice, using ammonia, are as follows. Ammonia is liquefied by compressing and cooling it. The pressure on the liquid ammonia is then released by passing it through an expansion valve, into coils of pipe immersed in brine. The liquid ammonia evaporates with the absorption of heat from the brine. Pure water, in containers which dip into the brine, is frozen. The gaseous ammonia is compressed and used again. This makes the process continuous. Other substances, such as sulfur dioxide, "Freon-12" compound (CCl_2F_2), and methyl chloride (CH_3Cl), are also used in refrigeration machines. See Figure 19-2.

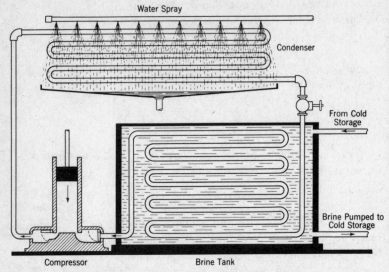

Fig. 19-2. Diagram of a refrigeration plant.

Fruits and vegetables can be kept fresh for several months or years by cooling them quickly to about $-50°F$. Special refrigerating machines are made for the purpose. Sudden cooling results in the formation of tiny ice crystals which do not pierce the cell walls. After the quick-cooling process has been completed, the fruits and vegetables are stored in rooms maintained at temperatures near $0°F$ until they are sent to the market.

Ammonium Compounds. Ammonium compounds contain the radical NH_4. It plays the role of a univalent metal.

Ammonium Hydroxide (NH_4OH). This compound is prepared by adding ammonia to water. It is a weak base, used to make ammonium salts and certain insoluble hydroxides.

$$NH_3 + H_2O \longrightarrow NH_4OH$$
$$2\,NH_4OH + H_2SO_4 \longrightarrow (NH_4)_2SO_4 + 2\,H_2O$$

Ammonium Sulfate [$(NH_4)_2SO_4$]. Ammonium sulfate is prepared by neutralizing sulfuric acid with ammonia water. It is used as a fertilizer.

AMMONIUM CHLORIDE (NH_4Cl). This salt is formed when hydrogen chloride fumes come in contact with ammonia. The salt is used in dry batteries and in soldering.

TEST FOR AMMONIA. Ammonium salts may be detected by mixing them with a base such as calcium hydroxide and heating. If ammonia is liberated, it can be detected by its odor or by its action on moist red litmus paper.

Nitrogen Oxides and Acids. Other important compounds of nitrogen besides ammonia are listed here.

NITROUS OXIDE (N_2O). Nitrous oxide, sometimes called *laughing gas*, is prepared by heating ammonium nitrate.

$$NH_4NO_3 \text{ (heat)} \longrightarrow N_2O + 2 H_2O$$

Nitrous oxide is a colorless gas with a faint odor and sweet taste. It is appreciably soluble in water and can be liquefied at 0°C by applying 30 atmospheres pressure. Its most important property is its ability to support combustion. The equation for the reaction is:

$$C + 2 N_2O \longrightarrow CO_2 + 2 N_2$$

The most important use of nitrous oxide is as an anaesthetic for minor operations. It is supplied to dentists and surgeons in liquid form in small cylinders.

NITRIC OXIDE (NO). This compound has been known since early times. Priestley made a careful study of its properties. When air is passed through an electric arc, a small percentage of the oxide is formed.

$$O_2 + N_2 \longrightarrow 2 NO$$

When ammonia is oxidized in the presence of a catalyst, nitric oxide is one product.

$$4 NH_3 + 5 O_2 \longrightarrow 4 NO + 6 H_2O$$

In the laboratory, dilute nitric acid reacts with copper turnings to give nitric oxide. The equation is:

$$3 Cu + 8 HNO_3 \longrightarrow 3 Cu(NO_3)_2 + 2 NO + 4 H_2O$$

In the presence of ferrous sulfate and dilute sulfuric acid, dilute nitric acid reacts to give pure nitric oxide.

$$6 FeSO_4 + 3 H_2SO_4 + 2 HNO_3 \longrightarrow 3 Fe_2(SO_4)_3 + 2 NO + 4 H_2O$$

Nitric oxide is slightly heavier than air, colorless, and only slightly soluble in water. Nitric oxide does not burn, but will support combustion at high temperatures. It combines readily with oxygen at room temperature, forming nitrogen dioxide.

$$2\,NO + O_2 \longrightarrow 2\,NO_2$$

Nitric oxide is used in making nitric acid and also in the manufacture of sulfuric acid by the Lead Chamber Process.

NITROGEN DIOXIDE (NO_2). This oxide is made by the action of oxygen on nitric oxide and by the action of heat on a heavy metal nitrate.

$$2\,NO + O_2 \longrightarrow 2\,NO_2$$
$$2\,Pb(NO_3)_2\ (heat) \longrightarrow 2\,PbO + O_2 + 4\,NO_2$$

It is a brown gas with a disagreeable odor, is soluble in water, and is easily liquefied. The liquid has the formula N_2O_4.

$$2\,NO_2 \rightleftharpoons N_2O_4$$

The two oxides, NO_2 and N_2O_4, are called *polymers* because they have the same percentage composition but different molecular weights.

Important chemical properties are: Nitrogen dioxide does not burn, but it will support the combustion of phosphorus and carbon. It is an oxygen carrier in the manufacture of sulfuric acid by the Chamber Process. It is a double acid anhydride. It reacts with water or a soluble base, forming nitrous and nitric acids or their salts.

$$2\,NO_2 + H_2O \longrightarrow HNO_3 + HNO_2$$
$$2\,NO_2 + 2\,KOH \longrightarrow KNO_3 + KNO_2 + H_2O$$

Nitrogen dioxide is used in making nitric and sulfuric acids.

NITROGEN PENTOXIDE (N_2O_5). If nitric acid is dehydrated by phosphorus pentoxide, nitrogen pentoxide is formed. The reaction is

$$P_2O_5 + 2\,HNO_3 \longrightarrow 2\,HPO_3 + N_2O_5$$

Nitrogen pentoxide is a colorless, crystalline solid which decomposes at room temperature. It is the acid anhydride of nitric acid.

There are no important uses of nitrogen pentoxide. However,

our knowledge of the mechanism of chemical reactions has been increased by a study of the decomposition of this oxide.

NITRIC ACID (HNO_3).   This acid was known to the alchemists, who called it *aqua fortis*.

Laboratory Preparation.   Nitric acid is prepared when a salt of the acid is treated with sulfuric acid. See Figure 19-3.

$$NaNO_3 + H_2SO_4 \longrightarrow NaHSO_4 + HNO_3$$

Commercial Preparation.   The action of sulfuric acid on Chile saltpeter has been an important commercial method of preparing the acid. Nitric acid is now made by two well-known methods: the Arc Process and the oxidation of ammonia (Ostwald Process).

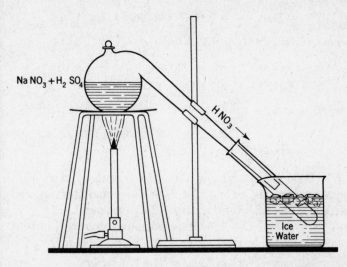

Fig. 19-3.   Preparation of nitric acid in the laboratory. Note that the apparatus is all glass.

In the Arc Process, air is passed through an electric arc where a small quantity of nitric oxide is formed. This oxide reacts with the oxygen present to give the dioxide. Nitrogen dioxide is passed into hot water to give nitric acid.

$$N_2 + O_2 \longrightarrow 2\,NO$$
$$2\,NO + O_2 \longrightarrow 2\,NO_2$$
$$3\,NO_2 + H_2O \longrightarrow 2\,HNO_3 + NO$$

In the Ostwald Process, a mixture of ammonia and air is passed over a catalyst (platinum gauze). The reactions are:

$$4 NH_3 + 5 O_2 \text{ (catalyst)} \longrightarrow 4 NO + 6 H_2O$$
$$2 NO + O_2 \longrightarrow 2 NO_2$$
$$3 NO_2 + H_2O \longrightarrow 2 HNO_3 + NO$$

Physical Properties. Nitric acid is a colorless liquid with a sharp choking odor. Its specific gravity is 1.52 and its boiling point is 86°C. It forms a constant boiling solution with water (68 per cent HNO_3). Commercial nitric acid is brown or yellowish-brown due to dissolved oxides of nitrogen.

Chemical Properties. The important chemical properties of nitric acid are: It is unstable when heated or exposed to light.

$$4 HNO_3 \longrightarrow 2 H_2O + 4 NO_2 + O_2$$

It is a powerful oxidizing agent.

$$3 Cu + 8 HNO_3 \longrightarrow 3 Cu(NO_3)_2 + 2 NO + 4 H_2O$$
$$C + 4 HNO_3 \longrightarrow 4 NO_2 + CO_2 + 2 H_2O$$

With water, it forms a strong acid because it is highly ionized.

$$HNO_3 \longrightarrow H^+ + NO_3^-$$
$$HNO_3 + H_2O \longrightarrow H_3O^+ + NO_3^-$$

Uses. Nitric acid is used in the manufacture of explosives, photographic supplies, celluloid, dyes, sulfuric acid, nitrates, and fertilizers. It is also an important laboratory reagent.

NITROUS ACID (HNO_2). A dilute, unstable solution of nitrous acid can be prepared by treating a nitrite with dilute H_2SO_4.

$$NaNO_2 + H_2SO_4 \longrightarrow NaHSO_4 + HNO_2$$

Sodium nitrite can be prepared by heating $NaNO_3$ with lead.

$$Pb + NaNO_3 \longrightarrow NaNO_2 + PbO$$

Nitrous acid is an unstable substance, the water solution of which is pale blue.

The solutions of the acid may act as an oxidizing or a reducing agent.

$$2 HI + 2 HNO_2 \longrightarrow 2 H_2O + 2 NO + I_2 \text{ (oxidizing agent)}$$

$$2 KMnO_4 + 3 H_2SO_4 + 5 HNO_2 \longrightarrow$$
$$K_2SO_4 + 2 MnSO_4 + 5 HNO_3 + 3 H_2O \text{ (reducing agent)}$$

Nitrous acid and its salts have a limited use in the chemical laboratory as reagents.

Fixation of Atmospheric Nitrogen.   Any process in which atmospheric nitrogen is converted into useful compounds is called *nitrogen fixation*. There are several ways in which this can be accomplished. The important ones are given here.

NATURAL FIXATION.   Nitrogen is fixed by such agencies as bacteria and lightning.

Nitrifying Bacteria.   The bacteria live on the roots of certain leguminous plants (peas, beans, clover, and alfalfa). These bacteria convert atmospheric nitrogen into nitrites and nitrates which may be assimilated by the plants.

Lightning.   Lightning causes the formation of small quantities of nitric oxide (NO) which is subsequently converted into nitric acid and nitrates.

FIXATION BY MAN.   Nitrogen is now fixed by a number of commercial methods. The following are important or interesting.

Arc Process.   Air is passed through an electric arc. A small quantity of nitric oxide is formed. This is subsequently converted into nitric acid and nitrates.

Cyanamide Process.   Calcium carbide (CaC_2) is heated in an atmosphere of nitrogen, forming calcium cyanamide ($CaCN_2$).

$$CaC_2 + N_2 \longrightarrow CaCN_2 + C$$

Ammonia is prepared by treating calcium cyanamide with superheated steam.

$$CaCN_2 + 3 H_2O \longrightarrow CaCO_3 + 2 NH_3$$

Nitride Process.   If certain metals such as magnesium or aluminum are heated in an atmosphere of nitrogen, a nitride is formed. The nitride then reacts with water to give ammonia.

$$2 Al + N_2 \longrightarrow 2 AlN$$
$$AlN + 3 H_2O \longrightarrow Al(OH)_3 + NH_3$$

Haber Process.   Nitrogen and hydrogen combine, under suitable temperature and pressure, in the presence of a catalyst. An important catalyst for this reaction is finely divided iron containing small amounts of certain oxides such as aluminum oxide and potassium oxide. The highest yields have been obtained at 500°C and 200 atmospheres pressure.

$$N_2 + 3 H_2 \text{ (catalyst)} \Longleftrightarrow 2 NH_3$$

See the paragraph on the Ostwald Process (p. 200). Ammonia may be converted into nitric acid by passing it, mixed with oxygen, over heated platinum gauze, which serves as a catalyst. The Haber (ammonia) Process is probably the most important method of fixing nitrogen.

The Nitrogen Cycle in Nature.   Nitrogen of the air cannot be assimilated by plants and animals as is the case with carbon dioxide. Certain nitrogen compounds, however, may be used by plants and animals. These compounds are synthesized from the free nitrogen by certain bacteria living on the roots of legumes. When plants and animals die, part of the protein nitrogen is set free. The remainder is converted into ammonia, nitrites, and nitrates by bacteria. These compounds are absorbed by the soil and are again available for plant use. The steps in the cycle are made clear by the diagram in Figure 19-4.

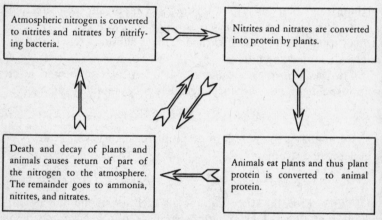

Fig. 19-4. The nitrogen cycle.

The Atmosphere.   The gaseous envelope surrounding the earth is called the *atmosphere*. Sometimes the word *air* is used in place of atmosphere. Ten thousand liters of dry air at 760 mm pressure contain approximately the following:

Nitrogen	7803 liters	Hydrogen	1	liter
Oxygen	2099 liters	Neon	0.123	liter
Argon	94 liters	Helium	0.04	liter
Carbon dioxide	3 liters	Krypton	0.005	liter
		Xenon	0.0006	liter

The atmosphere also contains water vapor, dust particles, microorganisms, and, in large cities, such gases as hydrogen sulfide, sulfur dioxide, and carbon monoxide. Also, the atmosphere in large cities may, under certain conditions, be contaminated with what is called *smog*. This word is a contraction of two words, *smoke* and *fog*. In addition, the atmosphere may contain what is called *radioactive fallout*. It is composed of dust particles, water vapor, radioactive isotopes, and fragments of soil, metals, and other materials. Radioactive fallout is caused by the explosion of atomic and hydrogen bombs near the surface of the earth. If the concentration of the radioactive material is high enough it may be injurious to human beings and animals. Recently air pollution has become the concern of all of us.

Essential Constituents. Oxygen, nitrogen, carbon dioxide, and water vapor are the essential constituents of the atmosphere. Oxygen supports life. It also supports combustion and aids in fermentation. Nitrogen is essential to plant growth and serves as a diluent for oxygen. This checks the speed of oxidation. Carbon dioxide is used as a food by plants. Water vapor prevents excessive evaporation from plants and animals and furnishes moisture for growth.

Nonessential Constituents. The nonessential constituents of the atmosphere include the inert gases, always present in air, and accidental substances such as hydrogen sulfide, sulfur dioxide, dust particles, traces of hydrogen peroxide, ozone, ammonia, and other pollutants.

Air as a Mixture. The following facts prove that the atmosphere is a mixture. Its composition varies. Dissolved air when expelled from water is about one-third oxygen. If liquid air is boiled, the nitrogen escapes first. Air can be made by mixing nitrogen and oxygen (and the other constituents) in the proper proportions. *There is no heat evolved or absorbed in the process.* This would not be true if a compound were formed.

Analysis of Air. The following is the method of analyzing the atmosphere for oxygen, nitrogen, carbon dioxide, and water vapor.

Oxygen and Nitrogen. Phosphorus is burned in a known volume of air (100 ml) confined over water. The phosphorus removes the oxygen by forming phosphorus pentoxide. When this substance dissolves, water rises in the tube to take the place of the oxygen. The remaining volume (79 ml) represents the nitrogen (plus a minute quantity of inert gases).

Carbon Dioxide. Pass a known volume of *dry* air through a weighed tube of potassium hydroxide. After the air has passed through, weigh the tube again. The increase in weight is due to the carbon dioxide which was removed from the air.

Water Vapor. Pass a known volume of air through a weighed drying tube containing phosphorus pentoxide or fused calcium chloride. Weigh the tube again. The increase in weight is due to the water vapor which was removed from the air.

HUMIDITY OF THE ATMOSPHERE. The atmosphere always contains moisture. The quantity present varies from time to time and from place to place. The following terms are used to express the amount of water vapor present. The *absolute humidity* is the number of grams of water vapor per unit volume of air. The *relative humidity* is the amount of water vapor in the air compared with the amount necessary for saturation under the given conditions of temperature and pressure. If, at a given temperature, air is saturated with water vapor, the humidity is 100 per cent. If the air is but half saturated, the relative humidity is 50 per cent.

The Inert Gases. The members of the zero group (or the VIIIA group in some periodic tables) are known as the *inert gases.* Helium, neon, argon, krypton, xenon, and radon make up this group. Because of the stable arrangement of the electrons in the outer shell of their atoms they are essentially chemically inert. However, a few compounds have been prepared, examples of which are given here. So far only the three heaviest members of the group are known to have entered into chemical combustion.

$$Xe_{(gas)} + F_{2(gas)} \xrightarrow{(-50°C)} XeF_{2(solid)}$$

Other fluorides of xenon are known: XeF_4, XeF_6, and also a xenon oxyfluoride $XeOF_{4(liquid)}$. Xenon trioxide (XeO_3) has been prepared by the action of water on XeF_6.

$$XeF_6 + 3 H_2O \longrightarrow XeO_3 + 6 HF$$

The XeO_3 is highly explosive. Figure 19-5 is a suggested structural formula for XeF_4.

PREPARATION. The inert gases can be prepared as follows:

Helium. Natural gas from certain wells in Texas contains about 1 per cent helium. If the gas mixture is cooled while under high pressure, all the constituents are liquefied except helium.

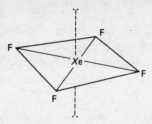

Fig. 19-5. A suggested structural formula for XeF_4.

Neon. Neon is prepared by the fractional distillation of argon.

Argon. This gas is prepared by the fractional distillation of liquid air.

Krypton and Xenon. These occur in such small quantities that they are not prepared for ordinary commercial use.

PHYSICAL PROPERTIES. The important physical properties of the inert gases are as follows:

Helium. Helium is, excluding hydrogen, the lightest gas known, one liter weighing 0.1785 g. It boils at $-268.9°C$ and freezes at $-272.2°C$. It was the last gas to be liquefied. Helium will not burn, is slightly soluble in water, and has but one atom per molecule.

Neon. One liter weighs 0.9002 g. It is slightly soluble in water, boils at $-245.9°C$ and freezes at $-248.7°C$.

Argon. One liter weighs 1.7823 g. It is appreciably soluble in water and is adsorbed by charcoal. It boils at $-186°C$ and freezes at $-189°C$.

USES. The important uses of the inert gases are: *Helium* is used to fill balloons and to prevent Caisson Disease. This disease is a disorder caused by the solution of nitrogen in the blood of men who work under increased air pressure as in deep-sea diving. As the high pressure is released when the diver returns to the surface, the nitrogen forms bubbles in the blood. These may cause death if they come in contact with the heart or brain cells. At the present time, helium is substituted for nitrogen because it is less soluble in the blood under the same conditions. *Neon* is used in electric signs. *Argon* is used in electric light bulbs and in electric signs. Special light bulbs (or discharge tubes) may contain *krypton* or *xenon*.

Phosphorus. Phosphorus was first prepared by the German alchemist Brand in 1669 while he was looking for the Philosopher's Stone. He heated a mixture of organic matter (rich in phosphorus) with sand, obtaining a substance that glowed in the dark. This was named phosphorus, meaning "light-bearer." Phosphorus is now prepared in the electric furnace (see Figure 19-6). The charge is a mixture of phosphate rock, silica, and carbon.

$$Ca_3(PO_4)_2 + 5\,C + 3\,SiO_2 \longrightarrow 3\,CaSiO_3 + 5\,CO + 2\,P$$

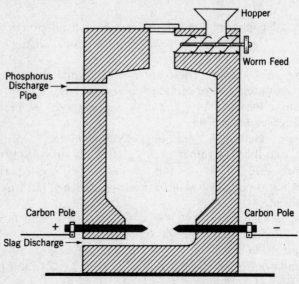

Fig. 19-6. An electric furnace for the preparation of phosphorus.

ALLOTROPIC MODIFICATIONS. Phosphorus prepared in the electric furnace is the white (or yellow) form. The other well-known form is red (or violet) phosphorus. Figure 19-7 shows the relation between the two modifications.

PHYSICAL PROPERTIES. White phosphorus is a translucent, waxy solid which melts at 44.1°C and boils at 280°C; has a density of 1.82; is insoluble in water; is soluble in carbon disulfide; is converted into the red modification by heating to 250°C in the absence of air; and is changed to the red form by light.

CHEMICAL PROPERTIES. White phosphorus has a low kindling temperature (60°C or lower), burns in oxygen and the halogens, and is very poisonous. It attacks the jaw bone.

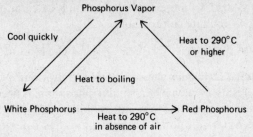

Fig. 19-7. The relation between phosphorus vapor, white phosphorus, and red phosphorus.

$$4 P + 5 O_2 \longrightarrow 2 P_2O_5$$
$$2 P + 5 Cl_2 \longrightarrow 2 PCl_5$$

USES. White phosphorus is used for smoke screens, for incendiaries, for rat poison, and for making P_4S_3, which is used in the match industry.

RED PHOSPHORUS. Red phosphorus is semicrystalline and vaporizes at 290°C or above, instead of melting. It is insoluble in water and in carbon disulfide. It has a kindling temperature of about 240°C and reacts with oxygen and the halogens. It is nonpoisonous.

$$4 P + 5 O_2 \longrightarrow 2 P_2O_5$$
$$2 P + 5 Br_2 \longrightarrow 2 PBr_5$$

Red phosphorus is used in the manufacture of matches.

Phosphorus Compounds. Phosphine (PH_3) is a hydride prepared by the action of water on calcium phosphide.

$$Ca_3P_2 + 6 H_2O \longrightarrow 3 Ca(OH)_2 + 2 PH_3$$

Another hydride, P_2H_4, is spontaneously combustible.

There are two oxides of phosphorus, P_2O_3 and P_2O_5. The latter is more important because it is an excellent drying agent and also because it is the acid anhydride of orthophosphoric acid.

$$4 P + 3 O_2 \longrightarrow 2 P_2O_3 \text{ (limited oxygen supply)}$$
$$4 P + 5 O_2 \longrightarrow 2 P_2O_5 \text{ (in excess of oxygen)}$$

Orthophosphoric acid (H_3PO_4) is a tribasic (triprotic) acid. It is made by the action of water on P_2O_5.

$$P_2O_5 + 3 H_2O \longrightarrow 2 H_3PO_4$$

Less pure acid is prepared by digesting calcium phosphate, $Ca_3(PO_4)_2$, with dilute sulfuric acid.

$$Ca_3(PO_4)_2 + 3 H_2SO_4 + 2 H_2O \longrightarrow 2 H_3PO_4 + 3 CaSO_4 \cdot 2 H_2O$$

Salts of Orthophosphoric Acid. Disodium phosphate (Na_2HPO_4) is used whenever a soluble phosphate is needed. It has a limited use in medicine. Trisodium phosphate (Na_3PO_4) gives an alkaline water solution (hydrolysis) and has detergent properties. It is used extensively for cleansing purposes. Microcosmic salt ($NaNH_4HPO_4 \cdot 4 H_2O$) is used in analytical chemistry. Superphosphate is a mixture of calcium sulfate and calcium acid phosphate, $Ca(H_2PO_4)_2$. It is used as a fertilizer.

Fertilizers. Plants require carbon, oxygen, hydrogen, nitrogen, phosphorus, potassium, calcium, magnesium, sulfur, and iron. In addition, plants also use small amounts of one or more of the following: manganese, aluminum, boron, zinc, copper, and silicon. All of these elements are found in the soil, but the plant probably obtains most of its carbon and oxygen from the atmosphere. Commercial fertilizers contain phosphates, nitrates or other nitrogen compounds, such as $(NH_4)_2SO_4$, and potash.

PHOSPHATES. Calcium phosphate, $Ca_3(PO_4)_2$, is the most important source of phosphorus. It must be made soluble before plants can make use of it. This is done by treating it with sulfuric acid or with phosphoric acid.

$$Ca_3(PO_4)_2 + 2 H_2SO_4 \longrightarrow Ca(H_2PO_4)_2 + 2 CaSO_4$$
$$Ca_3(PO_4)_2 + 4 H_3PO_4 \longrightarrow 3 Ca(H_2PO_4)_2$$

The mixture of $Ca(H_2PO_4)_2$ and $CaSO_4$ is called *superphosphate of lime*. The calcium acid phosphate made by using the phosphoric acid is called *triple superphosphate*.

NITROGEN COMPOUNDS. Plants obtain nitrogen from organic matter, ammonium sulfate, sodium nitrate, ammonium phosphate, and urea phosphate, $H_3PO_4 \cdot CO(NH_2)_2$. These are added to the soil to provide it with available nitrogen.

POTASH. Potassium is added to the soil as the chloride (KCl), the nitrate (KNO_3), or the sulfate (K_2SO_4). Balanced fertilizers containing phosphates, nitrogen compounds, and potash are on the market.

Matches. Safety matches will not ignite unless they are rubbed on a specially prepared surface. The composition of the match head is: antimony trisulfide (easily combustible material), potassium chlorate (the oxidizing agent), powdered glass (the filler), and glue (the binder). The special surface consists of: red phosphorus (low kindling), potassium chlorate or lead dioxide (oxidizing agent), powdered glass (to increase friction), and glue (the binder).

Any solid surface is suitable for the friction necessary to ignite "strike-anywhere" matches. The materials on the head of the match are: potassium chlorate (the oxidizing agent), sulfur or paraffin (easily combustible material), glue (the binder), ground glass (filler), and phosphorus sesquisulfide, P_4S_3 (nonpoisonous compound with a low kindling temperature). Because of the poisonous character of white phosphorus, laws have been passed preventing its use on the tips of matches.

Arsenic. Arsenic may be prepared by heating arsenopyrite.

$$FeAsS + (heat) \longrightarrow FeS + As$$

Another method is to roast a sulfide to the oxide and then reduce with carbon.

$$2\,As_2S_3 + 9\,O_2 \longrightarrow 2\,As_2O_3 + 6\,SO_2$$
$$As_2O_3 + 3\,C \longrightarrow 2\,As + 3\,CO$$

PHYSICAL PROPERTIES. There are at least two allotropic forms of arsenic—the yellow and the gray (semimetallic). Both modifications sublime upon heating. The gray form is a good conductor of heat but a poor conductor of electricity.

CHEMICAL PROPERTIES. Arsenic combines with oxygen, the halogens, and certain metals, and is very poisonous.

$$4\,As + 3\,O_2 \longrightarrow 2\,As_2O_3$$
$$2\,As + 3\,Cl_2 \longrightarrow 2\,AsCl_3$$
$$3\,Zn + 2\,As \longrightarrow Zn_3As_2$$

USES. There are very few uses for metallic arsenic. A small quantity is used in alloys, especially for hardening lead shot.

Arsenic Compounds. Arsenic forms two series of compounds: the arsenous, in which the valence (oxidation number) is $+3$, and the arsenic, in which the valence (oxidation number) is $+5$.

Arsenous oxide, As_2O_3 (white arsenic), is prepared by burning arsenic in air. It is a white, slightly soluble, finely divided crystalline substance. In alkaline solutions, it is a reducing agent. Large quantities are used in making insecticides.

Arsenic oxide (As_2O_5) is prepared by heating crystals of orthoarsenic acid.

$$2 H_3AsO_4 + (heat) \longrightarrow 3 H_2O + As_2O_5$$

Orthoarsenic acid (H_3AsO_4) is prepared by heating arsenous oxide in nitric acid. It is an oxidizing agent. The salts of this acid are quite similar to the salts of orthophosphoric acid.

Paris green, $Cu_3(AsO_3)_2 \cdot Cu(C_2H_3O_2)_2$, and lead arsenate, $Pb_3(AsO_4)_2$, are compounds used in insecticides.

Arsine (AsH_3) is a highly poisonous substance made by treating any arsenic compound, under suitable conditions, with an acid and a chemically active metal such as zinc. The acid and metal react to form hydrogen, which can then react with the arsenic compound.

$$As_2O_3 + 6 H_2 \longrightarrow 2 AsH_3 + 3 H_2O$$

In the extremely sensitive *Marsh test* for arsenic, the arsenic is first converted into arsine. This is subsequently decomposed by passing it through a hot tube. A black spot (mirror) of metallic arsenic, on the tube, indicates the presence of arsenic.

Antimony.   Antimony is prepared by heating the sulfide with iron.

$$Sb_2S_3 + 3 Fe \longrightarrow 3 FeS + 2 Sb$$

Antimony is a very brittle, shiny solid which can be volatilized. It expands slightly on solidifying and occurs in allotropic modifications similar to those of arsenic. One modification is called *explosive antimony* because it "explodes," with the liberation of energy, upon being struck with a hammer.

Antimony has metallic and nonmetallic properties. It is not acted upon by dilute acids, but reacts with strong nitric and concentrated sulfuric acids to form the nitrate and sulfate, respectively. It combines with oxygen and the halogens.

Antimony is an important constituent of such alloys as type metal and Babbitt metal. It is added to lead to harden it for use in storage batteries.

Compounds of Antimony.   The compounds of antimony, with the exception of the sulfides which are used in the match industry, are of little commercial importance. Antimony trichloride has had some use in medicine under the name of "butter of antimony."

Antimony hydroxide, $Sb(OH)_3$, is of interest since it can act basic or acidic. That is, it is *amphoteric*.

$$Sb(OH)_3 + 3 NaOH \longrightarrow Na_3SbO_3 + 3 H_2O \text{ (acidic)}$$
$$Sb(OH)_3 + 3 HCl \longrightarrow SbCl_3 + 3 H_2O \text{ (basic)}$$

Bismuth. Bismuth is obtained by heating the ore, containing native (free) bismuth, to about 300°C. This melts the bismuth, which can be poured into suitable containers. Bismuth sulfide ore is roasted to convert the sulfide into the oxide, which is subsequently reduced with carbon.

$$2 Bi_2S_3 + 9 O_2 \longrightarrow 2 Bi_2O_3 + 6 SO_2$$
$$Bi_2O_3 + 3 C \longrightarrow 2 Bi + 3 CO$$

Bismuth is a hard and brittle metal with a characteristic luster, is decidedly crystalline, has a low melting point and high boiling point, and expands on cooling. It is chemically inactive at ordinary temperatures, burns (at elevated temperatures) to the oxide, dissolves in hot concentrated nitric or sulfuric acids, and combines slowly with the halogens.

$$4 Bi + 3 O_2 \longrightarrow 2 Bi_2O_3$$
$$2 Bi + 3 Cl_2 \longrightarrow 2 BiCl_3$$

Bismuth is used in alloys, especially those with low melting points. Wood's metal, an alloy containing bismuth, melts at 65.5°C.

Compounds of Bismuth. The compounds are of little commercial importance. Because bismuth has a metallic character, there are no bismuth acids. The chlorides and nitrates hydrolyze, giving basic salts which have some use in medicine. Bismuth subcarbonate $(Bi_2O_3 \cdot CO_3 \cdot H_2O)$ is added to powders, tablets, and other preparations used to neutralize excess acidity in the stomach.

REVIEW QUESTIONS

1. What is the most important source of uncombined nitrogen for commercial use?
2. Describe a simple method for the preparation of a liter of nitrogen from the atmosphere. What impurities are present?
3. Describe a simple laboratory method for the preparation of a liter of pure nitrogen.
4. How is magnesium nitride prepared? Give the equations.

5. Describe a commercial method for synthesizing ammonia from the elements.

6. What is a general method of preparing ammonia in the laboratory?

7. List the oxides of nitrogen. How is nitrous oxide (N_2O) prepared?

8. Nitrogen dioxide and nitrogen tetraoxide are polymers. Explain.

9. Show that N_2O_5 is the acid anhydride of nitric acid.

10. Write equations illustrating the oxidizing property of nitric acid.

11. Give the approximate composition of ordinary air.

12. Distinguish between the essential and nonessential constituents of the atmosphere.

13. Prove that air is a mixture. How can one separate (by physical means) oxygen from the atmosphere?

14. How can one prove that what appears to be ordinary dry air contains moisture?

15. Name the inert gases found in the atmosphere.

16. Give the common uses of helium, neon, and argon.

17. XeF_2 is a solid. How may it be prepared?

18. In what ways do phosphorus and nitrogen resemble each other?

19. How does phosphorus occur in nature?

20. What is the common method of preparing phosphorus commercially?

21. How can one change phosphorus vapor into red phosphorus?

22. How is P_2O_5 prepared? What happens when it reacts with water?

23. What are safety matches? What role does phosphorus play in making safety matches?

24. What is a very sensitive test for arsenic? Give the equations.

25. Antimony hydroxide is said to be amphoteric. Explain.

26. What type of alloys are made using bismuth as one constituent?

27. Distinguish between microcosmic salt and Paris green.

28. Give an equation for converting $Ca_3(PO_4)_2$ into superphosphate.

Chapter 20
The Sulfur Family

Oxygen (8) $1s^2 2s^2 2p^4$
Sulfur (16) $1s^2 2s^2 2p^6 3s^2 3p^4$
Selenium (34) [   ] $4s^2 4p^4$
Tellurium (52) [   ] $5s^2 5p^4$

These elements belong to group VIA of the periodic table. Oxygen is a gas and was discussed in Chapter 9. Sulfur, selenium, and tellurium are solids. Table 20-1 tabulates the important physical properties of these elements. Polonium is not considered in this book although it is a member of group VIA.

Table 20-1

Physical Properties of the Sulfur Family

	Oxygen	Sulfur	Selenium	Tellurium
Color	colorless	yellow	gray or red	silver white
Physical State	gas	solid	solid	solid
Melting Point	$-218.4°C$	114.5°C	217°C	456°C
Boiling Point	$-183°C$	444.6°C	685°C	1390°C
Atomic radius	0.66Å	1.04Å	1.14Å	1.32Å
Ionization Energy	13.55 v	10.36 v	9.75 v	9.01 v
Heat of Fusion Kcal/Mole	53	293	1,250	4,280
Electro-negativity	3.5	2.5	2.4	2.1
Density g/cm^3	1.33×10^{-3}	2.07	4.36	6.24

Oxygen and sulfur are the two most important elements in Group VIA. This chapter deals largely with sulfur because of its importance. Selenium and tellurium are less common and not so important commercially. Because of the similarity of their electronic configurations, these two elements form compounds like those of sulfur.

SULFUR	SELENIUM	TELLURIUM
SO_2	SeO_2	TeO_2
SO_3	SeO_3	TeO_3
H_2SO_3	H_2SeO_3	H_2TeO_3
H_2SO_4	H_2SeO_4	H_2TeO_4
H_2S	H_2Se	H_2Te

Sulfur. This element has been known since early times. The Bible refers to sulfur as *brimstone.* Sulfur occupies the position next to oxygen in Group VIA of the periodic table. Each element has 6 electrons in its valence shell, but sulfur may exist in several different states of oxidation: -2, 0, $+2$, $+4$, and $+6$.

OCCURRENCE. Elemental (free) sulfur is found in Texas, Louisiana, Sicily, Japan, Spain, and Mexico. Sulfur is also found in a variety of minerals such as galena (PbS), zinc blende (ZnS), pyrite (FeS_2), gypsum ($CaSO_4 \cdot 2\ H_2O$), barite ($BaSO_4$), and epsom salt ($MgSO_4 \cdot 7\ H_2O$). Organic matter, such as protein found in egg yolk, also contains appreciable amounts of sulfur.

EXTRACTION OF FREE SULFUR.   Nearly 2,000,000 tons of sulfur are produced annually. The largest percentage comes from Texas and Louisiana deposits. The Frasch Method is the most important commercial process. Concentric pipes are sunk to the sulfur beds 500 to 800 feet below the surface. Water, heated to about 180°C, under pressure, is forced down between two of the pipes. This melts the sulfur. Hot air is forced down the innermost pipe. This forms a light, frothy mixture which is forced to the surface through the space between the other two pipes. The molten sulfur is run into settling tanks where it separates from the air and water. The product is 99.5 per cent pure.

In the Sicilian Method, crude sulfur, which is mixed with rock and earthy material, is heated. This melts the sulfur, which is drained off. This sulfur is purified by distilling in a retort, the exit tube of which opens into a brick cooling chamber. Sulfur vapor condenses on the walls of the chamber and is then called *flowers of sulfur.* When the walls of the cooling chamber get warm, some of the sulfur melts and runs into molds at the bottom. This is called *roll sulfur.*

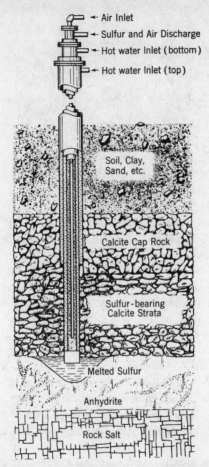

Air Inlet
Sulfur and Air Discharge
Hot water Inlet (bottom)
Hot water Inlet (top)

Soil, Clay, Sand, etc.

Calcite Cap Rock

Sulfur-bearing Calcite Strata

Melted Sulfur

Anhydrite

Rock Salt

Fig. 20-1. The Frasch Process for obtaining sulfur.

ALLOTROPIC FORMS.   Sulfur, like certain other elements, can exist in more than one physical form, usually crystalline. These different forms are said to be *allotropes* of that element. Sulfur may be crystallized as *rhombic* (*orthorhombic*) and also as *monoclinic* (*prismatic*). See Figure 20-2.

Sulfur may also exist in the *amorphous* (*plastic*) state. This is probably a supercooled liquid. The two crystalline forms of sulfur have different melting points. Rhombic melts at 112.8°C and monoclinic melts at 119.3°C.

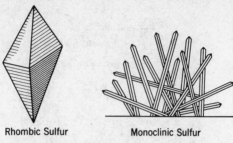

Rhombic Sulfur   Monoclinic Sulfur

Fig. 20-2.   Two well-known forms of sulfur.

Upon *slow* cooling, liquid sulfur freezes at 114.5°C. This is sometimes called the natural freezing point.

PREPARATION.   Rhombic sulfur is prepared by dissolving roll sulfur in carbon disulfide (CS_2) and allowing the liquid to evaporate. Crystals of yellow rhombic sulfur are formed. Monoclinic sulfur crystals are formed by melting sulfur and allowing the liquid to cool. This produces long needlelike crystals of monoclinic sulfur. If boiling sulfur is poured into cold water, so-called *plastic* sulfur is formed. It has a rubberlike quality, but on standing for several days will change to crystals of rhombic sulfur. When sulfur is first melted the pale yellow liquid flows freely. This sulfur is made up of 8-membered ring molecules (S_8). On heating the liquid sulfur to 160°C or higher the liquid becomes very viscous. This is caused by the rings breaking and becoming long chain *polymers*. The sulfur atoms on each end of the chain have unpaired electrons which can combine with each other to form entangled clusters of molecules. Consequently the viscosity of the liquid increases greatly. At about 250°C the long chains begin to break into short chains and, at the boiling point (445°C), the liquid flows freely again. Sulfur vapor just above the boiling point contains some S_8 molecules but as the temperature increases the molecules contain fewer sulfur atoms. At about 1900°C the molecules consist of single sulfur atoms. See Figure 20-3.

PROPERTIES AND USES.   At ordinary room temperature, sulfur is inactive, but hot sulfur will combine directly with oxygen, with many of the metals, and with certain nonmetals. The following equations are illustrative:

$$S + O_2 \longrightarrow SO_2$$
$$Hg + S \longrightarrow HgS$$
$$C + 2\,S\,(heat) \longrightarrow CS_2$$

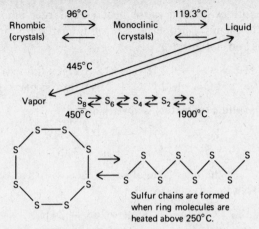

Fig. 20-3. Relationships between various forms of sulfur.

Sulfur is used in making sulfur dioxide (required in the manufacture of sulfuric acid), in vulcanizing rubber, in making sulfites and thiosulfates, in the preparation of carbon disulfide, in preparing insecticides, and in making a sulfur-sand cement. The action of sulfur on rubber (vulcanizing) is interesting and of considerable commercial value. Rubber tends to be brittle when it is cold and sticky when it is warm. However, if sulfur is mixed with the rubber and then heated to 140–145°C (vulcanized), sulfur atoms add on to the rubber molecules at their double bonds and extend the temperature range where rubber has desirable physical properties.

Compounds of Sulfur.   There are several important compounds of sulfur.

HYDROGEN SULFIDE (H_2S).   Hydrogen sulfide occurs in vapors issuing from volcanoes, in sulfur springs, and where organic matter is undergoing decay. It can be prepared in the laboratory by the action of a nonoxidizing acid on a metallic sulfide.

$$FeS + 2\,HCl \longrightarrow FeCl_2 + H_2S$$

Hydrogen sulfide can also be made by the direct combination of the elements. However, this method is impractical.

$$H_2 + S \longrightarrow H_2S$$

Hydrogen sulfide is a colorless gas with a foul odor, slightly heavier than air, moderately soluble in water, and very toxic.

Small quantities inhaled cause headache; larger amounts cause death. The chemical properties are summarized by the following equations and statements.

$$2 H_2S + O_2 \longrightarrow 2 H_2O + 2 S \text{ (oxygen limited)}$$
$$2 H_2S + 3 O_2 \longrightarrow 2 H_2O + 2 SO_2 \text{ (excess of oxygen)}$$

Hydrogen sulfide (in water) is a weak acid.

$$H_2S + H_2O \rightleftharpoons H_3O^{+1} + HS^{-1}$$
$$HS^- + H_2O \rightleftharpoons H_3O^{+1} + S^{-2}$$

Solutions of the heavy metal salts react with hydrogen sulfide to give insoluble sulfides.

$$CuCl_2 + H_2S \longrightarrow CuS + 2 HCl$$

Since water solutions of copper chloride and hydrogen sulfide contain copper ions (Cu^{+2}) and sulfide ions (S^{-2}) respectively, the reaction can be written:

$$Cu^{+2} + S^{-2} \longrightarrow CuS \text{ (insoluble)}$$

Hydrogen sulfide is also a strong reducing agent. For example, if it is bubbled through a nitric acid solution, a vigorous reaction takes place. Free sulfur is formed and, also, sulfur dioxide is formed.

$$3 H_2S + 2 HNO_3 \longrightarrow 4 H_2O + 2 NO + 3 S$$
$$3 H_2S + 6 HNO_3 \longrightarrow 6 H_2O + 6 NO + 3 SO_2$$

Hydrogen sulfide is used as a reducing agent and as a laboratory reagent in qualitative analysis.

SULFUR DIOXIDE (SO_2).   The most common oxide is sulfur dioxide. It is prepared by burning sulfur in oxygen, by the action of an acid on a sulfite, by roasting a metallic sulfide, and by the reduction of sulfuric acid or sulfates. Illustrative equations are:

$$S + O_2 \longrightarrow SO_2 \text{ (commercial)}$$
$$Na_2SO_3 + 2 HCl \longrightarrow 2 NaCl + H_2O + SO_2$$
$$2 ZnS + 3 O_2 \longrightarrow 2 ZnO + 2 SO_2$$
$$Cu + 2 H_2SO_4 \longrightarrow CuSO_4 + SO_2 + 2 H_2O$$
$$3 CaSO_4 + CaS \longrightarrow 4 CaO + 4 SO_2 \text{ (high temp. required)}$$

Sulfur dioxide is a colorless gas with a sharp penetrating odor, 2.2 times denser than air, easily liquefied (at $-10°C$), and soluble in water (80 ml in 1 ml water).

Sulfur dioxide reacts with water to form sulfurous acid, and reacts with oxygen in the presence of a catalyst to form sulfur trioxide. It also acts as a reducing agent.

$$SO_2 + H_2O \rightleftharpoons H_2SO_3$$
$$2\,SO_2 + O_2 \rightleftharpoons 2\,SO_3$$
$$3\,SO_2 + 2\,HNO_3 \longrightarrow H_2O + 2\,NO + 3\,SO_3 \text{ (reducing agent)}$$

Sulfur dioxide is used to prepare sulfuric acid, and to bleach straw, silk, or wool. Some refrigeration machines use sulfur dioxide. Sulfur dioxide is used to make calcium bisulfite, needed in the paper industry to convert wood pulp into paper pulp.

SULFUR TRIOXIDE (SO_3). This oxide is prepared by the catalytic oxidation of sulfur dioxide under suitable temperature and pressure conditions. Finely divided platinum, called platinum black, is the common catalyst for this reaction, but vanadium pentoxide (V_2O_5), combined with certain metallic oxides, has been used successfully. Up to 98 per cent conversion has been obtained.

$$2\,SO_2 + O_2 \text{ (catalyst) } \xrightarrow{\;400°C\;} 2\,SO_3$$

Sulfur trioxide may exist as a liquid and also in the solid state. There is good experimental evidence for the existence of at least three distinct solid forms of this oxide. Liquid SO_3 freezes at 15°C and boils at 46°C. If a trace of moisture is added to liquid SO_3, it changes to an asbestoslike solid. Sulfur trioxide is not readily soluble in water (probably because of a film of air around each particle), but it dissolves readily in sulfuric acid. The most important use of SO_3 is in making sulfuric acid.

SULFUROUS ACID (H_2SO_3). This is a rather weak and unstable acid made by dissolving SO_2 in water. It is a reducing agent and a bleaching agent. It has the power of reacting with colored compounds (found in organic material), changing them to colorless substances.

$$SO_2 + H_2O \rightleftharpoons H_2SO_3$$
$$H_2SO_3 + H_2O \rightleftharpoons H_3O^{+1} + HSO_3^{-1}$$
$$HSO_3^{-1} + H_2O \rightleftharpoons H_3O^{+1} + SO_3^{-2}$$

SULFURIC ACID (H_2SO_4). Sulfuric acid is probably used more extensively than any other manufactured chemical. More than 17,000,000 tons are used each year in the United States.

There are two methods of preparation, the *Contact Process*

and the *Lead Chamber Process*. The Contact Process is better for the preparation of 95–98 per cent acid, and the Lead Chamber Process is better for 66 per cent acid.

The steps in the Contact Process are: Burn sulfur to sulfur dioxide; oxidize sulfur dioxide to sulfur trioxide; and dissolve the trioxide in dilute sulfuric acid.

$$S + O_2 \longrightarrow SO_2$$
$$2\,SO_2 + O_2 \text{ (catalyst)} \longrightarrow 2\,SO_3$$
$$SO_3 + H_2SO_4 \longrightarrow H_2S_2O_7 \text{ (pyrosulfuric acid)}$$
$$H_2S_2O_7 + H_2O \longrightarrow 2\,H_2SO_4$$

The steps in the Lead Chamber Process (simplified) are: Sulfur is burned to the dioxide, which is subsequently oxidized to the trioxide by nitrogen dioxide. Steam reacts with sulfur trioxide, forming sulfuric acid.

$$S + O_2 \longrightarrow SO_2$$
$$SO_2 + NO_2 \longrightarrow SO_3 + NO$$
$$SO_3 + H_2O \longrightarrow H_2SO_4$$
$$2\,NO + O_2 \longrightarrow 2\,NO_2 \text{ (used again)}$$

Nitric oxide may be considered to be a catalyst or an oxygen carrier since it takes oxygen from the air and gives it to the sulfur dioxide.

Concentrated sulfuric acid is a heavy, oily liquid (specific gravity 1.84). Pure acid begins to decompose when it is heated to boiling (338°C).

Sulfuric acid is a strong acid, ionizing in two steps. Dilute solutions react with active metals to liberate hydrogen. The concentrated acid is an oxidizing agent. It is also a dehydrating agent, and since it has a high boiling point, it can be used to liberate volatile acids from their compounds.

$$H_2SO_4 + H_2O \rightleftharpoons H_3O^+ + HSO_4^-$$
$$HSO_4^- + H_2O \rightleftharpoons H_3O^+ + SO_4^{--} \text{ (acid property)}$$
$$H_2SO_4 + Zn \longrightarrow ZnSO_4 + H_2$$
$$2\,H_2SO_4 + Cu \longrightarrow CuSO_4 + 2\,H_2O + SO_2 \text{ (oxidizing)}$$
$$NaNO_3 + H_2SO_4 \longrightarrow NaHSO_4 + HNO_3$$

Table 20-2 shows some of the important uses of sulfuric acid.

Table 20-2

Uses of Sulfuric Acid

Where Used	Uses
Fertilizer industry	To make superphosphate and ammonium sulfate.
Petroleum refining	To purify petroleum products.
Chemical manufacturing	To make hydrochloric acid, nitric acid, and certain organic compounds.
Metallurgy	In electrolytic refining of metals.
Explosives	In making nitrated organic compounds.
Textile industry	In manufacture of rayon and similar products.

REVIEW QUESTIONS

1. Oxygen is a member of the sulfur family of elements. From the standpoint of atomic structure, why is this so?
2. Sulfur is found free in nature in underground deposits. How can it be recovered?
3. (a) Distinguish between rhombic and monoclinic sulfur. (b) What is the effect of heating sulfur vapor to a high temperature?
4. (a) How is hydrogen sulfide prepared in the laboratory? (b) How could one prove that bubbling hydrogen gas through molten sulfur produces hydrogen sulfide?
5. (a) Write an equation for a reaction in which hydrogen sulfide is a reducing agent. (b) Write an equation showing how one can make an insoluble sulfide.
6. Write an equation in which SO_2 is produced by roasting a metallic sulfide.
7. How can sulfuric acid be made from SO_2?
8. Write equations to show: (a) sulfuric acid as an oxidizing agent; (b) sulfuric acid as an acid; (c) the use of sulfuric acid in preparing a volatile acid from a salt of the acid desired.

Chapter 21
Group IVA Elements

Carbon (6) $1s^2 2s^2 2p^2$
Silicon (14) $1s^2 2s^2 2p^6 3s^2 3p^2$
Germanium (32) [   ] $4s^2 4p^2$
Tin (50) [   ] $5s^2 5p^2$
Lead (82) [   ] $6s^2 6p^2$

The first two members, carbon and silicon, are nonmetals. Tin and lead are classed as metals but germanium is sometimes spoken of as a metalloid. The common physical properties of this group are listed in Table 21-1.

Table 21-1

Physical Properties of Group IVA Elements

	Carbon	Silicon	Germanium	Tin	Lead
Color	gray-black	silvery gray	silvery gray	silvery white	silvery white
Physical State	solid	solid	solid	solid	solid
Melting Point	3730°C (graphite)	1410°C	937°C	232°C	327°C
Boiling Point	4830°C (graphite)	2680°C	2830°C	2270°C	1730°C
Atomic Radius	0.71Å	1.18Å	1.22Å	1.40Å	1.75Å
Ionization Energy	11.26 v	8.15 v	7.88 v	7.33 v	7.42 v
Heat of Fusion cal/g		337		14.5	6.3
Electro-negativity	2.5	1.8	1.8	1.8	1.8
Density g/cm³ at 20°C	2.26 (graphite)	2.33	5.32	7.30	11.4

Of the five elements in this family, silicon and germanium have become especially useful in making *semiconductors* and *transistors*. Silicon and germanium, when pure, are classed as non-conductors of the electric current, but if a few arsenic atoms (a group VA element) are added, say, to germanium its conductivity is greatly increased and it is now what is called an *n-type semiconductor*. If, on the other hand, a few aluminum atoms are added to germanium it becomes a *p-type semiconductor*. In the *n*-type semiconductor there is an excess of electrons present due to the fact that arsenic has 5 valence electrons whereas the germanium has only 4 valence electrons. This extra electron is free to move through the germanium. In the *p*-type semiconductor there is a deficiency of electrons because the aluminum added has only 3 valence electrons. Here we have what is called a "positive hole" and an electron can migrate from a nearby atom to fill the positive hole. But in so doing it leaves an electron deficiency in the atom it deserted. By this kind of electron movement we have a *p*-type of electrical conduction. See Figure 21-1.

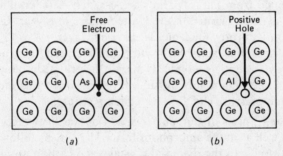

Fig. 21-1. Schematic representation of semiconductors. (*a*) *n*-type. (*b*) *p*-type.

As stated above, carbon and silicon are the first two elements of group IV of the periodic table. Carbon is the essential element in the organic (animal and vegetable) world; silicon is the important element in the inorganic (mineral) world.

Atomic Structure Similar. Atoms of carbon and silicon have four electrons in their outer energy level. This explains the similarity of many of their chemical and physical properties.

VALENCE. The valence (oxidation number) of these elements can be either −4 or +4. This accounts for their *amphoteric* nature.

In carbon dioxide and silicon dioxide each element has an oxidation number of +4. In methane and silicon hydride, the value is −4.

SELF-COMBINATION. The most outstanding property of carbon is its ability to combine with itself, atom to atom, forming stable long chain or ring compounds. Silicon has this property, but the chains are short and the compounds comparatively unstable.

STABILITY OF SILICATES. The chemistry of silicon is characterized by the stability and complexity of the many silicates known (compounds of silicon dioxide with metal oxides).

Occurrence of Carbon. Carbon is found free and in chemical combination. Free carbon is found as coal, graphite, and diamonds. The important compounds of carbon are: carbonates (limestone, dolomite, marble, etc.), petroleum and natural gas, organic compounds such as proteins and fats, and carbon dioxide (found in the atmosphere).

Commercial Source of Carbon. Diamonds, graphite, and coal are obtained from natural deposits. As much as 400 million tons of coal have been mined in the United States in a year. Graphite of high purity is now made in the electric furnace.

Small diamonds, suitable for industrial uses, can be made in the laboratory. Scientists of the General Electric Company made these synthetic diamonds by subjecting carbonaceous matter to a very high pressure and an elevated temperature.

Allotropic Forms of Carbon. Carbon occurs in two crystalline forms as well as in the amorphous form. They are as follows:

DIAMONDS. In the diamond crystal, each carbon atom is surrounded by four other atoms located at the corner of a regular tetrahedron. The physical properties of diamonds depend upon the crystal structure. Diamonds are found in South Africa, Brazil, and India. (The India deposits are now practically exhausted.)

Properties. A diamond is the hardest substance known and probably has the highest melting point of any substance. It refracts light brilliantly, is insoluble in all common solvents, and is a nonconductor of electricity. Its density is 3.51, and it burns in oxygen.

Uses. Diamonds are used in jewelry, for points for some phonograph needles, to polish other diamonds, and to point rock drills for cutting hard rock.

GRAPHITE. The graphite crystal is made up of planes of hexagonal rings. This substance is found in Siberia, in Ceylon, and in New York State. Large quantities of graphite are now made in the electric furnace.

Properties. Graphite is soft, smooth, and scaly, is a fair conductor of electricity, and withstands high temperatures without oxidizing. Its density is 2.25. The physical properties depend upon the crystal structure. When other forms of carbon are volatilized and then cooled, graphite is formed.

Uses. Graphite is used for electrodes, crucibles, pencils, and stove polish, as a lubricant, in black paint, and to prevent boiler scale.

AMORPHOUS CARBON. There are several forms of amorphous carbon. The most important are hard coal, coke, charcoal, boneblack, and lampblack. From X-ray studies and observations made with the electron microscope, there are indications that so-called amorphous carbon, such as charcoal and hard coal, is crystalline. The crystal patterns indicate that the structure is similar to that of graphite.

Anthracite coal (and other coals) were probably formed by the slow decomposition of vegetable matter in the absence of air and under high temperatures and pressures.

Hard Coal (Anthracite). Hard coal (85 per cent carbon) is found in Pennsylvania. It is a hard, black solid with a metallic appearance, possessing a high kindling temperature. It is used as a fuel, for making water gas, and in making graphite in the electric furnace.

Soft Coal (Bituminous). Soft coal contains less carbon and more volatile matter than anthracite coal. Bituminous coal is found in large deposits in various parts of the world. It is used as a fuel and for the manufacture of coke.

Destructive Distillation Products. When substances such as wood and soft coal are heated in the absence of air, they are decomposed. This process, carried out in retorts, is called *destructive distillation*. The volatile materials formed include coal gas, tarlike liquids, and ammonia. The nonvolatile residue is either coke or charcoal.

COKE. Coke contains about 90 per cent carbon. It is manufactured by heating soft coal in by-product ovens in the absence of air (destructive distillation). Coke is used as a reducing agent in the iron and steel industry, as a fuel, and in making water gas.

CHARCOAL.   Charcoal contains from 95 per cent to 98 per cent carbon. It is prepared by destructive distillation of wood, nut shells, sugar, or starch. Charcoal is used in gas masks to adsorb poisonous gases and as a reducing agent, especially in certain metallurgical processes. One cubic centimeter of active charcoal is said to have a surface of 1000 square meters. After being used for some time, charcoal loses its ability to adsorb gases. It regains its adsorptive property by heating and steaming. Charcoal, thus treated, is said to have been *activated*.

BONEBLACK (ANIMAL CHARCOAL).   Boneblack contains not more than 10 per cent carbon. The remainder is largely calcium phosphate..It is prepared by the destructive distillation of bones. Boneblack is used in the refining of sugar and for removing coloring matter from certain oils.

LAMPBLACK.   Lampblack is largely pure carbon, but it also contains some hydrocarbons. When an oil or gas flame, burning where there is an insufficient amount of air for complete combustion, is allowed to play against a cold surface, amorphous carbon, called lampblack, is deposited. This substance is used in making printers' ink, in the manufacture of certain black paints, in making shoe polish, and as a filler in certain rubber goods.

Electric Furnace Products.   At temperatures obtainable in the electric furnace, carbon can be made to combine with calcium, silicon, sulfur, and some of the metals.

CALCIUM CARBIDE (CaC_2).   When lime (CaO) and coke are heated in an electric furnace to 3000°C, calcium carbide is obtained.

$$3 \text{ C} + \text{CaO} \longrightarrow \text{CaC}_2 + \text{CO}$$

It is used in the making of acetylene and as the starting point in the fixation of nitrogen by the Cyanamide Process.

$$\text{CaC}_2 + 2 \text{ H}_2\text{O} \longrightarrow \text{Ca(OH)}_2 + \text{C}_2\text{H}_2 \text{ (acetylene)}$$
$$\text{CaC}_2 + \text{N}_2 \longrightarrow \text{CaCN}_2 + \text{C (calcium cyanamide)}$$

SILICON CARBIDE (CARBORUNDUM, SiC).   When sand, coke, and a little common salt are heated in a resistance-type electric furnace, silicon carbide is obtained. The reactions are:

$$\text{SiO}_2 + 2 \text{ C} \longrightarrow \text{Si} + 2 \text{ CO}$$
$$\text{Si} + \text{C} \longrightarrow \text{SiC}$$

Silicon carbide is used for polishing and grinding. See Figure 21-2.

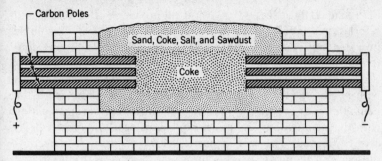

Fig. 21-2. A resistance-type electric furnace used in preparing carborundum.

CARBON DISULFIDE (CS_2). When sulfur vapors are passed over hot coke, in the absence of oxygen, carbon disulfide is formed.

$$C + 2S \longrightarrow CS_2$$

It is a heavy, colorless, and highly refractive liquid, which boils at 46.3°C and is highly flammable. Carbon disulfide is used as a solvent, as an insecticide, and in the manufacture of carbon tetrachloride.

$$CS_2 + 3Cl_2 \text{ (catalyst)} \longrightarrow CCl_4 + S_2Cl_2$$

Carbon Monoxide. There are two important oxides of carbon, carbon monoxide (CO) and carbon dioxide (CO_2). The monoxide was shown by Cruikshank in 1800 to be a compound of carbon and oxygen. It is prepared by reducing the dioxide of carbon; by burning carbon in a limited supply of air; by the decomposition of formic or oxalic acid; and by the decomposition (dehydration) of formic or oxalic acids with concentrated sulfuric acid. Equations for these four methods are:

1. $$CO_2 + C \longrightarrow 2CO$$
2. $$2C + O_2 \longrightarrow 2CO$$
3. $$CH_2O_2 + \text{(heat)} \longrightarrow H_2O + CO$$
   $$C_2H_2O_4 + \text{(heat)} \longrightarrow H_2O + CO + CO_2$$
4. $$C_2H_2O_4 (H_2SO_4) \longrightarrow H_2O(H_2SO_4) + CO + CO_2$$

Large quantities are produced by the incomplete combustion of gasoline in automobile engines. Since carbon monoxide is poisonous, it is dangerous to leave an automobile engine running in a closed garage.

PHYSICAL PROPERTIES. Carbon monoxide is a colorless and odorless gas, 0.967 times as dense as air. It is practically insoluble in water, but is quite soluble in cuprous chloride solutions containing ammonia.

CHEMICAL PROPERTIES. It is a powerful reducing agent, burns with a blue flame, and is very poisonous. It combines with the hemoglobin of the blood, thus preventing oxygen from reacting with the hemoglobin. In mine explosions, carbon monoxide is always formed. The rescue parties carry canaries to detect its presence. The death of the bird serves as a warning to the rescuers.

USES. Carbon monoxide is used as a reducing agent, as a fuel, and in the preparation of methyl alcohol by a catalytic process.

Carbon Dioxide. Carbon dioxide occurs in the atmosphere (4 parts per 10,000), in volcanic gases, in fissures of the earth, and in certain caves.

It is prepared by the complete combustion (oxidation) of any organic compound; by heating certain carbonates; by the action of an acid upon a carbonate; by fermentation of glucose; by reduction of metal oxides with carbon monoxide; and by adding water to baking powder. The equations are:

$$CH_4 + 2\,O_2 \longrightarrow 2\,H_2O + CO_2$$
$$CaCO_3 + heat \longrightarrow CaO + CO_2$$
$$MgCO_3 + 2\,HCl \longrightarrow MgCl_2 + H_2O + CO_2$$
$$C_6H_{12}O_6 \text{ (fermentation)} \longrightarrow 2\,C_2H_5OH + 2\,CO_2$$
$$CuO + CO \longrightarrow Cu + CO_2$$
$$Ca(H_2PO_4)_2 + 2\,NaHCO_3 \xrightarrow{\text{(water)}}$$
$$CaHPO_4 + Na_2HPO_4 + 2\,H_2O + 2\,CO_2$$

PHYSICAL PROPERTIES. Carbon dioxide is a colorless, odorless, and tasteless gas. It weighs 1.96 grams per liter and is moderately soluble in water; one hundred grams (100 g) of water will dissolve 0.145 g of the oxide at 25°C. It is easily liquefied and solidified. If the liquid is allowed to escape from a tank of the oxide, the heat of evaporation is sufficient to cool part of the oxide until it solidifies. This is called *carbon dioxide snow* (dry ice).

CHEMICAL PROPERTIES. Carbon dioxide does not burn or support combustion; it is very stable but can be decomposed at 2000°C or higher; and it reacts to a limited extent with water to form carbonic acid. In the presence of sunlight and a catalyst (chlorophyll), it combines with water to form starch or cellulose.

USES. Carbon dioxide is used as a leavening agent in bread-making; in fire extinguishers (air containing 15 per cent carbon dioxide will not support combustion); in making sodium bicarbonate in the Solvay Soda Process; in making soda water and such beverages; and as a refrigerant (carbon dioxide snow mixed with ether or acetone will give a temperature of $-80°C$). Cubes of carbon dioxide snow (dry ice) are used in refrigeration. See Figure 21-3.

Fig. 21-3. A common portable fire extinguisher.

The Carbon Dioxide Cycle in Nature. Carbon dioxide in the atmosphere is converted into plant tissue (starch and cellulose) by means of the sun's energy and the chlorophyll of the plant. This process is known as *photosynthesis*. The carbon dioxide, thus removed from the atmosphere, is returned by the decay of dead plants (carbon dioxide is one product of decomposition), and the decay of dead animals which have eaten the plants for food. See Figure 21-4, p. 230.

Occurrence of Silicon. About 27 per cent of the earth's crust is silicon. It ranks next to oxygen in abundance, but never occurs free in nature. There are many silicon minerals; a few of the more important are as follows:

FREE SILICA (SiO_2). In the crystalline form, this substance is called quartz. Sand, agate, and flint are also forms of silica.

CARBON DIOXIDE CYCLE IN NATURE

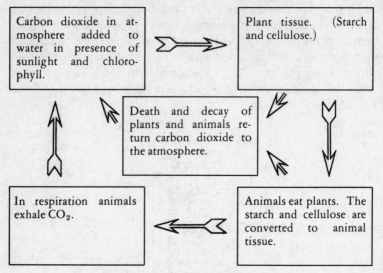

Fig. 21-4.   Carbon dioxide cycle in nature.

SILICATES.   These substances are salts of the silicic acids, and the names of some of them are: feldspar; orthoclase ($KAlSi_3O_8$) and anorthoclase ($NaAlSi_3O_8$); mica, an orthosilicate of potassium and aluminum [$H_2Al_3(SiO_4)_3$]; and clay, an impure form of kaolinite ($H_4Al_2Si_2O_9$).

Preparation and Properties of Silicon.   Silicon is prepared by the reduction of silicon dioxide with carbon in a resistance-type electric furnace.

$$SiO_2 + 2\,C \longrightarrow Si + 2\,CO$$

The metals magnesium and aluminum can be used in place of carbon for the reduction of silica, but are not used commercially because carbon is just as effective and is less expensive.

PHYSICAL PROPERTIES.   There are two allotropic forms, one crystalline and the other amorphous. The crystalline form, a gray, metallike substance, is a fair conductor of electricity.

CHEMICAL PROPERTIES.   Silicon burns in oxygen, combines with the halogens at suitable temperatures, and is not soluble in acids but dissolves readily in sodium hydroxide.

$$Si + 2\,NaOH + H_2O \longrightarrow Na_2SiO_3 + 2\,H_2$$

USES. Silicon is used for making alloys such as Duriron (containing 15 per cent silicon) and ferrosilicon (FeSi). It is also used as a "scavenger" (remover of dissolved gases) in the steel industry.

Compounds of Silicon. Silicon forms many compounds. Some of the most important are as follows:

SILICON DIOXIDE (SiO_2). There are several forms of silicon dioxide. The following are well known:

Quartz. Quartz is a hard, transparent crystalline substance with a low coefficient of expansion and a density of 2.65. It transmits ultraviolet light and is insoluble in water and acids. This substance is used for quartz glass. Quartz glass can be heated and then plunged into cold water without breaking. This is because the coefficient of expansion is very low. Quartz containing metal or metal oxide impurities is known under various names. The important forms are: agate (SiO_2 + Fe), amethyst (SiO_2 + Mn), and onyx (SiO_2 + C).

Sand. Sand (impure SiO_2) is insoluble in water and acids (except HF), has a low coefficient of expansion, and has a density of 2.65. It is used in making glass and mortar.

Amorphous Silicon Dioxide. This substance (silica gel) is used as a base in preparing catalysts and as an adsorbent.

SILICIC ACIDS. There are several silicic acids, all of which may be considered as having been derived from the ortho acid (H_4SiO_4) by the removal of water. The following equations illustrate the preparation of *orthosilicic acid* (H_4SiO_4).

$$4\,NaOH + SiO_2 \longrightarrow Na_4SiO_4 + 2\,H_2O$$
$$Na_4SiO_4 + 4\,HCl \longrightarrow H_4SiO_4 + 4\,NaCl$$

Metasilicic acid (H_2SiO_3) is prepared as follows:

$$Na_2CO_3 + SiO_2 \longrightarrow Na_2SiO_3 + CO_2$$
$$Na_2SiO_3 + 2\,HCl \longrightarrow H_2SiO_3 + 2\,NaCl$$

Both acids are white gelatinous precipitates which are easily dehydrated. They are easily converted into colloidal gels. Several salts of these acids are found in nature.

SILICON TETRACHLORIDE ($SiCl_4$). This is prepared by the action of carbon and chlorine upon silicon dioxide.

$$SiO_2 + 2\,C + 2\,Cl_2 \longrightarrow SiCl_4 + 2\,CO$$

It hydrolyzes easily to form silicic and hydrochloric acids.

$$SiCl_4 + 4\,H_2O \longrightarrow Si(OH)_4 + 4\,HCl$$

When sprayed into the air with ammonia, it forms an effective smoke screen since, in moist air, solid silicic acid and also ammonium chloride are formed.

SILICON HYDRIDES. The first six members of the series whose general formula is Si_nH_{2n+2} (corresponding to the paraffin hydrocarbon series) are known. They are very unstable, some taking fire upon exposure to the air.

FERROSILICON (FeSi). This substance is made by the reduction in an electric furnace, of an iron ore rich in silica. It is used in the steel industry as a deoxidizer and also in making acid-resisting alloys (Duriron).

SODIUM SILICATE. This is a mixture of sodium metasilicate and sodium orthosilicate. It is used for preserving eggs and as a cement or glue. Some is added to cheap soap. It is sometimes called *water glass* because of its solubility.

HYDROGENITE. This is a mixture of ferrosilicon and solid sodium hydroxide. When water is added, hydrogen is liberated. It is useful in filling balloons and as a convenient source of hydrogen.

Glass. A glass is a fused, usually transparent, noncrystalline mass composed of silica and basic oxides, such as sodium oxide and calcium oxide. Some glasses contain aluminum oxide, bismuth oxide, or lead oxide. Glass is spoken of as a supercooled liquid because it has no definite melting point.

SOFT GLASS (SODA GLASS). When sodium carbonate, calcium carbonate, and sand are mixed and melted, the fused mass obtained is soft glass. The probable equation is:

$$2\,Na_2CO_3 + CaCO_3 + 6\,SiO_2 \longrightarrow Na_4CaSi_6O_{15} + 3\,CO_2$$

Ordinary soft glass is used to make window panes and bottles.

Hard glass contains potassium oxide in the place of sodium oxide. *Flint glass* contains lead oxide.

Colored glasses are made by dissolving metals or metal oxides in ordinary glass. Ruby glass contains finely divided (colloidal) gold or colloidal selenium. Blue glass contains cobalt silicate.

PLATE GLASS. This is a soda glass, but the percentage composition is not exactly the same as ordinary soda glass. Plate glass is used for large windows and comes in sheets much thicker than ordinary window pane.

PYREX GLASS. Pyrex contains boric oxide (11–12 per cent) and also small amounts of iron oxide and potassium oxide. It is

Table 21-2

Composition of Glass

Type of Glass	Approximate Percentage Composition								
	SiO_2	Na_2O	K_2O	CaO	MgO	Fe_2O_3	PbO	Al_2O_3	B_2O_5
Soft glass (Soda glass)	75	15		8				2.0	
Plate glass	72	14		12.9		0.1		1.0	
Flint glass (Lead glass)	46	8.8	0.5			0.02	44.5		
Pyrex glass	80.8	3.8	0.8	0.2	0.2	0.1		2.0	11.9

used for chemical glassware because it is more resistant to chemicals, has greater mechanical strength, and is less likely to crack upon sudden temperature changes.

SAFETY GLASS. The glass used in automobile windshields and windows is made by placing a sheet of transparent plastic material between two sheets of glass and heating them under pressure until a good bond is formed.

NEW USES. During recent years, glass has been put to many new uses. Walls made of glass brick are durable and admit diffused light. *Glass fiber* is used in making dust filters and for heat and electrical insulators. Fabrics made of fiber glass are on the market and find a variety of uses.

The Silicones. These compounds are composed of chain molecules in which silicon and oxygen atoms are arranged alternately. Thus, methyl silicone is:

$$H_3C-\underset{\underset{CH_3}{|}}{\overset{\overset{CH_3}{|}}{Si}}-O-\underset{\underset{CH_3}{|}}{\overset{\overset{CH_3}{|}}{Si}}-O-\underset{\underset{CH_3}{|}}{\overset{\overset{CH_3}{|}}{Si}}-CH_3$$

Some silicones (through proper heating) acquire a ring structure. Thus:

PREPARATION.   Methyl chloride reacts with silicon which contains some copper as a catalyst.

$$2\,CH_3Cl\,+\,Si(Cu)\,\longrightarrow\,(CH_3)_2SiCl_2\,+\,Cu$$
$$\text{(dimethylsilicon dichloride)}$$

Hydrolysis gives:

$$(CH_3)_2SiCl_2\,+\,2\,H_2O\,\longrightarrow\,(CH_3)_2Si(OH)_2\,+\,2\,HCl$$

Condensation (by splitting out water) takes place between two or more molecules produced by hydrolysis. Thus:

$$
\begin{array}{ccccc}
CH_3 & & CH_3 & & CH_3 \quad CH_3 \\
| & & | & & | \qquad | \\
HO-Si-\overline{OH}\;\;\overline{H}O-Si-OH & \longrightarrow & HO-Si-O-Si-OH \\
| & & | & & | \qquad | \\
CH_3 & & CH_3 & & CH_3 \quad CH_3
\end{array}
$$

PROPERTIES AND USES.   Depending on molecular structure and molecular weight, silicones may be oily liquids, resins, or rubberlike substances. The liquids have low coefficients of viscosity, are excellent heat-resistors, and are chemically inert toward metals and many reagents. Silicones are used to make water-repellent surfaces, to make heat-resistant varnishes, for electrical insulation, and for liquids in hydraulic systems. A puttylike ball of silicone, plastic enough to be molded into sheets or drawn in threads, will bounce like a rubber ball if thrown against a hard wall.

Tin.   Tin is found in the Malay States, Bolivia, the East Indies, and China. Cassiterite (tinstone, SnO_2) is the most important compound of tin found in nature. Complex sulfides of tin are known, but they are of no commercial importance.

METALLURGY.   The oxide, SnO_2, is reduced with carbon in a reverberatory furnace. The common furnace is 12×30 feet and treats about 10 tons of charge per batch. There is a large tin loss in the slag, which is usually saved and retreated. The reaction is:

$$SnO_2\,+\,C\,\longrightarrow\,Sn\,+\,CO_2$$

If the tin ore contains sulfur or arsenic compounds, they must be removed by roasting before the oxide is reduced. Tin is also prepared by an electrolytic method.

PROPERTIES.   Tin exists in two well-defined allotropic forms, *gray* tin and *white* tin. The transition temperature is 18°C. Gray tin is brittle and easily crumbles into a powder. White tin is soft and malleable. It may be rolled into thin sheets. Tin resists oxidation, but reacts with the common acids. In the second reaction below, it will be noticed that the nitrate of tin is not a product.

$$Sn + 2 HCl \longrightarrow SnCl_2 + H_2$$
$$3 Sn + 4 HNO_3 \longrightarrow 2 H_2O + 4 NO + 3 SnO_2$$

USES. Tin is used as a protective coating for other metals (tin plate), for preparing alloys such as bronze, pewter, and solder, and for distilled water pipes. Tin foil has many uses.

Compounds of Tin. The important compounds of tin are as follows.

STANNOUS HYDROXIDE [$Sn(OH)_2$]. This compound is amphoteric. It reacts with acids and bases.

$$Sn(OH)_2 + 2 HCl \longrightarrow SnCl_2 + 2 H_2O$$
$$Sn(OH)_2 + 2 NaOH \longrightarrow Sn(ONa)_2 + 2 H_2O$$

$Sn(ONa)_2$ can also be written Na_2SnO_2.

STANNOUS CHLORIDE ($SnCl_2$). This compound, prepared by the action of hydrogen chloride on tin, is used as a mordant in dyeing and as a reducing agent in the laboratory. The mordant makes it possible to fix the dye to the cloth.

AMMONIUM CHLOROSTANNATE [$(NH_4)_2SnCl_6$]. This is used as a mordant in dyeing.

Lead. Lead is found in western United States, particularly in Missouri, Idaho, Utah, and Oklahoma. The important minerals include galena (PbS), cerussite ($PbCO_3$), and anglesite ($PbSO_4$). Low-grade sulfide ores are concentrated by milling and flotation.

METALLURGY. Today, most lead ores, especially the sulfides, are low grade and, therefore, must be concentrated. This is accomplished by the Flotation Method. The concentrate is treated by one of the following methods.

If the concentrate contains no silver or gold, it is placed in a Newman ore hearth where metallic lead is obtained by the following reactions:

$$2 PbS + 3 O_2 \longrightarrow 2 PbO + 2 SO_2$$
$$PbS + 2 PbO \longrightarrow 3 Pb + SO_2$$

If the concentrate contains silver and gold, it is roasted (and sintered) on a Dwight-Lloyd sintering machine. This removes most of the sulfur and makes a sinter which is easily handled in the blast furnace.

Blast Furnace Smelting. The sinter, mixed with proper fluxing materials, is charged to the lead blast furnace. The products obtained are: impure lead (lead bullion), some speiss (iron arsenide), and, perhaps, some leaded matte (iron, copper, and lead sulfides), and slag. The reactions are:

$$PbS + 2\,PbO \longrightarrow 3\,Pb + SO_2$$
$$PbS + Fe \longrightarrow Pb + FeS$$

Metallic (scrap) iron is usually charged to the furnace to help reduce PbS. Iron also combines with arsenic, if present, to form iron arsenide (speiss).

Refining (Softening) of Lead. The impure lead from the blast furnace is purified by:

1. *Drossing.* The molten blast furnace lead is placed in large kettles where air is blown through to oxidize impurities, and powdered sulfur is added to remove copper as copper sulfide.

2. *The Parkes Process.* Silver and gold are removed from the lead in the following way: About 1 per cent of zinc is added to the molten lead. This zinc combines with the gold and silver present and floats on the surface of the lead as a scum. This is skimmed off. The small amount of zinc remaining in the molten lead is oxidized and volatilized by blowing steam and air through it.

3. *The Betts Electrolytic Process.* Lead containing silver, gold, and such metals as bismuth may be purified by electrolysis. Impure lead (containing silver, gold, bismuth, etc.) is used for the anodes. Thin sheets of pure lead are the cathodes. The electrolyte is a solution of fluosilicic acid and lead fluosilicate.

PROPERTIES. Lead is a silvery metal which melts at 327.5°C, boils at 1620°C, and has a specific gravity of 11.34. It is a moderately good conductor of heat and electricity, is the softest of the heavy metals, and can be rolled into sheets and drawn into wire. Lead tarnishes readily in air and reacts with nitric acid.

$$3\,Pb + 8\,HNO_3 \longrightarrow 3\,Pb(NO_3)_2 + 4\,H_2O + 2\,NO$$

USES. Lead is used for making pipes, for lining acid tanks, for storage batteries, for plumbing fixtures, for making alloys such as solder, lead shot, type metal, and many others, for cable coverings, and for preparing white lead.

Compounds of Lead. The important compounds of lead include the oxides, acetate, carbonate, chromate, and arsenate.

OXIDES. Litharge, red lead (minium), and lead dioxide are the important oxides of lead.

Litharge (PbO) is an orange-yellow compound prepared by heating the metal in air at about 550°C. It is used in glazing pottery and making glass. When mixed with glycerine, it forms a useful cement.

Red lead (Pb_3O_4) is prepared by heating litharge at temperatures below 500°C. It is used in making flint glass and as a paint pigment.

Lead dioxide (PbO_2) is prepared by the oxidation of a lead salt in alkaline solutions, or by the action of nitric acid on red lead. It is used in lead storage batteries and as an oxidizing agent.

$$2\,PbCl_2 + Ca(OCl)_2 + 2\,H_2O \longrightarrow CaCl_2 + 2\,PbO_2 + 4\,HCl$$
$$Pb_3O_4 + 4\,HNO_3 \longrightarrow 2\,Pb(NO_3)_2 + PbO_2 + 2\,H_2O$$

LEAD ACETATE [SUGAR OF LEAD, $Pb(C_2H_3O_2)_2$]. This is prepared by dissolving litharge in acetic acid. It is used where a soluble lead salt is needed. It is very poisonous.

BASIC LEAD CARBONATE [WHITE LEAD, $PbCO_3 \cdot Pb(OH)_2$]. This is prepared by the action of air, carbon dioxide, and acetic acid on perforated lead plates. It is used as a paint pigment. White lead is also prepared by an electrolytic process called the *Sperry Process*.

LEAD CHROMATE (CHROME YELLOW, $PbCrO_4$). This is prepared by the action of a soluble lead salt on a soluble chromate.

$$Pb(C_2H_3O_2)_2 + K_2CrO_4 \longrightarrow PbCrO_4 + 2\,KC_2H_3O_2$$

LEAD ARSENATE [$Pb_3(AsO_4)_2$]. This is prepared by treating lead acetate with sodium arsenate. It is used as an insecticide.

LEAD TETRAETHYL. This is an important compound of lead used as an antiknock in gasoline. It is prepared by the action of ethyl chloride on a sodium lead alloy.

$$4\,C_2H_5Cl + Na_4Pb \longrightarrow Pb(C_2H_5)_4 + 4\,NaCl$$

Lead tetraethyl in gasoline contributes to the pollution of the atmosphere. Consequently lead-free gasolines are now on the market.

REVIEW QUESTIONS

1. Why are carbon and silicon placed in the same group in the periodic table? Give several reasons.
2. How does carbon occur in nature? How does silicon occur in nature?
3. How does silicon differ from carbon in its ability to combine with itself, atom to atom?
4. What are the important oxides of carbon? Of silicon? What can you say as to their ease of preparation?
5. What are the allotropic forms of carbon?
6. Contrast graphite and diamond in terms of physical properties.
7. Distinguish between: (a) boneblack and lampblack, (b) coke and charcoal.
8. (a) How is calcium carbide prepared? (b) How is silicon carbide prepared?
9. How can one prove that the ordinary atmosphere contains carbon dioxide?
10. Write a paragraph on "The Carbon Dioxide Cycle in Nature."
11. How can one make free silicon from SiO_2?
12. What is glass? Distinguish (chemically and physically) between quartz glass and ordinary soft glass.
13. What are the silicones? How are they made? What are their physical properties?
14. From a chemical standpoint, account for the fact that tin and lead were known to the early Romans.
15. Compare the two allotropic forms of tin.
16. What is the chemical action of concentrated nitric acid on metallic tin? Write the equation.
17. List the common uses of tin.
18. What are the names and formulas of the three important minerals of lead?
19. What happens to oxidized lead in a lead blast furnace?
20. How is silver removed from metallic lead? How is copper removed?
21. Distinguish between (a) litharge and white lead, (b) sugar of lead and lead tetraethyl.
22. Give several uses for metallic lead.
23. Write an equation for the preparation of lead arsenate.
24. How is lead peroxide (dioxide) used in car batteries? Write an equation for its preparation.

Chapter 22
Metallurgy, General Properties of Metals, Alloys

Metallurgy.   Metallurgy is the science (and art) of extracting metals from their naturally occurring compounds and adapting these metals to the use of mankind. Metallurgy, like chemistry, has several branches. These branches include hydrometallurgy, pyrometallurgy, electrometallurgy, and physical metallurgy. The first two will be considered later in this chapter.

Minerals.   Minerals are naturally occurring compounds of metals that are usually crystalline. They are found as oxides, sulfides, carbonates, chlorides, sulfates, phosphates, silicates, and others. Some metals such as gold, silver, and copper may be found in the uncombined (free) state in nature. Galena (PbS) is a mineral. It is a crystalline compound of lead and sulfur. Hematite (Fe_2O_3) is a mineral composed of iron and oxygen.

Ores.   The naturally occurring earthy substances (sometimes called *gangue*) containing one or more valuable minerals in sufficient quantity to make their recovery profitable are called *ores*. For example, a naturally occurring material contains copper minerals. The percentage of copper in this material is 25 per cent. This is considered to be a high-grade ore. On the other hand, material from the copper mine in Bingham, Utah, contains such minerals as chalcopyrite ($CuFeS_2$) and chalcocite (Cu_2S). This material is a low-grade ore because it contains less than 1 per cent copper.

Metallurgical Operations.   Once an ore is mined, metallurgical operations begin. The metallurgical steps involved in obtaining a metal from an ore, somewhat simplified, are listed below.

Concentration of the Ore.   Most ores, especially those of the nonferrous metals, are low-grade. That is, the metal content probably is less than 5 per cent. By concentration methods, a product containing, say, 30 per cent of the metal can be made. For economic reasons, the preparation of concentrates is necessary.

The first step in making a concentrate is to pulverize the ore. This process, called *milling*, is done by crushers, rolling mills, and ball mills. The mineral particles can be separated from the powdered ore by one or a combination of the following:

GRAVITY CONCENTRATION. Mineral particles usually have a higher specific gravity than particles of gangue. By passing a water suspension of the powdered ore over an inclined vibrating table (a Wilfley table), the mineral is separated from the gangue (waste material).

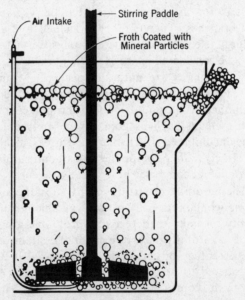

Fig. 22-1. A simplified laboratory model flotation cell.

FLOTATION. Mineral particles (especially sulfides) will attach themselves to the bubbles of a froth made by bubbling air up through a flotation cell containing a water suspension of the ore. A stable froth is made by adding a small amount of frothing agent (pine oil or similar substance). The mineral particles adhere to the froth better if a small quantity of collector (say, potassium ethyl xanthate) is present. The gangue is not present in the froth, and thus a separation is made. The froth is skimmed off the top of the cell, mixed with water, and run into a settling tank. After settling, the mineral matter is filtered and sent on to the next process. The preparation of an ore for smelting or other metallurgical treatment is often called *ore dressing* or *mineral dressing*. If an iron ore is being concentrated, the process is called *beneficiation*. Most nonferrous ores must be subjected to concentration before smelting. See Figure 22-1 for a simplified flotation cell.

Treatment of Concentrates.   There are several methods of obtaining a metal from a concentrate. Two methods will be considered here: pyrometallurgical and hydrometallurgical.

PYROMETALLURGICAL PROCESSES.   These are processes which take place at high temperatures. To obtain metals from their compounds, it is usually necessary to make use of chemical reactions which occur at high temperatures. Sometimes the term *smelting* is used as a general term to include the various processes listed here under pyrometallurgy. However, in some cases, smelting may be used in connection with reduction processes only. In the pyrometallurgical processes listed here, we shall assume that suitable ovens and/or furnaces are used to obtain the temperature required in each case.

Drying.   Sometimes it is necessary to remove the water (moisture) from a concentrate before it can be placed in a furnace.

Calcining.   This is a process in which carbonates are decomposed by heating.

$$CaCO_3 + heat \longrightarrow CaO + CO_2$$

Roasting.   This is essentially an oxidation process. Thus, sulfides are converted (roasted) into oxides by heating them in the presence of air or oxygen.

$$Cu_2S + 2O_2 \longrightarrow 2CuO + SO_2$$

Reduction.   In pyrometallurgical processes, a metal is produced by the action of a suitable reducing agent on a compound of the metal (usually an oxide, but not necessarily so). Often, blast furnaces are used in reduction processes. Many of the impurities (gangue) collect in what is called a *slag*. Ordinarily this is a complex mixture of fused silicates of calcium and aluminum in which the impurities are collected. The slag is subsequently discarded.

$$Fe_2O_3 + 3\,CO \longrightarrow 2\,Fe + 3\,CO_2$$
$$WO_3 + 3\,H_2 \longrightarrow W + 3\,H_2O$$
$$AlCl_3 + 3\,K \longrightarrow Al + 3\,KCl$$

Fire Refining. In this process, the impurities in a molten metal are removed by oxidation. Small amounts of impurities present in molten copper (such as iron, bismuth, or zinc) can be removed by blowing air through it. This oxidizes the impurities, which will be absorbed by the slag present or go off as a fume.

$$Cu(Zn) + (O) \longrightarrow Cu + ZnO$$

HYDROMETALLURGICAL PROCESSES. These are processes in which water (or a water solution) is used to dissolve (or extract) the metallic compound and thus separate it from the gangue. Metallurgists call such a process *leaching.* Dilute sulfuric acid is a good leaching agent for oxidized copper ores or concentrates. The reaction is:

$$CuO + H_2SO_4 \longrightarrow CuSO_4 + H_2O$$

The best leaching agent is one that has no solvent effect on the gangue in an ore or concentrate. The most efficient leaching process is one in which the leaching agent is regenerated in (one step of) the process.

Recovery of Metals. Two methods may be used to recover metals from solutions: electrowinning and replacement.

ELECTROWINNING. In this process, the metal ions (in the leaching solution) are plated out as metal by means of a direct current on suitable cathodes. Electrolytic copper is plated out from a copper sulfate solution. Electrolytic zinc is plated out from a zinc sulfate solution.

REPLACEMENT. A metal can be replaced from solution by adding to it a metal higher in the activity series (see Appendix XI). Copper can be replaced from copper sulfate solution by adding scrap iron to the solution.

$$CuSO_4 + Fe \longrightarrow Cu + FeSO_4$$

Gold can be replaced from a gold cyanide solution by the addition of metallic aluminum.

$$3\,NaAu(CN)_2 + 3\,NaOH + Al \longrightarrow 3\,Au + 6\,NaCN + Al(OH)_3$$

Extremely active metals, such as calcium, magnesium, and potassium, are prepared by the electrolysis of their fused salts. (This method will be explained in later chapters.)

General Properties of Metals. Elements which furnish simple cations when their compounds are dissolved in water are metals. The metals are those elements whose oxides or hydroxides are basic. From the standpoint of atomic structure, the metals are those elements whose outer electronic orbits are very incompletely filled. In chemical changes, the electrons present are surrendered (donated) to other elements or groups. About 70 per cent of the known elements are metals. (Review Chapter 2.)

PHYSICAL PROPERTIES. Metals are malleable and ductile, hard and tenacious (the alkali and alkaline earths are exceptions), and conductors of heat and electricity. They have a low specific heat. They reflect light and have luster. With the exception of copper and gold, all are silver white or bluish white in color.

CHEMICAL PROPERTIES. The metals are electropositive in character. This is due to their tendency to lose electrons. They combine directly with oxygen, the halogens, and sulfur. Some combine with nitrogen. Metals are present as cations when their salts are dissolved in water. The active metals (those high up on the activity series) replace the less active metals from their solutions. Metals below magnesium in the activity series are not "soluble" in water.

Alloys. With few exceptions, most of the metallic substances in use are alloys. In general, alloys are used in place of pure metals either because they have more desirable properties or because they are less expensive. Thus, stainless steel is used for certain purposes in spite of its cost whereas cheap jewelry from the "Five and Ten" is made from alloys which resemble, in a superficial way, the more expensive metals, gold, silver, and platinum.

An alloy is composed of two or more elements, usually metals. It may be a solid solution, an intimate mixture, a definite compound of its constituents, or a combination of these.

PREPARATION. Alloys are prepared by heating the constituents until they form a homogeneous fused mass. This is then allowed to cool.

PROPERTIES. The specific properties of alloys depend on their composition. In general, they have lower melting points, resist corrosion better, and can be machined, cast, rolled, forged, or welded more readily than their constituent elements. The alloys have metallic-like properties. They are crystalline, are fair to good conductors of heat and electricity, and have a metallic luster. The compositions given in the following table are not necessarily fixed, and the uses are merely suggestive.

Table 22-1

Alloy	Composition	Use
Gold coin	Au 90% Cu 10%	Coins
Sterling silver	Ag 92% Cu 8%	Silverware
German silver	Cu 50% Ni 30% Zn 20%	Silver imitation
Brass	Cu 70% Zn 30%	Castings, fixtures
Bearing metal	Cu 3% Sn 90% Sb 7%	Antifriction alloy
Solder	Sn 50% Pb 50%	Joining metals
Bronze	Cu 92% Zn 2% Sn 6%	Castings, statuary
Wood's metal	Sn 12.5% Pb 25% Cd 12.5% Bi 50%	Low-melting alloy used for sprinklers
Monel metal	Cu 33% Fe 6.5% Ni 60% Al 0.5%	Rust resisting and high tensile strength
Alnico	Fe 63% Ni 20% Al 12% Co 5%	Permanent magnets

EQUILIBRIUM DIAGRAM. The diagram in Figure 22-2, often called a *phase diagram*, represents the relationship between temperature and composition of alloy systems. A phase diagram may be prepared for a pure substance (a metal) but since we are interested here only in alloys, we will not consider what is called a *one-component system*. An alloy of two metals is a two-component alloy; one of three metals is a three-component alloy. Figure 22-2 is for the tin and lead system and is, therefore, a two-component system. Usually, phase diagrams for alloys are not concerned with a vapor phase. Thus, for the lead-tin phase diagram, we are interested only in liquid and solid phases.

Point *A* represents the melting point (327°C) of lead. Point *C* is the melting point (232°C) of tin. If tin is added to lead, the melting point is lowered. If lead is added to tin, the melting point is lowered. At point *E* the *lowest* melting mixture (183°C) of tin and lead is obtained. This comes at approximately 61.9 per cent tin, and is known as the *eutectic point*. According to the diagram, a 50 per cent lead, 50 per cent tin alloy will melt at about 212°C.

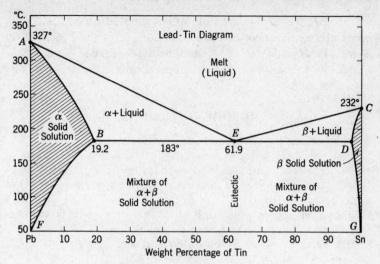

Fig. 22-2. An equilibrium diagram.

A . Melting point of pure lead, 327°C.
B . Maximum solubility of tin in lead, 19.2%.
C . Melting point of pure tin, 232°C.
D . Maximum solubility of lead in tin, 2.5%.
E . Melting point of lead-tin eutectic, 183°C.
E . Eutectic composition, 61.9% tin.

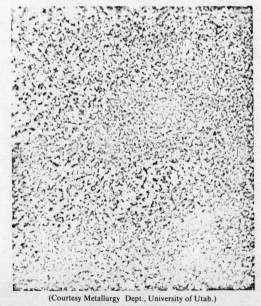

(Courtesy Metallurgy Dept., University of Utah.)

Fig. 22-3. Photomicrograph of lead-tin eutectic, 400×.

Figure 22-3 is a photomicrograph of the lead-tin eutectic. This material is a mixture of two solid solutions shown in Figure 22-2, alpha (α) shaded area *ABF*, and beta (β) shaded area *CDG*. Careful examination of this photomicrograph indicates that it is heterogeneous.

REVIEW QUESTIONS

1. What is metallurgy? Why is it both an art and a science?
2. What is the relationship of minerals to metallurgy?
3. Distinguish between a high-grade and a low-grade ore.
4. Most ores containing the common metals are low-grade. How are these concentrated?
5. Write a sentence in which each of the following words or terms is used correctly: ore dressing, mineral dressing, beneficiation.
6. Distinguish between pyrometallurgical and hydrometallurgical processes.
7. What are some of the common (general) properties of metals?
8. What is an alloy? How does it differ from a pure substance?
9. In Figure 22-2, what is the composition of the eutectic? How does it differ from the alpha solid solution?

Chapter 23
The Alkali Metals
(Group IA)

Lithium (3) $1s^2 2s^1$
Sodium (11) $1s^2 2s^2 2p^6 3s^1$
Potassium (19) [   ] $4s^1$
Rubidium (37) [   ] $5s^1$
Cesium (55) [   ] $6s^1$
Francium (87) [   ] $7s^1$

General Properties.   The free metals of group IA are so chemically active that they have, with an exception or two, few uses. Many of their compounds, on the other hand, are very important from a commercial standpoint. In this chapter we shall be chiefly concerned with sodium and potassium and their compounds. The physical properties of the group are listed in Table 23-1, p. 248.

PHYSICAL PROPERTIES.   These alkali metals are soft and silvery white in color, and possess a bright metallic luster. Their spectra are easily excited; a platinum wire moistened with a solution of the chloride of the group will impart a characteristic color to a flame (lithium, red; sodium, yellow; potassium, violet). Practically all the compounds of these metals are white solids that are soluble in water.

CHEMICAL PROPERTIES.   All are very active, electropositive metals that never occur free in nature. They are univalent. From the standpoint of their atomic structures, this is due to the fact that each metal has but one electron in the outer energy level of its atom. They decompose water vigorously, and all are good reducing agents.

Because of their relative unimportance, the chemistry of lithium, rubidium, and cesium is omitted from this book. Francium was first recognized in 1939. It is an unstable radioactive metal that has not been found in nature.

Sodium.   Metallic sodium was first prepared by electrolyzing

Table 23-1

Physical Properties of the Alkali Metals*

	Lithium	Sodium	Potassium	Rubidium	Cesium
Color	silvery white	silvery white	silvery white	silvery white	silvery white
Physical State	solid	solid	solid	solid	solid
Melting Point	179°C	98°C	64°C	39°C	28°C
Boiling Point	1336°C	883°C	758°C	700°C	670°C
Atomic Radius	1.22Å	1.57Å	2.02Å	2.16Å	2.35Å
Ionization Energy	5.39 v	5.14 v	4.34 v	4.18 v	3.89 v
Heat of Fusion Kcal/Mole	0.70	0.63	0.57	0.53	0.51
Electronegativity	1.0	0.9	0.8	0.8	0.7
Density g/cm^3 at 20°C	0.53	0.97	0.86	1.53	1.90

*Probably all the metals in this group crystallize in the body-centered cubic system. See Figure 8-6b.

fused sodium hydroxide (Sir Humphrey Davy, in 1807). It is now prepared commercially by electrolyzing fused sodium chloride in a Downs cell. The reaction within the cell is:

$$2\ NaCl + (electric\ current) \longrightarrow 2\ Na + Cl_2$$

The properties of sodium are illustrated by the following reactions: It combines directly with the halogens.

$$2\ Na + Cl_2 \longrightarrow 2\ NaCl$$

It has great affinity for oxygen.

$$2\ Na + O_2 \longrightarrow Na_2O_2$$

It is a good reducing agent. (Water must be absent.)

$$3\ Na + CrCl_3 \longrightarrow Cr + 3\ NaCl$$

Sodium will replace hydrogen from water or alcohol.

$$2\ Na + 2\ C_2H_5OH \longrightarrow 2\ C_2H_5ONa + H_2$$

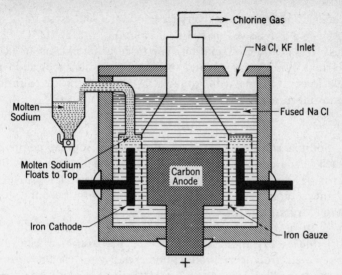

Fig. 23-1. The Downs cell for the preparation of metallic sodium.

Sodium metal is used as vapor in sodium vapor street lamps. It is used as a coolant and as a heat transfer agent in nuclear reactors. Sodium is also used as a starting substance in making some sodium compounds and as a reducing agent in preparing certain metals from their chlorides.

$$ZrCl_4 + 4\,Na \longrightarrow Zr + 4\,NaCl$$

Compounds of Sodium. The more important compounds are discussed here.

SODIUM CHLORIDE. Sodium chloride is the most important compound of sodium. Practically all other sodium compounds can be prepared from the chloride. Millions of tons are produced annually in the United States. This compound occurs in beds of rock salt, in salt lakes, such as Great Salt Lake, and in the oceans. Sodium chloride is obtained by the mining of rock salt; by the evaporation of brine solutions by either natural or artificial heat; and in cold countries by concentration of brines by freezing. After concentrating, the salt is recovered from the brine by evaporation by means of artificial heat. Sodium chloride can be prepared in the laboratory on a small scale by neutralizing sodium hydroxide with hydrogen chloride.

Sodium chloride crystallizes from water solutions as anhydrous cubes. Temperature has very little effect upon its solubility. At 0°C, 26.3 g dissolves in 100 g of solution, and at 100°C, 28.1 g dissolves in 100 g of solution. The fused salt conducts an electric current. It is present in water solutions as sodium ions and chloride ions. It reacts with nonvolatile acids, such as sulfuric or phosphoric acids, to give hydrogen chloride.

$$NaCl + H_2SO_4 \longrightarrow NaHSO_4 + HCl$$

Sodium chloride is used to make other sodium compounds such as sodium carbonate, sodium bicarbonate, and sodium hydroxide; in preparing chlorine and hydrogen chloride; in the manufacture of soap; as a food preservative; in seasoning foods; and in freezing mixtures.

SODIUM HYDROXIDE (NaOH) (CAUSTIC SODA). Sodium hydroxide is a by-product obtained in the preparation of chlorine by by the electrolytic method. Commercially, this is the most important method. More than 1,500,000 tons are produced yearly. It is also prepared by treating sodium carbonate with slaked lime.

$$Na_2CO_3 + Ca(OH)_2 \longrightarrow CaCO_3 + 2\,NaOH$$

Sodium hydroxide is a white solid, very soluble in water, and corrosive. Its solutions dissolve wool, and its water solution is strongly basic to indicators. Sodium hydroxide is deliquescent, and is a good absorbent for carbon dioxide.

This compound is used in making soap, in refining oil, in mercerizing cotton, in the paper industry, in rubber reclaiming, in making rayon, and in neutralizing acids to form salts.

SODIUM BICARBONATE (NaHCO₃). To prepare this compound, a solution of sodium chloride is saturated first with ammonia and then with carbon dioxide. Upon cooling this solution, the sodium bicarbonate, which is slightly soluble under these conditions, separates out. The ammonia can be recovered. This is the basis of the commercial *Solvay Process.*

$$Na^+ + Cl^- + NH_3 + H_2O + CO_2 \longrightarrow NaHCO_3 + NH_4^+ + Cl^-$$

The normal carbonate can be prepared by heating the bicarbonate.

$$2\,NaHCO_3 \text{ (heat)} \longrightarrow Na_2CO_3 + H_2O + CO_2$$

Sometimes, $Na_2CO_3 \cdot 10\,H_2O$ is called *washing soda* or *sal soda,* and $NaHCO_3$ is called *baking soda*. Anhydrous sodium carbonate (Na_2CO_3) is called *soda ash*.

Sodium bicarbonate decomposes on heating (see above) and reacts with bases.

$$NaOH + NaHCO_3 \longrightarrow Na_2CO_3 + H_2O$$

Acids (even weak acids) liberate CO_2. Thus,

$$NaHCO_3 + HCl \longrightarrow NaCl + H_2O + CO_2$$
$$NaHCO_3 + H \cdot C_2H_3O_2 \longrightarrow NaC_2H_3O_2 + H_2O + CO_2$$

Sodium bicarbonate is used in making baking powder, in fire extinguishers, and in the manufacture of effervescent salts.

SODIUM CARBONATE (Na_2CO_3) (SODA ASH). This is prepared by heating sodium bicarbonate (made by the Solvay Process). The LeBlanc Process, now obsolete, prepared sodium carbonate by the reduction of sodium sulfate (by carbon) to sodium sulfide, which subsequently was treated with calcium carbonate.

If a mixture of sodium carbonate and a silicate is heated, soluble sodium silicate is formed. This is a good way to decompose a silicate for chemical analysis. Weak acids, such as acetic acid, will liberate CO_2 from sodium carbonate. It will hydrolize in water to give an alkaline reaction.

$$2\,Na^+ + CO_3^{--} + H_2O \rightleftharpoons 2\,Na^+ + OH^- + HCO_3^-$$

Sodium carbonate is used in making glass, in the manufacture of sodium hydroxide, in soap making, in oil refining, in the manufacture of paper, in the preparation of certain chemicals, and for water softening. Millions of tons of sodium carbonate are used yearly in the United States.

SODIUM NITRATE ($NaNO_3$) (CHILE SALTPETER). Impure sodium nitrate is found in large deposits in Chile and Peru. It was probably formed by the decomposition of organic matter in the presence of sodium salts. The crude nitrate is freed from impurities by repeated crystallizations from water solutions. The crystals are slightly hygroscopic and are very soluble in water. They decompose upon heating.

$$2\,NaNO_3 + (heat) \longrightarrow 2\,NaNO_2 + O_2$$

It should be pointed out that a substance which takes up water (moisture), but does not dissolve in it, is *hygroscopic*.

Sodium nitrate is used in making nitric acid, in the manufacture of explosives, and in the preparation of fertilizers. Sodium nitrate can be prepared in the laboratory by neutralizing sodium hydroxide with nitric acid. This, however, is not an economical method.

OTHER SODIUM COMPOUNDS. Sodium sulfate, $Na_2SO_4 \cdot 10$ H_2O (*Glauber salt*), is used in the manufacture of glass and in medicine. Sodium tetraborate, $Na_2B_4O_7 \cdot 10$ H_2O (*borax*), is used in the manufacturing of soaps, glass, and fluxes. It is also used for softening water. Sodium thiosulfate, $Na_2S_2O_3 \cdot 5$ H_2O (*hypo*), finds use in photography and in chemical analysis. Disodium phosphate, Na_2HPO_4, is used in medicine and in chemical work whenever a soluble phosphate is needed. Sodium hypochlorite, $NaClO$, is used in bleaching cotton goods and as a germicide.

Potassium. Compounds of potassium occur widely distributed in nature. The Stassfurt salts in Germany contain sylvite (KCl), carnallite ($KCl \cdot MgCl_2 \cdot 6$ H_2O), and kainite ($MgSO_4 \cdot KCl \cdot 3$ H_2O). Searles Lake in California and Great Salt Lake contain potassium salts. Ash from plants and dust from cement kilns and blast furnaces also contain some potassium compounds.

Potassium is prepared by the electrolysis of the fused hydroxide or chloride. The common properties and uses of metallic potassium are very similar to those of sodium and, therefore, will not be considered here.

Compounds of Potassium. The more important compounds of potassium include the following.

POTASSIUM CHLORIDE (KCl). This is used in making other potassium compounds.

POTASSIUM HYDROXIDE (KOH) (CAUSTIC POTASH). Potassium hydroxide is made by the electrolysis of potassium chloride solutions. It is used in making soft soaps and also in the Edison storage battery.

POTASSIUM NITRATE (KNO_3) (SALTPETER). This is prepared by the action of a hot solution of sodium nitrate on potassium chloride. The sodium chloride separates out first. The nitrate is used in making gunpowder.

$$NaNO_3 + KCl \longrightarrow KNO_3 + NaCl$$

POTASSIUM CHLORATE ($KClO_3$). This salt is made by the action of chlorine on hot potassium hydroxide solution and also by the electrolysis of potassium chloride under suitable conditions. Potassium chlorate is used in making matches, explosives, and fireworks, and for the laboratory preparation of oxygen.

Baking Powders. A baking powder is a mixture of sodium bicarbonate with certain salts (with an acid reaction in water) and some inert filler such as starch to take up the moisture.

When moisture is added to a baking powder, carbon dioxide is set free. Heat aids in the escape of the gas, which leaves the material light and spongy.

Reactions for three typical baking powders follow. It is assumed that water is present in each case.

CREAM OF TARTAR BAKING POWDER.

$$KHC_4H_4O_6 + NaHCO_3 \longrightarrow KNaC_4H_4O_6 + H_2O + CO_2$$

ALUM BAKING POWDER.

$$NaAl(SO_4)_2 + 3\,NaHCO_3 \longrightarrow 2\,Na_2SO_4 + Al(OH)_3 + 3\,CO_2$$

PHOSPHATE BAKING POWDER.

$$Ca(H_2PO_4)_2 + 2\,NaHCO_3 \longrightarrow CaHPO_4 + Na_2HPO_4 + 2\,H_2O + 2\,CO_2$$

REVIEW QUESTIONS

1. List the four most useful alkali metals. Tell why sodium is used more than francium.
2. Draw the atomic (electronic) structure of sodium.
3. Which is the best way to store metallic potassium, (a) in a bottle of water or (b) in a bottle of kerosene? Why?
4. Write four chemical equations illustrating four different chemical properties of sodium.
5. How is metallic potassium prepared?
6. Distinguish between (a) caustic potash and caustic soda, (b) soda ash and washing soda, (c) baking soda and baking powder.
7. Describe the method of making potassium chlorate in the laboratory. Write the equations.
8. Sodium carbonate hydrolyzes when dissolved in water. Explain what happens. Write an equation to show this.

Chapter 24
The Alkaline Earth Metals
Group IIA

Beryllium (4) $1s^2 2s^2$
Magnesium (12) $1s^2 2s^2 2p^6 3s^2$
Calcium (20) $1s^2 2s^2 2p^6 3s^2 3p^6 4s^2$
Strontium (38) [   ] $5s^2$
Barium (56) [   ] $6s^2$
Radium (88) [   ] $7s^2$

Of these six elements, beryllium and radium are of special interest. Radium is strongly radioactive and because of this property it finds many special uses. See Chapter 14 for more on radium. One or two per cent beryllium in certain copper alloys is used to increase their hardness. Metallic beryllium is used for windows in X-ray tubes because of its permeability to X rays. Beryllium is also used to make nonsparking alloy tools.

Calcium, strontium, and barium are usually considered to be the alkaline earth metals. Of these three, calcium is much more important, commercially, than the other two. Normally one does not think of magnesium as an alkaline earth metal but its atomic structure places it in this subgroup and therefore it will be considered here. Table 24-1 lists the important physical properties of the group.

Group Characteristics. These metals belong to Family A of Group II of the periodic table. The group characteristics of these elements are due to the arrangement of the electrons in the highest energy level of their atoms.

PHYSICAL PROPERTIES. The metals of this group are grayish white in color, with a bright metallic luster, malleable and ductile, and good conductors of heat and electricity. The spectral colors are: calcium, orange; strontium, red; barium, green.

CHEMICAL PROPERTIES. The alkaline earth metals have a valence of +2. This is because the atoms have two valence elec-

Table 24-1

Physical Properties of Group IIA Metals

	Beryllium	Magnesium	Calcium	Strontium	Barium
Color	grayish white	silver	silver white	silver white	silver white
Physical State	solid	solid	solid	solid	solid
Melting Point	1280°C	651°C	842°C	800°C	725°C
Boiling Point	2970°C	1110°C	1487°C	1384°C	1140°C
Atomic Radius	1.12Å	1.60Å	1.97Å	2.15Å	2.17Å
Ionization Energy	9.32 v	7.64 v	6.11 v	5.69 v	5.21 v
Heat of Fusion Kcal/Mole	2.8	2.1	2.2	2.2	1.8
Electro- negativity	1.5	1.2	1.0	1.0	0.9
Density g/cm^3 at 20°C	1.85	1.74	1.55	2.60	3.50

NOTE.   In keeping with the practice of most textbook writers, radium is not included in this table. Its only importance is due to its radio-active properties.

Table 24-2

Common Minerals of Group IIA Metals

Beryllium	Beryl	$3 BeO \cdot Al_2O_3 \cdot 6 SiO_2$
Magnesium	Magnesite	$MgCO_3$
	Dolomite	$MgCO_3 \cdot CaCO_3$
	Carnallite	$MgCl_2 \cdot KCl \cdot 6 H_2O$
Calcium	Calcite	$CaCO_3$
	Gypsum	$CaSO_4 \cdot 2 H_2O$
	Fluorite (Fluorspar)	CaF_2
Strontium	Strontianite	$SrCO_3$
Barium	Barite	$BaSO_4$
	Witherite	$BaCO_3$
Radium	Minute quantities found with carnotite	$K_2O(UO_3)_2V_2O_5 \cdot 3 H_2O$

trons. They are very active, tarnish quickly in air, burn brilliantly, and react with water, forming hydroxides and liberating hydrogen. They combine directly with the halogens and nitrogen. The hydroxides of these metals increase in solubility with increase in atomic weight of the metal. The sulfates decrease in solubility with increase in atomic weight of the metal.

Magnesium. During World War II, magnesium became very important and was used to make a number of light alloys for use in airplanes. Because of its chemical activity, magnesium must be made by special processes. Two of the important methods are given here.

1. *Electrolysis of the Fused Chloride.* A mixture of magnesium chloride and sodium chloride (to lower the fusion temperature) is heated to about 700°C where it is fused, and then an electric current is passed through. The anode is graphite and the cathode is steel. Magnesium collects at the cathode, and chlorine is liberated at the anode. Precautions are taken to prevent the chlorine gas from coming in contact with the freshly formed magnesium. In the Dow Sea-Water Process, magnesium ion is removed from sea water by precipitating it as $Mg(OH)_2$ by the use of milk of lime. The magnesium hydroxide is then converted to the chloride and treated as outlined above.

$$Mg^{++} + Ca(OH)_2 \longrightarrow Mg(OH)_2 + Ca^{++}$$
$$Mg(OH)_2 + 2\,HCl \longrightarrow MgCl_2 + 2\,H_2O$$

2. *The Pidgeon Ferrosilicon Process.* Calcined dolomite and ferrosilicon (75 per cent grade) are ground, mixed, and briquetted. They are then charged into horizontally placed chromium-nickel steel retorts. The retorts are evacuated and then heated (by gas) to 1150°C. Approximately 32 pounds of metallic magnesium is obtained from 235 pounds of briquettes. The reaction is:

$$2\,MgO \cdot CaO + Si(Fe) \longrightarrow 2\,Mg + (CaO)_2SiO_2 + Fe$$

PROPERTIES. Magnesium is a light, silvery metal. Its density is 1.74 as compared with 2.70 for aluminum and 7.86 for iron. When exposed to the atmosphere, it becomes covered with a thin, closely adhering oxide film. This prevents further oxidation unless the temperature is raised. Magnesium burns with a brilliant flame which is rich in ultraviolet light. It is a reducing agent. The following equations illustrate some of the properties:

$$2\,Mg + O_2 \longrightarrow 2\,MgO \text{ (bright flame)}$$
$$Mg + Cl_2 \longrightarrow MgCl_2$$
$$Mg + 2\,H_2O \longrightarrow H_2 + Mg(OH)_2$$
$$3\,Mg + Cr_2O_3 \longrightarrow 3\,MgO + 2\,Cr$$
$$Mg + CO_2 \longrightarrow MgO + CO \text{ (high temperature)}$$
$$Mg + H_2SO_4 \longrightarrow MgSO_4 + H_2 \text{ (dilute acid)}$$

USES. Most of the magnesium produced is used in alloys. These are characterized by being light and strong. They are easily forged, rolled, and drawn into rods or wire. A typical high-strength alloy contains 95 per cent magnesium, 3.5 per cent aluminum, 0.2 per cent manganese, and 1.3 per cent zinc. Magnesium is also used in flashlight powder, in fireworks, and in incendiary bombs.

Compounds of Magnesium. The most important compounds are discussed here.

MAGNESIUM OXIDE (MAGNESIA, MgO). This compound is prepared by heating the carbonate.

$$MgCO_3 + \text{(heat)} \longrightarrow MgO + CO_2$$

The oxide is a white powder that does not melt or decompose at high temperatures. It is a poor conductor of heat and is but slightly soluble in water. The important uses include insulation for covering water and steam pipes, and making bricks which are used to line electric furnaces.

MAGNESIUM CHLORIDE ($MgCL_2$). This compound is found in natural water and in certain salt deposits. When mixed with the oxide and water, an oxychloride, $(MgO)_3MgCl_2 \cdot 10\ H_2O$, is formed, which is used as a cement.

MAGNESIUM SULFATE ($MgSO_4$). This salt is found in natural waters and in salt beds. It is used in the dye industry, in tanning, in the manufacture of paints, and in soap making. The hydrate, $MgSO_4 \cdot 7\ H_2O$ (*Epsom salt*), is used in medicine.

MAGNESIUM SILICATES. These compounds (serpentine, asbestos, talc, and meerschaum) are found in the complex silicates widely distributed in nature. Asbestos is used for fireproofing curtains, making brake-band linings, pipe insulation, and the manufacture of shingles and cement. Talc is used for a paper filler and as a base for toilet articles.

Calcium. Metallic calcium is prepared by the electrolysis of fused calcium chloride. Calcium fluoride may be added to lower the fusing temperature of the bath. An unsatisfactory product is obtained unless the calcium is removed from the bath as soon as it is deposited. This is accomplished by having a movable iron cathode which is raised as the metal is deposited on the cathode end. Calcium is a silvery white metal with a bright luster and is malleable and ductile at moderate temperatures. It resists oxidation in dry air, is chemically active (its position is high in the activity series), and burns with a brilliant white light at higher temperatures. Calcium has a limited use as a reducing agent. In the laboratory, it is used as a drying agent for certain organic compounds.

Compounds of Calcium. Certain calcium compounds are important industrially. Important ones will be considered here.

CALCIUM CARBONATE (MARBLE, LIMESTONE, CHALK, CALCITE, ICELAND SPAR, $CaCO_3$). Calcium carbonate is the most abundant and most widely distributed compound of calcium. It is a white or transparent solid (depending on whether it is in the amorphous or crystalline form). It is insoluble in water but "dissolves" in most acids. Carbon dioxide is liberated when an acid attacks the salt.

$$CaCO_3 + 2\,HCl \longrightarrow CaCl_2 + H_2O + CO_2$$
$$CaCO_3 + 2\,H \cdot C_2H_3O_2 \longrightarrow Ca(C_2H_3O_2)_2 + H_2O + CO_2$$
$$CaCO_3 + H_2CO_3 \longrightarrow Ca(HCO_3)_2$$

Calcium carbonate is used for making other calcium compounds, for making glass and cement, for neutralizing soil acidity, as building material, and as a flux in the iron blast furnace. Freshly precipitated chalk is used in tooth pastes (dental creams). A test for a carbonate is to add an acid to the substance and pass the liberated carbon dioxide into lime water, $Ca(OH)_2$. A white precipitate of calcium carbonate is formed.

$$MgCO_3 + 2\,HCl \longrightarrow MgCl_2 + H_2O + CO_2$$
$$CO_2 + Ca(OH)_2 \longrightarrow CaCO_3 + H_2O$$

CALCIUM OXIDE (QUICKLIME, CaO). This compound is prepared by heating calcium carbonate. Since the reaction is reversible, it is necessary to remove the carbon dioxide as it is formed.

$$CaCO_3 \text{ (heat)} \rightleftharpoons CaO + CO_2$$

A high temperature favors the decomposition, but care must be taken to prevent the melting of the siliceous material that may be present; otherwise, the calcium oxide will be coated and rendered unfit for use. See Figure 24-1.

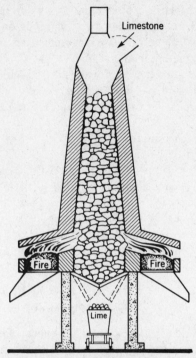

Fig. 24-1. A vertical lime kiln.

Calcium oxide is a white solid with a very high melting point. It reacts with water, liberating considerable heat. Moisture in the air causes the reaction to take place slowly. The product is called *slaked lime.*

Calcium oxide is used in making mortar, plaster, slaked lime, and bleaching powder.

CALCIUM HYDROXIDE [SLAKED LIME, $Ca(OH)_2$]. This compound is formed by allowing calcium oxide to stand in moist air, or by adding water to it.

$$CaO + H_2O \longrightarrow Ca(OH)_2$$

Calcium hydroxide is slightly soluble in water, reacts basic (in water) to indicators, and combines with carbon dioxide to form calcium carbonate.

Calcium hydroxide is used to soften water, in dehairing hides before tanning, for neutralizing acid soils, and in making glass. Whitewash is a suspension of calcium hydroxide in water that is sometimes called *milk of lime*. Lime water is a saturated but dilute solution of calcium hydroxide used in testing for carbon dioxide. It is also used in medicine.

Slaked lime (with an excess of water) mixed with three or four times its bulk of sand is called *mortar*. When mortar sets, the following steps occur: The water evaporates. The calcium hydroxide (in the mortar) reacts with the carbon dioxide of the air, forming calcium carbonate. This explains why old mortar is harder than that freshly made.

CALCIUM SULFATE (GYPSUM, $CaSO_4 \cdot 2H_2O$).   Large quantities of this substance occur widely distributed in nature. Millions of tons are mined yearly in the United States. In its natural form, it is used as a fertilizer. When the water of hydration is partially removed, it is called *plaster of Paris*.

When gypsum is heated to about 125°C, the following reaction takes place:

$$2\,(CaSO_4 \cdot 2\,H_2O) + (heat) \longrightarrow (CaSO_4)_2 \cdot H_2O + 3\,H_2O$$
$$\text{Plaster of Paris}$$

When water is added, hydration takes place. This causes the plaster to set. Upon setting, the plaster expands slightly. Upon this fact depends its usefulness in making casts.

Plaster of Paris is used for making casts, for stucco, for wall plaster, and for plaster board.

BLEACHING POWDER (CHLORIDE OF LIME, $CaOCl_2$).   This substance is prepared by passing chlorine over slaked lime. Thousands of tons are made in the United States every year.

$$Ca(OH)_2 + Cl_2 \longrightarrow CaOCl_2 + H_2O$$

Bleaching powder is an unstable white powder. The commercial product has no definite composition. It reacts with acids, liberating chlorine. Moist bleaching powder is decomposed by the carbon dioxide of the air.

$$CaOCl_2 + 2\,HCl \longrightarrow CaCl_2 + H_2O + Cl_2$$
$$2\,CaOCl_2 + CO_2 + H_2O \longrightarrow CaCl_2 + CaCO_3 + 2\,HClO$$

True calcium hypochlorite, $Ca(OCl)_2$, is now on the market under the trade name H.T.H. Bleaching powder is used for bleaching and as a disinfectant.

CALCIUM CARBIDE (CaC_2). This compound is prepared on a large scale by heating a mixture of coke and lime in an electric furnace.

$$CaO + 3 C \longrightarrow CaC_2 + CO$$

Calcium carbide is a transparent crystalline substance (when pure), is insoluble in all common solvents, reacts with water to form acetylene (C_2H_2), and reacts with nitrogen to form calcium cyanamide ($CaCN_2$).

$$CaC_2 + 2 H_2O \longrightarrow C_2H_2 + Ca(OH)_2$$
$$CaC_2 + N_2 \longrightarrow CaCN_2 + C$$

Calcium carbide is used for making acetylene and also calcium cyanamide, which is used as a fertilizer.

CALCIUM PHOSPHATE (PHOSPHATE ROCK), $Ca_3(PO_4)_2$. Calcium phosphate is found in deposits in South Carolina, Tennessee, and the Rocky Mountain states. It is also found in bones. It is used as a source of phosphorus and also for the preparation of superphosphate of lime. This is an important fertilizer made by treating phosphate rock with sulfuric acid (see Chapter 19).

$$Ca_3(PO_4)_2 + 2 H_2SO_4 \longrightarrow CaH_4(PO_4)_2 + CaSO_4$$

Water Softening. Water containing dissolved calcium and magnesium salts is called *hard water*. For industrial and household uses, it is desirable to remove these impurities. Hard water forms deposits on the inside of steam boiler pipes which destroy their usefulness. If hard water is used in the home, considerable soap must be added to the water to soften it.

Temporarily hard water contains dissolved calcium and magnesium bicarbonates. These salts can be removed by boiling the water. The normal carbonates are precipitated.

$$Ca(HCO_3)_2 + (heat) \longrightarrow CaCO_3 + H_2O + CO_2$$
$$Mg(HCO_3)_2 + (heat) \longrightarrow MgCO_3 + H_2O + CO_2$$

Slaked lime will soften temporarily hard water.

Permanently hard water contains dissolved calcium and magnesium sulfates and chlorides. It cannot be softened by boiling. The methods of softening are as follows.

ADDING WASHING SODA (Na_2CO_3).   This precipitates calcium and magnesium carbonates.

$$CaSO_4 + Na_2CO_3 \longrightarrow CaCO_3 + Na_2SO_4$$
$$MgCl_2 + Na_2CO_3 \longrightarrow MgCO_3 + NaCl$$

PASSING WATER THROUGH A PERMUTIT SOFTENER.   This is a tank containing a complex sodium aluminum silicate. When water is passed through the vessel, calcium and magnesium ions exchange places with the sodium in the silicate. The "spent" silicate can be reactivated by forcing a saturated brine solution through it.

To soften water:

$$Na_2O \cdot Al_2O_3 \cdot 2\ SiO_2 + CaSO_4 \longrightarrow$$
$$CaO \cdot Al_2O_3 \cdot 2\ SiO_2 + Na_2SO_4$$

To reactive the Permutit:

$$CaO \cdot Al_2O_3 \cdot 2\ SiO_2 + 2\ NaCl \longrightarrow$$
$$Na_2O \cdot Al_2O_3 \cdot 2\ SiO_2 + CaCl_2$$

ADDING SOAP.   This is the household method of softening hard water. Ordinary laundry soap is chiefly sodium stearate ($C_{17}H_{35}COONa$). This reacts with the calcium and magnesium salts in the water.

$$2\ C_{17}H_{35}COONa + MgSO_4 \longrightarrow (C_{17}H_{35}COO)_2Mg + Na_2SO_4$$

In each of the examples given above, it will be noticed that a sodium salt remains in the water. This is not considered to be detrimental because of its low concentration.

USING ION-EXCHANGE RESINS.   These are synthetic resins made from phenol and formaldehyde to give *cationic resins*, or from aromatic amines and certain aldehydes to give *anionic resins*.

Cationic resins will remove (by adsorption) cations such as Na^+, Mg^{++}, K^+, Ca^{++}, Cu^{++}, and Ag^+ from solution.

Anionic resins will remove (by adsorption) anions such as Cl^-, SO_4^{--}, PO_4^{---}, and NO_3^- from solution.

Cationic resins may be regenerated by passing hydrochloric or sulfuric acid solutions through them. Anionic resins are usually regenerated by passing sodium hydroxide or sodium carbonate solutions through them. Water containing various dissolved salts can be purified by passing it through a cationic resin and then through an anionic resin. Such water is as pure as ordinary distilled water. See Chapter 10 for additional information on this topic.

REVIEW QUESTIONS

1. From the standpoint of atomic structure, what do beryllium and magnesium have in common?
2. How does magnesium occur in sea water? How can magnesium be recovered from sea water?
3. Describe the preparation of magnesium by an electrolytic process.
4. Write five equations illustrating five different properties of magnesium.
5. How can quicklime be prepared from calcium carbonate?
6. Compare the solubilities of calcium, strontium, and barium sulfates in water.
7. How can gypsum be changed into plaster of Paris? Write the equation.
8. Calcium carbide is made in an electric furnace. What materials are used? What reactions take place?
9. Distinguish between (a) hard and soft water, (b) temporarily hard and permanently hard water.
10. How does a Permutit water softener remove hardness from water?

Chapter 25
Group IIIA Elements

Boron (5) $1s^2 2s^2 2p^1$
Aluminum (13) $1s^2 2s^2 2p^6 3s^2 3p^1$
Gallium (31) [   ] $4s^2 4p^1$
Indium (49) [   ] $5s^2 5p^1$
Thallium (81) [   ] $6s^2 6p^1$

From a commercial point of view boron and aluminum are most important. Gallium, indium, and thallium are less important. Gallium has a low melting point, 29.8°C, which is just above room temperature. Indium and gallium form an alloy that melts at approximately 12°C. The physical properties of group IIIA are shown in Table 25-1.

Boron is a nonmetal; the other members of this subgroup are metals. Gallium is one of the elements whose existence and properties were predicted by Mendeleev (of periodic law fame) years before it was actually discovered. Liquid gallium is corrosive. Thallium compounds are poisonous.

Boron. The earth's crust contains about 0.0001 per cent boron. It never occurs free. Large quantities of borax ($Na_2B_4O_7 \cdot 10\ H_2O$) and colemanite ($Ca_2B_6O_{11} \cdot 5\ H_2O$) are found in California. Some boric acid is found in hot springs in Italy. Tourmaline, $H_2MgNa_9Al_3(B_2O_3)_2Si_4O_{20}$, is a complex silicate found widely distributed in rocks. Boron is prepared by the reduction of the oxide (B_2O_3) with powdered magnesium. The product of reduction contains amorphous boron mixed with magnesium boride (Mg_3B_2).

$$B_2O_3 + 3\ Mg \longrightarrow 3\ MgO + 2\ B$$

An electrolytic method has been developed for the preparation of boron. Here, a fused mixture of B_2O_3, KBF_4, KF, and KCl is electrolyzed, using a graphite anode and an iron cathode.

Boron is a brown amorphous powder that can be fused to a

Table 25-1

Physical Properties of Group IIIA Metals

	Boron	Aluminum	Gallium	Indium	Thallium
Color	brown powder	silver white	silver white	silver white	grayish white
Physical State	solid	solid	solid below 29.8°C	solid	solid
Melting Point	2300°C	660°C	29.8°C	156°C	303°C
Boiling Point		2467°C	2403°C	2000°C	1457°C
Atomic Radius	0.79Å	1.43Å	1.35Å	1.62Å	1.70Å
Ionization Energy	8.3 v	5.98 v	6.0 v	5.78 v	6.11 v
Heat of Fusion cal/g		94.6	19.2		7.2
Electro-negativity	2.0	1.5	1.6	1.7	1.8
Density g/cm³ at 20°C	2.34	2.7	5.91	7.31	11.85

very hard and brittle mass. It is a poor conductor of heat and electricity.

Boron combines slowly with oxygen at 100°C. At higher temperatures, it burns with a green flame.

$$4 \, B + 3 \, O_2 \longrightarrow 2 \, B_2O_3$$

Boron combines with the halogens at higher temperatures.

$$2 \, B + 3 \, Cl_2 \longrightarrow 2 \, BCl_3$$

It combines with metals to form borides.

$$3 \, Zn + 2 \, B \longrightarrow Zn_3B_2$$

Nitric acid oxidizes boron to boric acid.

$$B + HNO_3 + H_2O \longrightarrow H_3BO_3 + NO$$

Nitrogen combines directly with boron.

$$2 \, B + N_2 \longrightarrow 2 \, BN$$

A small amount of boron is used in certain alloy steels. A very low percentage in steel has a remarkable effect in hardening it. Boron has also been used as a moderator in uranium-graphite piles.

Compounds of Boron. Two important boron compounds are discussed here.

BORIC ACID (H_3BO_3). This is a white crystalline substance obtained by recrystallizing the condensed vapors of certain hot springs in Tuscany or by treatment of borax with an acid.

When borax is treated with sulfuric acid, boric acid is formed.

$$Na_2B_4O_7 + 5\,H_2O + H_2SO_4 \longrightarrow Na_2SO_4 + 4\,H_3BO_3$$

Boric acid is a white crystalline substance. It is sparingly soluble in water and has weak acidic properties.

It is used as a mild antiseptic, especially as an eyewash.

BORAX ($Na_2B_4O_7$). This is a white crystalline salt that is obtained by recrystallization of the crude salt found in nature.

If colemanite is treated with sodium carbonate, borax is formed.

$$2\,Ca_2B_6O_{11} + 4\,Na_2CO_3 + H_2O \longrightarrow 4\,CaCO_3 +$$
Colemanite

$$3\,Na_2B_4O_7 + 2\,NaOH$$
Borax

Below 60°C borax crystallizes from water solutions to give $Na_2B_4O_7 \cdot 10\,H_2O$. Above 60°C, the crystals have the composition $Na_2B_4O_7 \cdot 5\,H_2O$. It is sparingly soluble in water at 0°C but is very soluble at 60°C. Borax hydrolyzes in water to give a basic reaction. It dissolves metallic oxides.

Borax is used in soldering and welding, to remove oxide film, for softening water, for making soaps, and for reagent use in the laboratory. Borax beads are frequently used in the analytical laboratory. Large quantities of borax are used in making glass, pottery, and enamels.

Aluminum. The earth's crust is about 7 per cent aluminum. It is the most abundant metal. Due to its chemical activity, it never occurs free. The important minerals are bauxite ($Al_2O_3 \cdot 2\,H_2O$) and many silicates such as feldspar ($KAlSi_3O_8$) and mica ($KH_2Al_3(SiO_4)_3$).

METALLURGY. Aluminum can be prepared by two important

methods, *chemical* reduction and *electrolytic* reduction. The action of metallic sodium on anhydrous aluminum chloride is an example of chemical reduction.

$$AlCl_3 + 3\ Na \longrightarrow Al + 3\ NaCl$$

In electrolytic reduction (the Hall Process) the furnace (or electrolytic cell) is an iron box, lined with carbon. This serves as the cathode. Carbon rods are suspended in the box and serve as anodes. The furnace is charged with fused cryolite (natural or synthetic) which contains dissolved, chemically pure aluminum oxide. (This oxide is made by the purification of bauxite by the Bayer or similar process). When a suitable direct electric current is passed through this fused solution, aluminum is plated out on the cathode and oxygen is liberated at the anodes. The aluminum prepared in this manner is about 99.5 per cent pure. See Figure 25-1.

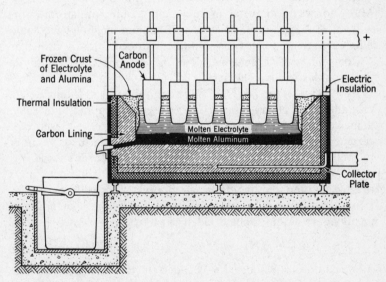

Fig. 25-1. The Hall Process for preparation of aluminum.

PROPERTIES AND USES. Aluminum is a light metal, is twice as good a conductor of electricity as copper (weight for weight), is somewhat malleable and ductile, and may be cast and welded. Aluminum is an active metal, but it does not normally corrode because of the formation of a protective oxide film on the surface. It reacts with acids and bases such as HCl and NaOH. In each case, hydrogen is one product.

$$2 Al + 6 HCl \longrightarrow 2 AlCl_3 + 3 H_2$$
$$2 Al + 6 NaOH \longrightarrow 2 Na_3AlO_3 + 3 H_2$$

Aluminum is a good reducing agent (see paragraph on thermite, p. 269), and it burns with a brilliant light in oxygen.

$$4 Al + 3 O_2 \longrightarrow 2 Al_2O_3$$

Aluminum is used for making household utensils, for electric transmission lines, for airplane alloys, for foil in place of tin, for making paints, for removing dissolved oxygen from steel castings, and for flashlight bulbs used in indoor photography. Thousands of tons of aluminum are used yearly to make alloys for washing machines. More than a million miles of steel-reinforced aluminum cable are used in transmission lines. Considerable aluminum is used in making automobiles and also in the building industries. Many houses now have aluminum siding, aluminum storm doors and windows, and roofs covered with aluminum shingles. Aluminum cans are used widely as containers for beer and soft drinks.

Compounds of Aluminum. The more important aluminum compounds are discussed here.

ALUMINUM OXIDE (Al_2O_3). The natural forms of the oxide are corundum and emery, used in abrasives; and sapphire and ruby, valued as gems. The color of these gems is due to the presence of certain dissolved oxides. Gems can be prepared in the laboratory by melting aluminum oxide and adding the appropriate oxide.

ALUMINUM HYDROXIDE [$Al(OH)_3$]. This is a white gelatinous precipitate prepared by adding ammonium hydroxide to a solution of an aluminum salt. It is used as a mordant in dyeing and in purifying water and, when dried, as a catalyst.

$$AlCl_3 + 3 NH_4OH \longrightarrow 3 NH_4Cl + Al(OH)_3$$

A mordant is a substance that is added to cloth before it can take a dye.

Recently, aluminum hydroxide has been used in neutralizing excess acid in the stomach. Since $Al(OH)_3$ is amphoteric, it will react with a base such as NaOH or an acid such as HCl.

ALUMINUM SULFATE [$Al_2(SO_4)_3$]. This is prepared by the action of sulfuric acid on bauxite. It is used in the manufacture of alum and paper, and to purify water.

$$Al_2O_3 + 3 H_2SO_4 \longrightarrow Al_2(SO_4)_3 + 3 H_2O$$

Aluminum sulfate hydrolyzes in water, forming aluminum hydroxide. This settles to the bottom of the water tank, carrying suspended matter and bacteria with it.

ALUMS. An alum is a double sulfate of a monovalent (M′) and trivalent (M‴) metal containing a definite amount of water of hydration (crystallization). The general formula is:

$$M'M'''(SO_4)_2 \cdot 12 H_2O$$

or

$$M'_2SO_4 \cdot M_2'''(SO_4)_3 \cdot 24 H_2O$$

Potassium alum, $KAl(SO_4)_2 \cdot 12 H_2O$, and ammonium alum, $NH_4Al(SO_4)_2 \cdot 12 H_2O$, are the most widely used. Alums are used in paper manufacture, in water purification, and in the preparation of baking powders.

ALUMINUM CHLORIDE ($AlCl_3$). This compound is prepared by heating aluminum with hydrogen chloride or chlorine. It is used for refining mineral oils and as a catalyst in organic chemistry and in the "cracking" of petroleum.

$$2 Al + 6 HCl \longrightarrow 2 AlCl_3 + 3 H_2$$
$$2 Al + 3 Cl_2 \longrightarrow 2 AlCl_3$$

KAOLIN (PURE CLAY) $[H_2Al_2(SiO_4)_2 \cdot H_2O]$. This is found in certain natural deposits. It is used in making pottery—ornaments, utensils, dishes, and similar objects made from clay.

THERMITE. This is a mixture of powdered aluminum and a metallic oxide (Fe_2O_3). When it is ignited, the heat of the reaction is sufficient to raise the temperature of the mixture to approximately 3000°C. At this temperature, the reduced metal is fluid and can be poured into molds or cracks in defective machinery. Thermite is used in welding rails and repairing heavy machinery.

$$Fe_2O_3 + 2 Al \longrightarrow Al_2O_3 + 2 Fe + 185,000 \text{ calories}$$

Metals such as chromium can be prepared by this method. The above reaction is sometimes called the *Goldschmidt Reaction.*

PORTLAND CEMENT. This is a complex calcium aluminum silicate which possesses the property of hardening (setting) in water as well as in air.

Iron blast furnace slag (impure calcium aluminum silicate) or similar material is ground to a fine powder and then heated in a furnace until it vitrifies. This material is called *clinker*. It is pulverized and gypsum is added, which retards the setting of the cement when water is added.

The following compounds have been identified: Ca_2SiO_4, Ca_3SiO_5, and $Ca_3(AlO_3)_2$. The percentage composition of ordinary cement is about as follows: CaO—58 to 67 per cent; Al_2O_3—4 to 11 per cent; MgO—0 to 5 per cent; SiO_2—19 to 26 per cent; Fe_2O_3—2 to 5 per cent; SO_3—0 to 2.5 per cent; Na_2O and K_2O—0 to 3 per cent. When mixed with sand and crushed stone (and water), it forms concrete, which is used for highway pavement, for foundations of buildings, and for dams. The setting of cement is due to the hydration of the calcium silicate present. This is especially true of the tricalcium silicate (Ca_3SiO_5). The setting of cement which occurs within the first 3 to 6 hours may be due to the following reaction:

$$Ca_3(AlO_3)_2 + 6\,H_2O \longrightarrow 3\,Ca(OH)_2 + 2\,Al(OH)_3$$

Note. The setting of cement is probably much more complicated than suggested here.

REVIEW QUESTIONS

1. Why are aluminum and boron placed in the same group of the periodic table?
2. Aluminum is abundant in nature, yet it is more expensive than iron. Why?
3. Describe the Hall Process for preparing aluminum.
4. Give an equation for the preparation of boron.
5. Give the equation for the preparation of boric acid from borax.
6. From a commercial point of view, what are the interesting properties of aluminum?
7. Write two equations to show (*a*) the action of hydrochloric acid on aluminum hydroxide, and (*b*) the action of sodium hydroxide on aluminum hydroxide.
8. What is an alum? Give a general formula.
9. What is thermite? Describe how it is used.

Chapter 26
Group IB Metals

Copper (29) [   ] $4s^1$
Silver (47) [   ] $5s^1$
Gold (79) [   ] $6s^1$

These three metals belong to group IB of the periodic table. Unlike the metals in group IA, which have high chemical activity, these three are relatively inactive. All three are found free in nature and are sometimes called *noble metals*. Copper and gold have oxidation numbers of +1 or more, but silver has an oxidation number of +1 only. All three metals were known long before the time of Christ. Table 26-1 gives the physical properties of these three metals.

Table 26-1
Physical Properties of Group IB Metals

	Copper	Silver	Gold
Color	reddish yellow	silver white	yellow
Physical State	solid	solid	solid
Melting Point	1083°C	961°C	1063°C
Boiling Point	2595°C	2212°C	2966°C
Atomic Radius	1.28Å	1.44Å	1.44Å
Ionization Energy	7.72 v	7.57 v	9.22 v
Heat of Fusion cal/g	50.6	25	16.1
Electronegativity	1.8	1.9	2.4
Density g/cm^3 at 20°C	8.96	10.5	19.3

There is a difference in chemical activity of group IA and group IB metals. If a group IA metal is stripped of its valence electron, the next underlying shell has the stable structure of a noble gas (zero group) element. Group IA elements have an oxidation number of +1 only. On the other hand, if a group IB metal is stripped of its valence electron, to give it an oxidation number of +1, the underlying shell does not have a zero group structure and therefore it may lose one or more electrons. Thus copper may lose its valence electron and one electron from the M shell. Thus:

	K	L	M	N	
Potassium	$1s^2$	$2s^2 2p^6$	$3s^3 3p^6$	$4s^1$	$K \longrightarrow K^{+1} + e$
Copper	$1s^2$	$2s^2 2p^6$	$3s^2 3p^6 3d^{10}$	$4s^1$	$Cu \longrightarrow Cu^{+2} + 2e$

The M shell for potassium is the same as the M shell for argon. But the M shell for copper is not the same as for argon. Copper can lose 2 electrons, one from the N shell and one of the d electrons in the M shell. Both copper and gold have oxidation numbers greater than +1. Copper may have an oxidation number of +2 and gold may have an oxidation number of +3. Silver is the exception; it has an oxidation number of +1 only. In this respect it is like the alkali metals.

Copper. The important naturally occurring compounds of copper are sulfides such as chalcocite (Cu_2S) and chalcopyrite ($CuFeS_2$), and oxidized minerals such as malachite [$CuCO_3 \cdot Cu(OH)_2$] and cuprite (Cu_2O). The ores of copper are usually low-grade (less than 2 per cent copper) and, therefore, must be concentrated in some suitable manner. The sulfides are concentrated by *flotation*, and the oxidized minerals are usually *leached*. The sulfide concentrate, made by the flotation process, is put through a smelting process.

METALLURGY. The important methods of recovering copper from its ores are as follows.

Flotation Process. The ore is passed through a mill where it is crushed and ground to fine particles. To a water suspension of this powdered ore is added a *frother* and a *collector*, and the suspension is passed through a *flotation cell* (see Chapter 22). The copper sulfide minerals adhere to the froth and rise to the surface, where they are skimmed off. The gangue (worthless minerals) is removed through the bottom of the cell. The froth is added to a stream of water (to break down the froth) and then sent to a settling tank (*Dorr thickener*) where the sulfide particles settle out. Finally, the "mineral mud" is filtered on a revolving American type of filter or on an Oliver filter. It is then called the *concentrate*. It is possible to remove 90 per cent or more of the copper sulfide from an ore containing as little as 1 per cent copper sulfide.

Smelting. This general term includes various pyrometallurgical processes such as calcining, roasting, melting, converting, and reduction in high-temperature furnaces. After the concentrate of copper sulfides has been obtained, it is treated (smelted) as follows:

Roasting. The concentrate is passed through a multiple-hearth roaster. This oxidizes sulfides of arsenic and antimony to volatile oxides such as arsenic oxide and antimony oxide and some of the sulfur to the dioxide (SO_2). Most of the copper sulfide is unchanged and makes up the greater part of the discharge—called *calcines*.

Formation of Matte. The product (calcine) from the roaster is mixed with suitable fluxing material (silica and/or calcium carbonate) and charged into a copper reverberatory furnace. In this furnace a fused mass called copper matte ($FeS + Cu_2S$) is formed. Also, a slag is formed which is removed and sent to the slag dump. Any gold and silver in the concentrate goes into the matte.

Blister Copper. Molten matte is transferred to a copper converter where air is blown through it. The FeS is oxidized to FeO and becomes part of the slag. The copper sulfide is then decomposed to give SO_2 and blister copper, which is approximately 98 per cent copper. The reactions are:

$$2\,FeS + 3\,O_2 \longrightarrow 2\,FeO + 2\,SO_2$$
$$x\,FeO + y\,SiO_2 \longrightarrow (FeO)_x \cdot (SiO_2)_y \text{ (slag)}$$
$$2\,Cu_2S + 3\,O_2 \longrightarrow 2\,Cu_2O + 2\,SO_2$$
$$Cu_2S + 2\,Cu_2O \longrightarrow 6\,Cu + SO_2$$

REFINING OF COPPER. Blister copper is refined (purified) by two methods: electrolytic refining, and fire refining.

Electrolytic Refining. A high percentage of copper is refined by this method. The essential steps are: Copper plates weighing about 700 pounds each, cast from fire-refined copper, are made the anodes in electrolyzing tanks containing a solution of copper sulfate and dilute sulfuric acid. Thin sheets of pure copper serve as the cathode. When a direct electric current is passed through, the impure copper from the anode goes into solution and pure copper plates out on the cathode.

The inactive impurities such as gold, silver, and platinum do not dissolve. They settle to the bottom of the tank and become part of what is called *anode mud*. Ions of the more active metals do not plate out. See Figure 26-1.

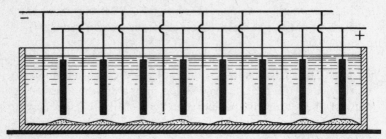

Fig. 26-1. Refining copper by the electrolytic method. The impure copper (+ charged anodes) shown as heavy rods goes into solution as copper ion (Cu^{++}) and subsequently plates out on the thin (− charged) cathodes.

Fire Refining. Copper ingots (or pigs) made of blister copper are melted in a reverberatory furnace. Molten copper is "flapped" by striking the surface with a paddle, or more likely air is blown through it. This oxidizes some of the impurities and gives "set" copper which contains 0.6 per cent to 0.9 per cent oxygen.

The "set" copper is *poled* by forcing green poles into the molten copper. This removes the excess oxygen. After poling, it is called *tough pitch copper*. It is used commercially, or further refined by electrolysis.

PROPERTIES AND USES.   Copper is a heavy metal with a characteristic color. Next to silver, copper is the best conductor of an electric current (for wires of equal cross section and length). Copper is not attacked appreciably by nonoxidizing acid such as hydrochloric acid or dilute sulfuric acid. When heated, copper combines directly with oxygen, sulfur, and the halogens.

$$2\,Cu + O_2 \longrightarrow 2\,CuO$$
$$Cu + S \longrightarrow CuS$$
$$Cu + Cl_2 \longrightarrow CuCl_2$$

It reacts with nitric acid and hot concentrated sulfuric acids.

$$3\,Cu + 8\,HNO_3 \longrightarrow 3\,Cu(NO_3)_2 + 2\,NO + 4\,H_2O$$
$$Cu + 2\,H_2SO_4 \longrightarrow CuSO_4 + SO_2 + 2\,H_2O$$

Moist air containing carbon dioxide attacks copper, giving the basic carbonate.

$$2\,Cu + O_2 + CO_2 + H_2O \longrightarrow CuCO_3 \cdot Cu(OH)_2$$

Copper is used for making wire bars, and for alloys such as bronze and brass. This metal is also used for tubing, roofing, electrical instruments, water stills, and many similar products.

Compounds of Copper. Copper forms two series of compounds, those where its oxidation number is $+1$ and those where it is $+2$.

COPPER(I) OXIDE (Cu_2O). This is a red oxide, formed by heating copper(II) oxide or by reducing a copper(II) salt in an alkaline solution.

COPPER(I) IODIDE (CuI). This is prepared by treating a copper(II) salt with an iodide.

$$2\,CuSO_4 + 4\,KI \longrightarrow 2\,CuI + I_2 + 2\,K_2SO_4$$

It is interesting to note that a quantative method for the determination of copper (depending on the formation of copper(I) iodide with the simultaneous liberation of iodine) is based on this reaction.

The following are important copper(II) compounds.

COPPER(II) OXIDE. This compound is prepared by heating metallic copper in air. It is a black solid, sometimes used as an oxidizing agent.

$$2\,Cu + O_2 \longrightarrow 2\,CuO$$

COPPER(II) SULFIDE (CuS). This is prepared by treating a copper(II) salt with hydrogen sulfide. It is of special importance in qualitative analysis.

$$CuCl_2 + H_2S \longrightarrow CuS + 2\,HCl$$

COPPER SULFATE ($CuSO_4$). This substance is prepared when sulfuric acid and oxygen act on metallic copper. The acid (dilute) is allowed to drip on copper shavings in the presence of air. The blue pentahydrate, $CuSO_4 \cdot 5\,H_2O$, crystallizes from a water solution. The anhydrous salt is made by heating the blue crystals. Copper sulfate is used in gravity batteries (Daniell cells), in copper refining, and in electrotyping and calico printing. It is sometimes thrown into lakes and reservoirs to kill organisms such as algae, and *Bordeaux Mixture*, copper sulfate mixed with calcium hydroxide ($CuSO_4$ + $Ca(OH)_2$), is used as a fungicide. Copper sulfate is also used in preparing Fehling's solution. Anhydrous copper sulfate is used as a test for moisture in kerosene and similar substances. In the presence of moisture, the white powder turns blue.

Silver. The symbol for silver is Ag. It is derived from the Latin word *argentum*. Silver is found in many areas of the world. The United States and Mexico are large producers. Some free (native) silver is found in nature, but most of it is found as the sulfide (Ag_2S, argentite). There is also some chloride ($AgCl$, cerargyrite, sometimes called *horn silver*). Usually, silver is associated with copper and lead ores and may be considered as a by-product metal of the copper and lead industries. In copper metallurgy, silver follows the copper until the latter is electrolytically refined. Here the silver leaves the copper and goes into the *anode mud* from which the silver can be recovered. Likewise, in lead smelting, the silver follows the lead until molten *lead bullion* (containing the silver) is treated with a limited supply of zinc (the Parkes Process). The silver concentrates in the zinc layer and is skimmed off. The zinc silver alloy is then distilled to remove the zinc.

Silver is a soft white metal with a bright luster. It is very malleable and ductile and is the best known conductor of heat and electricity. Silver is not attacked by nonoxidizing acids such as hydrochloric acid. Hydrogen sulfide and certain sulfur-containing compounds will tarnish silver. Metallic silver is soluble in nitric acid and in hot concentrated sulfuric acid.

$$3\,Ag + 4\,HNO_3 \longrightarrow 3\,AgNO_3 + 2\,H_2O + NO$$
$$2\,Ag + 2\,H_2SO_4 \longrightarrow Ag_2SO_4 + 2\,H_2O + SO_2$$

Silver is used for coins, ornaments, silver plating, mirrors, electrical instruments, and certain silver compounds used in photography.

Compounds of Silver. The important compounds of silver are as follows.

SILVER NITRATE. Silver nitrate, $AgNO_3$, called *lunar caustic*, is made by the action of nitric acid on pure silver. This is the substance used in the manufacture of other silver compounds. It is also used in medicine and for making indelible inks.

SILVER CHLORIDE AND SILVER BROMIDE. These compounds are used in photography. The following equation illustrates their preparation. The silver salt is insoluble.

$$NaCl + AgNO_3 \text{ (in water)} \longrightarrow AgCl + NaNO_3$$
$$NaBr + AgNO_3 \text{ (in water)} \longrightarrow AgBr + NaNO_3$$

Photography. The art (and science) of photography is based upon the fact that silver halides are reduced by light, or made sensitive to the action of reducing (developing) agents. Light causes reduction in proportion to its intensity and time of exposure. The chemical change can be written:

$$2\,AgBr + \text{(light)} \longrightarrow 2\,Ag + Br_2$$

EFFECT OF LIGHT. The effect of light of different wave lengths is shown in the following table.

Table 26-2
Time for Equal Reduction

Seconds	Kind of Light
15	Violet (shortest wave length)
29	Blue
37	Green
330	Yellow
660	Red (longest wave length)

This table explains why the developing of photographic plates or films is carried out in a room where only red light is used. The "color blindness" of silver halides is overcome in panchromatic emulsions by the use of dye sensitizers.

PLATE OR FILM. A glass plate or a cellulose acetate film is covered with a thin layer of gelatin impregnated with colloidal silver bromide. This film must be kept in the absence of light until the photographic exposure is made.

EXPOSURE. The plate (or film) is placed in the camera, and the image of the object to be photographed is focused on the plate by opening the shutter on the camera lens. There is no visible effect, but the silver bromide struck by the light can be reduced more easily than that not exposed. This is sometimes called the *latent image*. The effect is proportional to the intensity of the light and the time of exposure.

DEVELOPING. The exposed plate is placed in a reducing bath. The reducing agent is a crystalloid which diffuses into the gelatin and reacts with the silver bromide. The exposed silver bromide is reduced more readily than the unexposed bromide. This reduction produces metallic silver. If ferrous oxalate is the reducing agent, the equation is:

$$3\ FeC_2O_4\ +\ 3\ AgBr\ \longrightarrow\ Fe_2(C_2O_4)_3\ +\ FeBr_3\ +\ 3\ Ag$$

FIXING. The process of developing is stopped (fixed) when the complete picture is obtained. This is accomplished by dipping the plate into a solution of sodium thiosulfate ($Na_2S_2O_3$) called *hypo*. The fixer dissolves the unreduced silver bromide. The reaction is:

$$AgBr\ +\ Na_2S_2O_3\ \longrightarrow\ NaAgS_2O_3\ +\ NaBr$$

The image on the plate is called the *negative*. The entire process is illustrated in Figure 26-2.

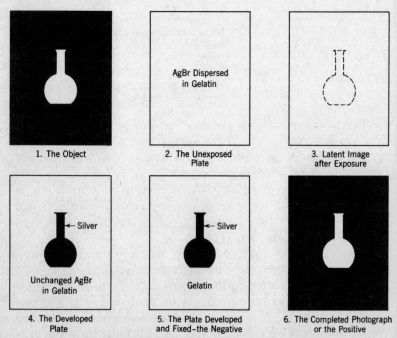

1. The Object

2. The Unexposed Plate

AgBr Dispersed in Gelatin

3. Latent Image after Exposure

4. The Developed Plate

← Silver

Unchanged AgBr in Gelatin

5. The Plate Developed and Fixed—the Negative

← Silver

Gelatin

6. The Completed Photograph or the Positive

Fig. 26-2. The steps in making a photograph.

PRINTING. Paper is coated with a gelatin containing a suspension of silver chloride or silver bromide. The negative is placed on the "light-sensitive" side of the paper and exposed to the light. This gives a "negative of the negative" or a *positive*. After exposure, the positive must be fixed in the usual way.

Gold. Gold, like silver, has been known since pre-Christian times. The name *aurum* was given to the metal by the early Romans, who prized it for its value in making ornaments and coins.

Gold is widely distributed in nature. The world's supply comes from South Africa, South America, Alaska, western United States, and Canada. Native (free) gold is found in quartz veins and in alluvial gravel. It is also alloyed with certain metals, such as in sylvanite $(AuAg)Te_2$, and is present in sea water (3 to 4 milligrams per ton of water). The latter may be a commercial source in the future.

METALLURGY. The important metallurgical processes for gold are the following.

Cyanide Process. In this process, the powdered ore is treated with a dilute sodium or potassium cyanide solution. This dissolves the gold. It can then be recovered from this solution either by electrolysis or by displacement with zinc. The equation for the solution of gold is:

$$4\,Au + 8\,NaCN + 2\,H_2O + O_2 \longrightarrow 4\,NaAu(CN)_2 + 4\,NaOH$$

The deposition of gold from the cyanide solution is:

$$2\,NaAu(CN)_2 + Zn \longrightarrow Na_2Zn(CN)_4 + 2\,Au$$

Amalgamation Process. In the amalgamation process, finely divided gold ore, mixed with water, is run over copper plates coated with mercury. The gold *amalgamates* (forms an alloy) with the mercury. The mercury is then distilled off, leaving the gold.

PROPERTIES AND USES. Gold is a soft, heavy, bright yellow metal. It is a good conductor of heat and electricity, and is more malleable and ductile than any other metal. Gold sheets can be made so thin that more than 100,000 of them are necessary to make a pile one inch high. Gold is very inactive and will not react with any common acid. It is soluble in *aqua regia* (mixture of HCl and HNO_3) and in sodium cyanide in the presence of oxygen. Gold is also corroded by fused alkalies.

It is used for jewelry and coins, and in dentistry and gold plating. It is the standard of value on which some of the currency systems of the world are based.

REVIEW QUESTIONS

1. What are the reasons for grouping copper, silver, and gold in the same group of the periodic table?
2. Give the names and formulas of three important minerals of copper and two of silver. How is gold distributed in nature?
3. Low-grade copper sulfide ore must be concentrated before it can be sent through the smelting operation. Explain how this is done.
4. Make a simple diagram (flow sheet) for smelting copper sulfide concentrate.
5. Describe the Parkes Process for separating silver from lead.
6. What are the four steps in making a photograph?
7. Gold is soluble in a sodium cyanide solution containing dissolved oxygen. Explain what happens. Write the chemical equation.
8. Describe the amalgamation process for recovery of gold from a siliceous ore.
9. Write an equation that shows what happens when a clean copper wire is placed in a solution of silver nitrate. What happens if a silver wire is placed in a copper nitrate solution?

Chapter 27
Group IIB Metals

Zinc (30) [   ] $4s^2$
Cadmium (48) [   ] $5s^2$
Mercury (80) [   ] $6s^2$

These three metals have oxidation numbers of $+2$. Mercury also has an oxidation number of $+1$. Zinc has a tendency to form complex ions. The physical properties are summarized in Table 27-1.

Table 27-1
Physical Properties of Group IIB

	Zinc	Cadmium	Mercury
Color	silver white	silver white	silver gray (liquid)
Physical State	solid	solid	liquid
Melting Point	419°C	321°C	−38°C*
Boiling Point	906°C	765°C	357°C
Atomic Radius	1.37Å	1.52Å	1.55Å
Ionization Energy	9.39 v	8.99 v	10.43 v
Heat of Fusion cal/g	24.09	13.2	2.7
Electronegativity	1.6	1.7	1.9
Density g/cm³ at 20°C	7.14	8.65	13.6

*Mercury is the only metal that is liquid at temperatures below 0°C.

Zinc. Zinc does not occur free in nature. The important zinc minerals are sphalerite (zinc blende, ZnS), zincite (ZnO), smithsonite (zinc carbonate, $ZnCO_3$), franklinite ($ZnFe_2O_4$), and willemite (Zn_2SiO_4). The important zinc-producing states are

Oklahoma, New Jersey, Montana, and Kansas. About 500,000 metric tons are produced in one year in this country. (A metric ton is 1,000 kilograms.)

METALLURGY. The important zinc ores contain sphalerite (ZnS). Like those of copper, they are low-grade and require concentration. The methods used are gravity concentration and flotation, or a combination of these. The concentrate is treated as follows.

Roasting. Depending on the subsequent treatment planned, the concentrate is roasted in multiple-hearth roasters to convert the ZnS either to ZnO or to a mixture of ZnO and $ZnSO_4$. The reactions are:

$$2\,ZnS + 3\,O_2 \longrightarrow 2\,ZnO + 2\,SO_2$$
$$ZnS + 2\,O_2 \longrightarrow ZnSO_4$$

If the calcines are to be leached, the roasting operation is regulated so that some zinc sulfate will be formed. If the calcines are to be "distilled," all of the ZnS is converted to ZnO.

$$ZnO + H_2SO_4 \longrightarrow ZnSO_4 + H_2O$$

This makes the process cyclic. However, the calcines should contain about 3 per cent $ZnSO_4$ in order to make enough extra acid to replace that lost because of the formation of insoluble sulfates and unavoidable mechanical losses.

Batch Retort Process. The calcines (zinc as ZnO) are mixed with crushed anthracite coal and placed in a fire-clay retort. This retort is placed in a horizontal position in a retort furnace and heated to about 1250°C, which reduces ZnO to metallic zinc. At this temperature, the zinc vapor distills into an attached condenser. The zinc obtained, usually called *spelter*, is not pure since it contains ZnO, cadmium, lead, etc. The New Jersey Zinc Company has developed a continuous vertical retort for preparing zinc. The retort is about 25 feet high and has a rectangular cross section of 1 × 6 feet. It is said to produce 30 pounds of zinc per 24 hours per square foot of heated wall surface. The British have recently developed a blast furnace for smelting zinc concentrates. This new furnace operates rather efficiently, and it appears that additional zinc blast furnaces will be built in the future.

PROPERTIES AND USES. Zinc is a bluish-white metal. It is malleable and ductile between 100°C and 150°C and crystalline, hard, and brittle (if allowed to solidify from the liquid state).

Zinc tarnishes readily (a basic carbonate film is formed which prevents its oxidation), burns with a blue flame, reacts readily with acids (pure zinc reacts very slowly due to the formation of a film of hydrogen on the surface), and reacts with soluble bases, forming zincates.

$$Zn + 2\,HCl \longrightarrow ZnCl_2 + H_2$$
$$Zn + 2\,NaOH \longrightarrow Na_2ZnO_2 + H_2$$

Zinc is used for making galvanized iron, for making brass, bronze, German silver, and other alloys, and for making dry batteries.

Compounds of Zinc. The more important compounds of zinc are as follows. All zinc compounds are somewhat poisonous.

ZINC OXIDE (ZINC WHITE, ZnO). This compound is made by burning zinc vapor in oxygen. It is a white powder used in paints, as a filler for rubber, and in the manufacture of oilcloth. Zinc oxide, unlike white lead, is not darkened on exposure to hydrogen sulfide fumes or other sulfur compounds.

ZINC CHLORIDE ($ZnCl_2$). This is used as a caustic, as a wood preservative, and as a soldering fluid.

ZINC SULFATE ($ZnSO_4$). This is used in printing and dyeing cloth.

ZINC SULFIDE (ZnS). This is a white compound used in the manufacture of paint (lithopone).

Paints. An oil-based paint consists of three essential ingredients:

THE VEHICLE (LIQUID MEDIUM). This must be a rapid-drying oil that forms a flexible, hornlike film. Linseed oil is the best oil for this purpose. The "drying" of an oil is due to its oxidation. Manganese dioxide catalyzes the reaction.

THE BODY (BASE). A solid material suspended in the oil, which gives a smooth surface as the vehicle dries, is called the *body* of a paint. It must have great covering power. White lead lithopone (a mixture of zinc sulfide and barium sulfate) are important for this purpose.

THE PIGMENT. Metallic oxides, certain dyes, and certain salts are added to give the desired color. White paints contain only the vehicle and the body.

Mercury. This metal, sometimes called *quicksilver*, is found in Spain, Italy, and California. Unlike zinc, it occurs free as well as in chemical combination. The most important ore of mercury contains the red mineral known as *cinnabar* (HgS).

METALLURGY.   The sulfide (HgS) is roasted to decompose it. A rotary-type roasting furnace is used in the United States. The mercury vapor is condensed in a separate container. The sulfur is burned to the dioxide and escapes. Sometimes, iron (or lime) is added to help decompose the sulfide.

$$HgS + O_2 \longrightarrow Hg + SO_2$$
$$HgS + Fe \longrightarrow Hg + FeS$$
$$4\,HgS + 4\,CaO \longrightarrow 4\,Hg + 3\,CaS + CaSO_4$$

PROPERTIES.   Mercury is a mixture of at least six isotopes, is a fair conductor of heat and electricity, expands uniformly over a wide temperature range, does not wet glass, and has a low vapor pressure at room temperature.

Mercury combines with oxygen, the halogens, and sulfur.

$$2\,Hg + O_2 \Longrightarrow 2\,HgO$$
$$Hg + Cl_2 \longrightarrow HgCl_2$$
$$Hg + S \longrightarrow HgS$$

Mercury also reacts with nitric acid and with hot concentrated sulfuric acid.

$$3\,Hg + 8\,HNO_3 \longrightarrow 3\,Hg(NO_3)_2 + 4\,H_2O + 2\,NO$$
$$Hg + 2\,H_2SO_4 \longrightarrow HgSO_4 + SO_2 + 2\,H_2O$$

It amalgamates with metals such as gold and silver. (An amalgam is an alloy containing mercury.)

USES.   Mercury is used in drugs and chemicals, in the manufacture of barometers, thermometers, mercury vapor arc lamps, and certain types of vacuum pumps, as a cathode in some electrolytic processes, and for physical and chemical research.

Compounds of Mercury.   Mercury forms two series of compounds, those where mercury has an oxidation number of $+1$ and those where it is $+2$. Some of the important compounds are: mercury(II) chloride ($HgCl_2$), mercury(I) chloride ($HgCl$), mercury oxide (HgO), mercury sulfide (HgS), and mercury(I) sulfate (Hg_2SO_4).

$HgCl_2$ is sometimes called *corrosive sublimate*. It is very poisonous. Mercury(I) chloride ($HgCl$ or Hg_2Cl_2) is called *calomel* and is used in medicine.

REVIEW QUESTIONS

1. Structurally, what do zinc and mercury atoms have in common?
2. How does zinc occur in nature? Write an equation for the preparation of metallic zinc from ZnO.
3. Zinc sulfide is sometimes converted to a mixture of oxide and sulfate. Write equations to show what happens.
4. How can zinc oxide be separated from lead oxide? Give two methods.
5. Write equations for the conversion of HgS into metallic mercury.
6. Describe the retort process for preparing zinc from ZnO.
7. For use in paints, what advantage does ZnO have over white lead?
8. List the important physical properties of metallic mercury.
9. What is meant by the statement, "mercury is sometimes called a *noble metal*"?
10. How is mercury prepared commercially?

Chapter 28
Group VIII Metals—Other Industrial Metals

Iron (26) [	] $4s^2$	Palladium (46) [	]
Cobalt (27) [	] $4s^2$	Osmium (76) [	] $6s^2$
Nickel (28) [	] $4s^2$	Iridium (77) [	] $6s^2$
Ruthenium (44) [	] $5s^1$	Platinum (78) [	] $6s^1$
Rhodium (45) [	] $5s^1$		

These nine metals belong to group VIII. (Note. In some periodic tables group VIII is considered to be group VIIIB.) Iron, cobalt, and nickel are usually studied as a separate subgroup (triad). The remaining 6 elements are known as the platinum metals, the first 3 being known as the light platinum metals and the remaining 3 the heavy platinum metals.

In this chapter only iron will be considered in detail because of its great industrial importance. In the tabulated material at the end of the chapter a limited amount of information concerning the preparation, properties, and use of cobalt, nickel, and platinum will be found.

Iron. Iron was known and used for tools and weapons by the early Egyptians and Assyrians. The Romans exploited the mines of Spain for their supply. The early use of iron depended upon the fact that iron oxide can easily be reduced by charcoal. Iron occurs free and as alloy (in meteorites) and chemically combined as oxides, carbonates, silicates, and sulfides. The important minerals of iron are hematite (Fe_2O_3), magnetite (Fe_3O_4), limonite ($2 Fe_2O_3 \cdot 3 H_2O$), and siderite ($FeCO_3$). Iron pyrites (FeS_2) occurs in large quantities, but at present is of no commercial importance as a source of iron.

METALLURGY.   Iron oxide is reduced to the metallic state in a blast furnace by means of carbon. The materials used in the blast furnace are:

Iron Ore.   An oxide is preferable. In the United States, hematite is generally used.

Table 28-1

Physical Properties of Group VIII Metals

	Iron	Cobalt	Nickel	Ruthenium	Rhodium	Palladium	Osmium	Iridium	Platinum
Color	silvery white	silvery white reddish tinge	silvery white yellowish tinge	iron gray	silvery white	silvery white	iron gray	silvery white	silvery white
Physical State	solid	solid	solid	solid	solid	solid	solid	solid	solid
Melting Point	1535°C	1490°C	1452°C	2400°C	1965°C	1554°C	2700°C	2450°C	1774°C
Boiling Point	3000°C	2900°C	2732°C	3900°C	3727°C	2927°C	5000°C	4527°C	3827°C
Atomic Radius	1.165Å	1.157Å	1.149Å	1.241Å	1.247Å	1.287Å	1.255Å	1.260Å	1.290Å
Ionization Energy	7.9 v	7.86 v	7.63 v	7.36 v	7.46 v	8.33 v	8.7 v	9.2 v	9.0 v
Electronegativity	1.8	1.8	1.9	2.2	2.2	2.2	1.9	2.2	2.2
Density g/cm^3 at 25°C	7.87	8.9	8.9	12.41	12.4	12.02	22.57	22.4	21.45

Carbon.   Coke is used both as a fuel and as a reducing agent.

Hot Air (Oxygen).   This helps to keep the temperature high enough to insure active combustion of the fuel. If commercial oxygen is used in place of air, the process takes less time and the product is superior.

Flux.   If the ore is rich in silica, calcium carbonate is added. If it is rich in calcium carbonate, silica is added. The flux removes the earthy matter present in the ore by forming a fused mass which has a comparatively low melting point. This becomes part of the *slag.*

Slag.   Slag is a glasslike material obtained by a combination of calcium carbonate, silica, alumina, and other impurities of the ore. Being less dense than molten iron and not soluble in it, slag forms a liquid layer on the surface of the molten iron, which prevents its reoxidation.

THE BLAST FURNACE.   The structure is made of sheet steel, lined with firebrick. The average furnace is more than 20 feet in diameter and 100 feet high. Iron ore, coke, and flux (limestone and/or silica) are added through the top of the furnace, and hot air is forced in through the *tuyeres.* After the reactions are completed, slag is drawn off through the slag tap-hole and the pig iron (hot metal) is drawn off through the iron tap-hole. The resulting flue gases (called blast furnace gas), consisting chiefly of CO, CO_2, and N_2, are drawn out through the top, and may be used as a fuel due to the CO present. See Figure 28-1.

Some of the modern blast furnaces produce more than 1500 tons of pig iron per 24 hours.

The reactions taking place in the furnace are:

$$3\ Fe_2O_3 + CO \longrightarrow 2\ Fe_3O_4 + CO_2\ (450°C)$$
$$Fe_3O_4 + CO \longrightarrow 3\ FeO + CO_2\ (600°C)$$
$$CO_2 + C \longrightarrow 2\ CO\ (800°C)$$
$$CaCO_3 \longrightarrow CaO + CO_2\ (800°C-1000°C)$$
$$FeO + CO \longrightarrow Fe + CO_2\ (800°C-1000°C)$$
$$CaO + SiO_2 \longrightarrow CaSiO_3\ (1200°C)$$
$$2\ C + O_2 \longrightarrow 2\ CO\ (1300°C)$$

Iron saturated with carbon melts at temperatures ranging from 1150°C to 1250°C. This is 300°C or more below the melting point of pure iron. The hottest part of the furnace is just above the tuyeres, the coolest at the top near the flue.

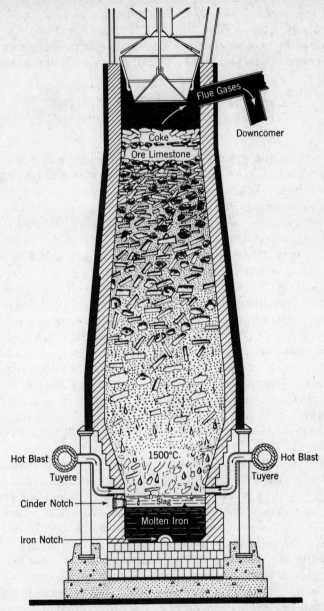

Fig. 28-1. A schematic cross-sectional view of an iron blast furnace.

CAST IRON. This is iron cast from pig iron (often called *hot metal*) as it comes from the blast furnace. Rapid cooling gives *white cast iron*; a slow rate of solidification and cooling produces the commercially important *gray cast iron*. Cast iron contains the following substances:

Carbon (2–4%)	Sulfur (0.1–0.3%)
Silicon (0.7–3%)	Phosphorus (0–0.3%)
Manganese (0.2–1%)	

White cast iron is hard, brittle, and crystalline, and melts at or near 1150°C. Gray cast iron is available in a wide range of hardness values, is of limited ductility, possesses a gray fracture, and melts at or near 1150°C. The separation of graphite from cast iron on solidification offsets normal contraction to a marked degree and increases castability.

Cast iron is used for making castings, for pipes, and for radiators used in heating buildings.

WROUGHT IRON. If pig iron is melted in a small hand-operated reverberatory (puddling) furnace in an oxidizing atmosphere, wrought iron is obtained. In this process, the charge melts at about 1150°C, but as the impurities are burned out, the melting point rises and lumps of iron are formed. These are raked together into lumps of 75 to 100 pounds each and then removed from the furnace. The enclosed slag is then squeezed out by hammering or passing the lumps of iron through suitable rolls. However, after cooling, a microscopic examination of the metal shows that some slag is still present, usually in the form of fibers or threads.

The Aston Process. The A. M. Byers Company of Pittsburgh, using the Aston Process, makes wrought iron on a rather large scale. The steps in the process are: Pig iron is melted in a cupola furnace and then transferred to a Bessemer converter where it is purified. An iron silicate slag is made in an open-hearth furnace. The molten metal (from the Bessemer) at a temperature of 1500°C is poured as a steady stream into slag maintained at 1300°C. The metal settles to the bottom in the form of a spongy mass because it contains trapped fibrous slag. This spongy mass is passed through a squeezer where the excess slag is removed. The product is then ready for the rolling mills.

Properties and Uses. The properties of wrought iron are due largely to the low carbon content (less than 0.3 per cent) and the presence of a small quantity of enclosed slag. It has a high melting point (near 1500°C), is malleable and ductile (i.e., it can be worked), has a fibrous structure due to slag content and rolling, and does not rust easily.

Wrought iron is used for pipes, horseshoes, chains, and bolts, and in making crucible steel.

Steel. When iron containing from 0.3 per cent to 1.7 per cent carbon is heated to a suitable temperature and then quenched, it is hardened and toughened. This product is called *steel*. By variations in the heating and quenching treatments, steels possessing various properties are produced. Most steels are characterized by their hardness, toughness, and high tensile strength. More than 100 million tons of the various types of steel have been manufactured in the United States in a single year. Steels are classified as follows:

PLAIN CARBON STEELS. The properties of these steels depend upon the percentage of carbon present and upon the heat treatment applied.

ALLOY STEELS. In addition to carbon, the iron contains varying amounts of certain metals such as nickel, vanadium, manganese, chromium, cobalt, and tungsten. The added metals have a remarkable effect upon the properties of the steel, especially when it has been given a suitable heat treatment. See Table 28-3, p. 294.

BESSEMER STEEL. Molten pig iron is poured into an egg-shaped furnace which can be tilted. This is the converter. A blast of air entering through tuyeres in the bottom is forced through the charge, and the carbon, silicon, and certain other impurities are burned out. After the process is completed (about 10 to 15 minutes per charge), spiegeleisen, an alloy of iron, manganese, and carbon, is added to furnish the desired amount of carbon and manganese. See Figure 28-2, p. 292.

In this country, most Bessemer converters are lined with silica and are acid-type furnaces. Consequently, they do not remove phosphorus and sulfur. Usually, pig iron made from iron ores low in phosphorus and sulfur is charged into the Bessemer converter.

BASIC OPEN-HEARTH STEEL. Pig iron, scrap iron, and some iron oxide are heated in an open-hearth furnace (40 × 12 × 2 feet or larger) lined with magnesia brick. The heat for smelting the charge comes from a gas flame on either end of the furnace, which is reflected from the low-arched ceiling onto the charge. Several hours are required for each charge, as determined by

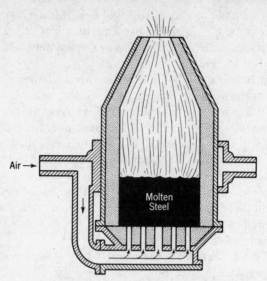

Fig 28-2.   A Bessemer converter.

analysis of samples taken from the furnace. For many uses, open-hearth steel is superior to Bessemer steel because its composition is more uniform. Also, a basic open-hearth furnace will remove most of the phosphorus (and a major part of the sulfur) from the metal placed in it. An acid open-hearth (silica-lined) furnace may be used on metals low in phosphorus and sulfur. The use of commercial oxygen (in place of air) greatly speeds up the process, and the product is of good quality.

CRUCIBLE STEEL.   Wrought iron is melted in graphite crucibles. Clean charcoal is added to give the desired carbon content. The molten mass is poured into molds, forming ingots. These ingots are retreated and worked into tools.

ELECTRIC-FURNACE STEELS.   Various types of electric furnaces are used in making steel. The electricity is used as a source of heat. One popular type of furnace is the Heroult arc furnace, in which an arc is maintained between graphite electrodes, furnishing the desired heat. Electric furnaces produce superior steel because of the superior charge added and also because of the temperature control possible.

BASIC OXYGEN STEEL.   Recently oxygen gas has been used in preparing high-quality steel. A furnace, shaped somewhat like a

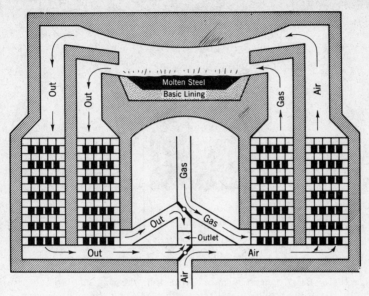

Fig. 28-3. A diagram of an open-hearth furnace.

Bessemer, but without tuyeres, is provided with an oxygen-supplying tube with nozzle which produces a jet of oxygen (sometimes called an *oxygen lance*) that strikes the top surface of the molten metal (pig iron) contained in the furnace. This brings about the necessary steel-making chemical reactions. Millions of tons of steel are now made yearly in the United States by this time-saving and interesting process.

In the year 1969 as much basic oxygen steel was produced in the United States as open-hearth steel. See Figure 28-4, p. 294.

Table 28-2
Uses of Plain Carbon Steels

Kind of Steel	Uses
Bessemer	Rails, wire bars, axles, structural units
Open-hearth	Wire, nails, axles, rails, structural units
Basic oxygen	Wire, nails, axles, rails, structural units
Crucible	Tools, bars, castings
Electric furnace	Tools, bars, castings, springs

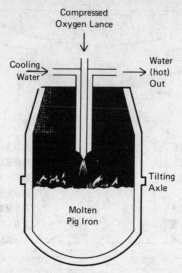

Fig. 28-4. Basic oxygen furnace. Note that the oxygen impinges on the surface of the molten metal rather than entering through tuyeres in the bottom as in the Bessemer converter.

Table 28-3

Composition, Properties, and Uses of Alloy Steels

Type	Composition	Properties	Uses
Low alloy	Many contain low percentages (2% or less) of 2, 3, or 4 alloying metals such as Mn, W, V, Mo, Ni, and .4–.8% carbon.	Respond to heat treatment more favorably than plain carbon steels	Automobiles, railroad cars, and wherever special steels are required
Special alloy steels	(a) High chromium steel (13–20% Cr)	Stainless (does not corrode under many industrial uses)	Cutlery, heat- and corrosion-resistant equipment
	(b) High nickel (Invar) steel, 36% Ni	Low coefficient of expansion	Watch springs
	(c) 20% Ni, 5% Co, 12% Al	Permanent magnet	In the electronic industries
High-speed tool steels	18% W, 4% Cr, 1% V, .25% Mn, .7% carbon	Hard, strong up to 650°C	Cutting tools, such as lathes

HEAT TREATMENT OF STEEL. Physical properties (hardness, tensile strength, ductility) of plain carbon steel depend on its carbon content and on the heating and cooling treatment it receives. Heat treating is influenced by the following factors:

1. Iron exists in three allotropic modifications:

$$\text{alpha} \underset{}{\overset{910°C}{\rightleftarrows}} \text{gamma} \underset{}{\overset{1400°C}{\rightleftarrows}} \text{delta} \underset{}{\overset{1535°C}{\rightleftarrows}} \text{liquid iron}$$

2. Carbon and iron combine chemically to give iron carbide (Fe_3C), called *cementite*.

3. Carbon dissolves in gamma iron up to 1.7 per cent but only very slightly in alpha iron. The solid solution formed is called *austenite*. On cooling to room temperature, this solution decomposes into iron (called *ferrite*) and Fe_3C (cementite).

4. A mixture consisting of alternate layers of ferrite and cementite (visible under a metallographic microscope) is called *pearlite*. See Figure 28-5, p. 296.

Table 28-4 summarizes the results obtained by heat-treating three samples of plain carbon steel, each of which contained .8 per cent carbon at a temperature where austenite is stable.

Table 28-4

Heat Treating Plain Carbon Steel

Sample Number	Composition	Description of Cooling Rate	Product	Hardness	Tensile Strength
1	.8% C	Cooled very slowly	Coarse pearlite	Relatively soft	Low
2	.8% C	Medium rate of cooling	Medium pearlite	Hard	High
3	.8% C	Quenched in water	Very fine pearlite	Very hard	Very high

Note 1. Sample No. 3 will be very brittle. If reheated to 180°C (*tempered*), it will become less brittle but the hardness and tensile strength will remain high.

Note 2. If sample No. 3 were reheated to about 650°C and then cooled slowly, it would be soft like sample No. 1. This process is called *annealing*.

Pure Iron. The commercial methods now used for preparing iron and the alloys of iron are not satisfactory for the preparation of chemically pure iron. Chemists who studied this problem found that high-purity iron can be prepared by the hydrogen reduction

(Courtesy H. E. Flanders, University of Utah.)

Fig. 28-5. Photomicrograph of pearlite, $600\times$.

of hematite at a suitable temperature or by the electrolysis of purified ferrous sulfate solution. These methods are expensive, so only small amounts of iron are produced by them. Pure iron, produced by one of these methods, is a silvery white metal that is soft, malleable, and ductile. Its density is 7.68 and it melts at 1535°C.

Compounds of Iron. Iron has oxidation numbers of $+2$ and $+3$. Those compounds where iron is $+2$ are known as iron(II) compounds (formerly ferrous compounds). The $+3$ compounds are known as iron(III) compounds (formerly ferric compounds). In general the iron(II) compounds are rather easily oxidized to iron(III) compounds. Some of the common iron(II) compounds are:

Iron (II) oxide (FeO) is prepared by heating iron(II) oxalate.

$$FeC_2O_4 + heat \longrightarrow FeO + CO_2 + CO$$

Iron(II) hydroxide [Fe(OH)$_2$] is prepared by adding a soluble base such as NaOH to an iron(II) sulfate (FeSO$_4$) solution.

$$FeSO_4 + 2\,NaOH \longrightarrow Fe(OH)_2 + Na_2SO_4$$

Iron(II) chloride (FeCl$_2$) is prepared by treating metallic iron with hydrochloric acid *in the absence of air.*

$$Fe + 2\,HCl \longrightarrow FeCl_2 + H_2$$

or

$$Fe + 2\,H^{+1} + 2\,Cl^{-1} \longrightarrow FeCl_2 + H_2$$

Iron(II) sulfate (FeSO$_4$) is prepared by adding sulfuric acid to metallic iron. With water of crystallization this salt is green and is called *green vitriol.*

$$Fe + H_2SO_4 \longrightarrow FeSO_4 + H_2$$

Water solutions of this salt oxidize readily to give
Iron(III) sulfate [Fe$_2$(SO$_4$)$_3$].

$$2\,FeSO_4 + H_2SO_4 + \tfrac{1}{2}\,O_2 \longrightarrow Fe_2(SO_4)_3 + H_2O$$

Iron(II) sulfide (FeS) is easily prepared by heating a mixture of iron filings and powdered sulfur.

$$Fe + S \longrightarrow FeS$$

Iron(II) ammonium sulfate, generally called ferrous ammonium sulfate, [(NH$_4$)$_2$SO$_4$·FeSO$_4$·6 H$_2$O] is known as *Mohr's salt.* It is prepared by adding the two salts, iron(II) sulfate and ammonium sulfate, in molecular proportions, to water.

$$(NH_4)_2SO_4 + FeSO_4 + 6\,H_2O \longrightarrow (NH_4)_2SO_4 \cdot FeSO_4 \cdot 6\,H_2O$$
$$\text{(green crystals)}$$

Important iron(III) compounds are:

Iron(III) oxide (Fe$_2$O$_3$) is found in nature and can be prepared in the laboratory by dehydrating Fe(OH)$_3$ or by controlled roasting of FeS$_2$ (iron pyrites). Fe$_2$O$_3$ is used as a pigment and known as *Venetian red.*

$$2\,Fe(OH)_3 + heat \longrightarrow Fe_2O_3 + 3\,H_2O$$
$$2\,FeS_2 + 11/2\,O_2 \xrightarrow{\text{(heat)}} Fe_2O_3 + 4\,SO_2$$

Iron(III) chloride (FeCl$_3$) is prepared by burning iron wire in chlorine. It is very soluble in water from which it crystallizes as the hexahydrate ($FeCl_3 \cdot 6 H_2O$). Dissolved in alcohol it is known as *tincture of iron*.

$$2 Fe + 3 Cl_2 \longrightarrow 2 FeCl_3$$
$$FeCl_3 + 6 H_2O \longrightarrow FeCl_3 \cdot 6 H_2O \text{ (crystals)}$$

Complex Cyanides. Iron(II) and iron(III) compounds form complex cyanides. The following are important:

Potassium Ferrocyanide [$K_4Fe (CN)_6$]. In this compound iron has an oxidation number of +2. It is prepared by heating together iron, potassium carbonate, and waste organic matter containing nitrogen. This salt is used as a test for iron(III) ion (Fe^{+3}). It gives a dark blue precipitate called *Prussian blue*.

$$4 Fe^{+3} + 3 [Fe(Cu)_6]^{-4} \longrightarrow Fe_4 [Fe (CN)_6]_3$$
Prussian blue

Potassium Ferricyanide [$K_3F_e(CN)_6$]. In this compound iron has an oxidation number of +3. It is prepared by treating potassium ferrocyanide with an oxidizing agent such as chlorine. This salt is used as a test for iron(II) ions. The resulting compound is *Turnbull's blue*.

$$3 Fe^{+2} + 2 [Fe (CN)_6]^{-3} \longrightarrow Fe_3 [Fe (CN)_6]_2$$
Turnbull's blue

Blueprint Paper. This paper is made by treating white paper (in the dark) with a solution of ammonium ferricyanide and iron(III) citrate. Exposure to light causes the formation of Turnbull's blue. The print is fixed by washing out the unchanged mixture.

REVIEW QUESTIONS

1. Why was iron prepared long before aluminum?
2. What are the important minerals of iron?
3. Distinguish between: (*a*) the product of the blast furnace and the product of the open-hearth furnace, (*b*) wrought iron and crucible steel, (*c*) Bessemer steel and open-hearth steel.
4. What is meant by each of the following? (*a*) slag, (*b*) flux, (*c*) tuyeres, (*d*) hot metal, (*e*) white cast iron, (*f*) blast furnace gas.
5. Why is electric-furnace steel considered to be superior to open-hearth steel?

6. In general, what advantages do alloy steels have over plain carbon steels?

7. Discuss the importance of the allotropic forms of iron in relation to heat treating of steel.

8. What is stainless steel? Compare stainless steel with ordinary carbon steel.

9. How can one make coarse pearlite? How is the coarseness (or fineness) of pearlite controlled?

OTHER METALS OF

METAL	MINERALS	PREPARATION
Beryllium	Beryl $((BeO)_3 \cdot Al_2O_3 \cdot 6 SiO_2)$	Electrolysis of fused barium and sodium fluorides (to which have been added beryllium oxide) at about 1300°C. The metal separates out on an iron cathode.
Cadmium	Greenockite (CdS) (Not important as a source of cadmium)	Usually obtained as a by-product: (1) in lead and zinc smelting, or (2) from electrolytic zinc plants.
Chromium	Chromite $(FeCr_2O_4)$	(1) By Goldschmidt Process: $Cr_2O_3 + 2 Al \longrightarrow 2 Cr + Al_2O_3$ (2) Chromium can be plated out from a solution of chromic acid using a suitable electric current. (3) $Cr_2O_3 + 3 H_2 \longrightarrow 2 Cr + 3 H_2O$ (Very pure metal)
Cobalt	Smaltite $(CoAS_2)$ Cobaltite (CoAsS)	(1) The sulfide is roasted to give an oxide. After the impurities are removed, the oxide is reduced with charcoal or coke in an electric furnace. (2) Oxidized cobalt ores can be leached with sulfuric acid. The cobalt is removed from this solution by electrolysis.
Manganese	Pyrolusite (MnO_2) Rhodochrosite $(MnCO_3)$	(1) By Goldschmidt Process: $3 MnO_2 + 4 Al \longrightarrow 3 Mn + 2 Al_2O_3$ (2) By reduction of the oxide in a blast furnace. This gives ferromanganese or spiegeleisen. (3) By electrolysis of purified manganese sulfate solution. This gives very pure metal.

INDUSTRIAL IMPORTANCE

Properties	Uses
(1) Silver-colored metal. (2) Specific gravity 1.85. (3) High melting point 1285°C. (4) It combines with copper to give an age-hardening alloy.	(1) To harden copper. (2) To make nonsparking alloys for tools. (3) To make light-weight alloys.
(1) Silver-white metal. (2) Specific gravity 8.65. (3) Melting point 321°C. (4) Resembles zinc somewhat but is not malleable.	(1) As a protective coating for steel. (2) In making bearing metals. (3) For low-melting alloys. (4) In atomic piles.
(1) Silver-white metal. (2) Specific gravity 7.14. (3) Melting point 1615°C. (4) Is very brittle. (5) Has high resistance to corrosion. (6) Forms a number of interesting colored compounds such as: K_2CrO_4, Cr_2O_3, $K_2Cr_2O_7$, etc.	(1) To make ferrochrome (an alloy of Fe—Cr—C) used in the steel industry. (2) To make various alloys. (3) For chromium plating of steel. Note. Stainless steel may contain up to 20 per cent of chromium.
(1) Silver colored with reddish tinge. (2) Specific gravity 8.9. (3) Melting point 1490°C (4) Somewhat malleable and ductile.	(1) For making high-speed alloys used for cutting tools. (2) For making magnetic alloys. (3) For electrical resistance alloys. (4) For the preparation of cemented carbide alloys. Here cobalt is the matrix.
(1) Silver colored with reddish tinge. (2) Specific gravity 7.44. (3) Melting point 1240°C. (4) Soft and ductile if pure. (5) Forms a number of interesting colored compounds such as: $KMnO_4$ (purple) and K_2MnO_4 (green).	(1) Used as a deoxidizer in iron and steel manufacture. (2) To make alloy steels such as manganese steel which contains 11–14% manganese. These steels are hard, tough, and wear-resistant.

OTHER METALS OF

METAL	MINERALS	PREPARATION
Molybdenum	Molybdenite (MoS_2) Wulfenite ($PbMoO_4$)	The sulfide is roasted to give the oxide (MoO_3). This is heated in a current of hydrogen: $$MoO_3 + 3H_2 \longrightarrow Mo + 3H_2O$$ The metal is in powder form. It must be pressed into suitable form and sintered to give rods and wires.
Nickel	Garnierite ($NiOMgO$ $SiO_2 n H_2O$) Also complex sulfides of nickel, copper, iron, and arsenic.	"Top and Bottom Process" A nickel matte is made by roasting the ore in a reverberatory furnace, then in a converter. To this matte is added sodium sulfate and coke. The smelting action causes two layers to form. The top layer contains most of the copper as sulfide; the bottom, the nickel as sulfide. This is purified, oxidized, and then dissolved in H_2SO_4 and finally electrolyzed.
Platinum *Note.* The other metals of the platinum group are: palladium, iridium, osmium, rhodium, and ruthenium.	(1) Found as free (or native) metal as grains and nuggets in alluvial sands. (2) Cupperite (PtS) Bragite ($Pd, Pt, Ni)S$	Ore containing platinum is treated with with aqua regia to dissolve the metal. To this solution is added NH_4Cl which precipitates ammonium chloroplatinate [$(NH_4)_2PtCl_6$]. This is ignited to give the metal.
Titanium	Rutile (TiO_2) Ilmenite ($FeTiO_3$)	(1) Reduction of TiO_2 in an electric furnace: $$TiO_2 + 2C \longrightarrow Ti + 2CO$$ (2) Reduction of $TiCl_4$ with sodium: $$TiCl_4 + 4Na \longrightarrow Ti + 4NaCl$$ Kroll Process. Sponge titanium is produced by reduction of $TiCl_4$ with molten magnesium in an inert atmosphere: $$TiCl_4 + 2Mg \longrightarrow Ti + 2MgCl_2$$

INDUSTRIAL IMPORTANCE (continued)

PROPERTIES	USES
(1) Silver-colored metal. (2) Specific gravity 10.2. (3) Melting point 2620°C. (4) The most interesting compound is: $(NH_4)_2Mo_7O_{24}4H_2O$ which is an analytical reagent for the determination of phosphates.	(1) Used in incandescent lamps and radio tubes. (2) In ribbon form as a heating element in electric furnaces. (3) Used in making alloy steels.
(1) Silver-colored metal. (2) Specific gravity 8.9. (3) Melting point 1452°C. (4) If pure is malleable and ductile. (5) Forms many interesting salts. (6) Resistant to corrosion or oxidation.	(1) For various alloys such as nickel steel, monel metal, and nichrome. (2) For electroplating steel. (3) For high-alloy steels.
(1) Silver-colored metal. (2) Specific gravity 21.4. (3) Melting point 1754°C. (4) Malleable and ductile. (5) Is chemically inactive. Does not react with ordinary chemicals. (6) Does not tarnish. Difficult to oxidize.	(1) Platinum laboratory ware for handling corrosives. (2) In jewelry. (3) As a catalyst for making sulfuric acid. (4) In photography.
(1) Silver white (2) Specific gravity 4.5. (3) Melting point about 1800°. (4) Resists corrosion. (5) Alloyed with iron (ferrotitanium) it is a deoxidizer. (6) Can be forged at red heat.	(1) As a deoxidizer. (2) Titanium alloys make permanent magnets. (3) As a corrosion-resistant metal.

OTHER METALS OF

METAL	MINERALS	PREPARATION
Tungsten	Scheelite ($CaWO_4$) Wolframite [$(Fe, Mn)WO_4$]	The tungsten ore is fused with soda ash (Na_2CO_3) and the Na_2WO_4 leached out. This is converted to H_2WO_4 with HCl. Heating strongly gives WO_3 which is subsequently reduced with hydrogen to give tungsten powder.
Uranium	Pitchblende (U_3O_8) Carnotite ($K_2(UO_2)_2 \cdot (VO_4)_2 \cdot 8 H_2O$)	Crushed and ground ore is leached in a sulfuric acid solution. This is treated with a kerosene-based solvent containing tributylphosphate. This solvent extracts the uranium. Next, the solvent is treated with a 10% sodium carbonate solution to extract the uranium again. Finally, the carbonate solution is treated with ammonia to give yellow cake (largely U_3O_8) which in turn is converted to the chloride (UCl_4). This is reduced with metallic sodium or calcium to give metallic uranium.
Zirconium	Zircon ($ZrSiO_4$)	(1) $ZrSiO_4 + 2 C \longrightarrow ZrC + \cdots$ (2) $ZrC + 2 Cl_2 \longrightarrow ZrCl_4 + C$ (3) $ZrCl_4 + 4 Na \longrightarrow Zr + 4 NaCl$

INDUSTRIAL IMPORTANCE (continued)

PROPERTIES	USES
(1) Steel-gray metal. (2) Melting point 3350°C. (3) Specific gravity 19.6. Due to the high melting point, tungsten powder is compressed into rods, then sintered and swaged. Finally, it is drawn into wire. It is also made into sheets.	(1) To make high-speed tool steel and other alloys. (2) For filaments used in X-ray tubes, radio tubes, and incandescent lamps.
(1) Metallic uranium is silver white but turns brown on oxidation in air. (2) Specific gravity 19.05. (3) Melting point 1133°C. (4) Boiling point 3927°C. (5) Powdered uranium is spontaneously combustible in air. (6) Uranium is radioactive.	(1) The most important uses for uranium are in nuclear devices, such as reactors and atomic bombs. (2) Alloys of uranium have been made but are of little industrial importance.
(1) Silver-colored metal. (2) Specific gravity 6.4. (3) Melting point 1930°C. (4) Combines with iron to form ferro-zirconium.	(1) Very limited use in the steel industry. (2) ZrO_2 is used as a refractory. (3) Very useful in atomic energy work.

Chapter 29
Organic Chemistry

Organic chemistry is that branch of chemistry which deals with the compounds of carbon. More than 1,000,000 carbon compounds have been identified. Many of these occur in nature; the others have been synthesized by chemists in the laboratory. The important classes of organic compounds include the *hydrocarbons* and the *derivatives of hydrocarbons*. The first group of compounds contain carbon and hydrogen only; the second group in addition to these two elements may contain oxygen, nitrogen, sulfur, phosphorus, and other elements. At one time, it was thought that natural products could not be synthesized outside the living organism. But in 1828 Wöhler succeeded in making urea (a compound formed in the body) from the inorganic substance ammonium cyanate, thus:

$$NH_4OCN \longrightarrow NH_2CONH_2$$
(Urea)

This success in producing an organic substance synthetically was the stimulus which has resulted in the preparation of thousands of compounds. Even some of the vitamins and hormones are now made in the laboratory. The unusual thing about carbon is its ability to combine with itself, atom to atom (C—C—C, etc.), and these combinations may be in the form of straight chains, branched chains, or rings.

Hydrocarbons. Hydrocarbons are compounds containing only carbon and hydrogen. Table 29-1 gives the various classes of these compounds. A general formula is given at the head of each column. The letter *n* represents the number of carbon atoms in each molecule. Open-chain compounds, such as propane (C_3H_8) and ethyl acetate ($CH_3COOC_2H_5$), are known as *aliphatic* compounds.

Compounds whose compositions can be represented by a general formula belong to a *homologous series*. In the alkane series (sometimes called the *methane series* because methane is the first member), the carbon atoms are held together by single valence bonds. For this reason, these compounds are said to be *saturated*.

Table 29-1

Classes of Hydrocarbons

Alkane C_nH_{2n+2}	Alkene C_nH_{2n}	Alkyne C_nH_{2n-2}	Benzene C_nH_{2n-6}
Methane...CH_4	Ethene.....C_2H_4	Ethyne ....C_2H_2	Benzene ...C_6H_6
Ethane ....C_2H_6	Propene ...C_3H_6	Propyne ...C_3H_4	Toluene ...C_7H_8
Propane ...C_3H_8	Butene.....C_4H_8	Butyne ....C_4H_6	Xylene.....C_8H_{10}
Butane ....C_4H_{10}	Pentene....C_5H_{10}	Pentyne....C_5H_8	

In the other series, some carbon atoms are bound together by double bonds (as in the alkene series) or by triple bonds (as in the alkyne series). These compounds are said to be *unsaturated*. From the standpoint of atomic structure, the carbon atoms are held together by an electron pair (C:C) as in the alkane compounds, or by two electron pairs (C::C) as in the alkene compounds, or by three pairs (C:::C) as in the alkyne compounds.

Structural Formulas. It is customary for chemists to write structural formulas of organic compounds. In this way, the properties and relationships for the various compounds are more readily understood. In a majority of cases, the (C:C) bond is represented by (C—C). This notation will be used in this chapter. Likewise, the (C::C) bond is represented by (C=C) and (C:::C) by (C≡C).

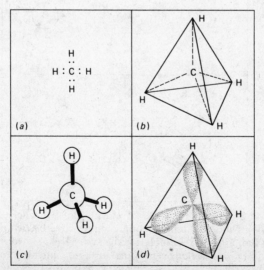

Fig. 29-1. Four structural formulas for methane.

Organic molecules, like inorganic molecules, occupy three-dimensional space. Molecules do not exist in one plane even though one-plane representations are simpler for the elementary student and more convenient for the advanced student. Methane (CH_4) the simplest hydrocarbon (compound of hydrogen and carbon) can be written in several ways to emphasize structural features. These are shown in Figure 29-1. Figure 29-1(a) is the usual way of writing the structural formula. It shows the 4 covalent bonds between the 4 hydrogen atoms and the carbon atom. Figure 29-1(b) depicts the carbon atom in the center of a tetrahedron with each of the 4 hydrogen atoms occupying a corner of the tetrahedron. Figure 29-1(c) shows the special ball and stick model of methane, and Figure 29-1(d) is a more elaborate representation of the methane molecule showing the sp^3 orbitals.

Figure 29-2 shows the structural formulas of ethene (ethylene) and benzene. Figure (a) shows the two covalent bonds between the two carbon atoms and (b) shows the ball and stick formula for the same substance. Figures (c) and (d) show two ways of writing the benzene structural formula. Figure (c) is the formula

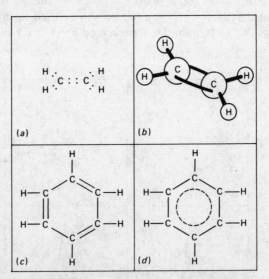

Fig. 29-2. Structural formulas of ethene and benzene. (a) Covalent formula of ethene. The double bond is represented by two pairs of electrons. (b) Ball and stick formula of ethene. (c) Kekule's formula of benzene. (d) Modified Kekule's benzene formula, intended to emphasize that all six bonds in the benzene ring have the same value.

as Kekule wrote it and (*d*) is the modified form intended to emphasize the fact that all six carbon atoms in benzene have the same bond value. The structural formulas of some of the compounds in Table 29-1 follow.

ALKANE SERIES COMPOUNDS:

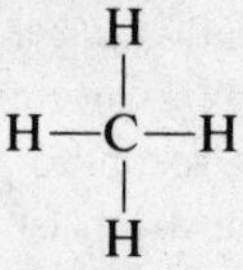

Methane 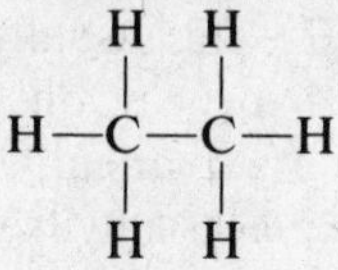   Ethane 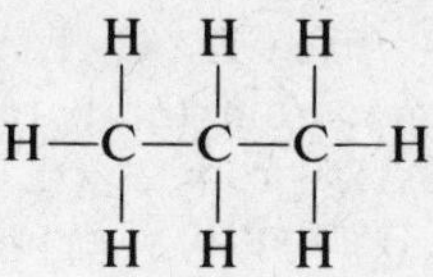   Propane

ALKENE SERIES:

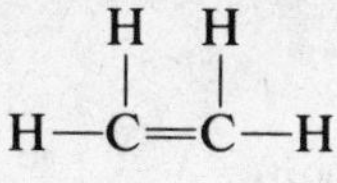

Ethene 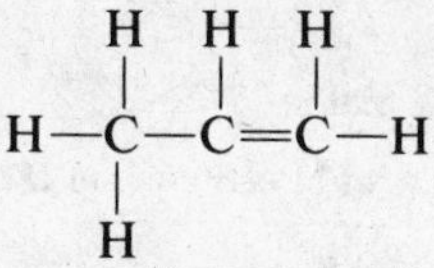   Propene 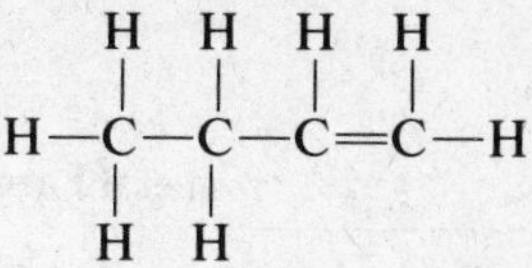   Butene

ALKYNE SERIES:

Ethyne
(Acetylene)   Propyne 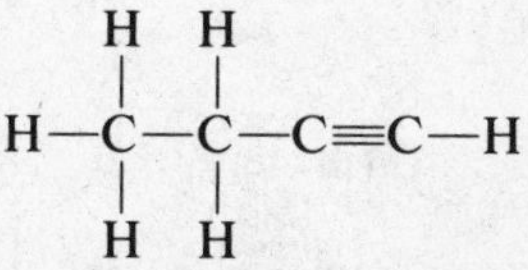   Butyne

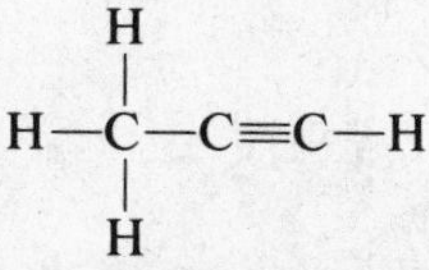

BENZENE SERIES:

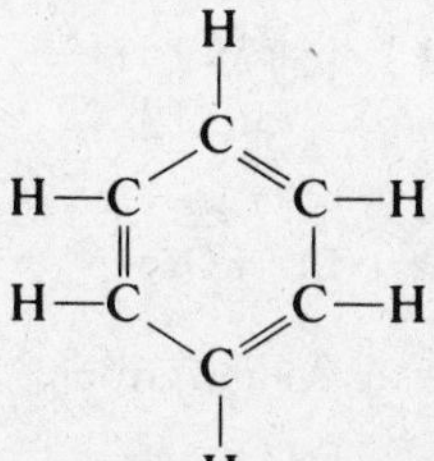

Benzene 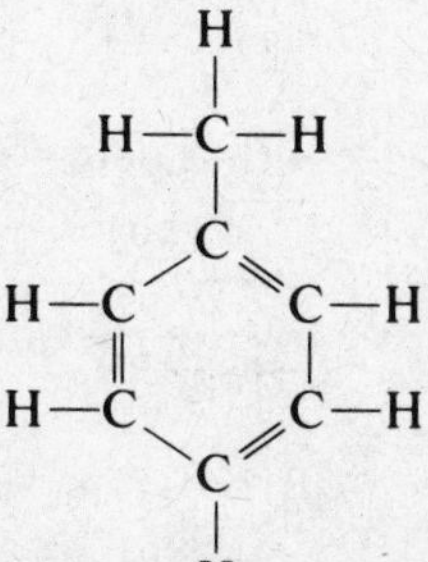   Toluene 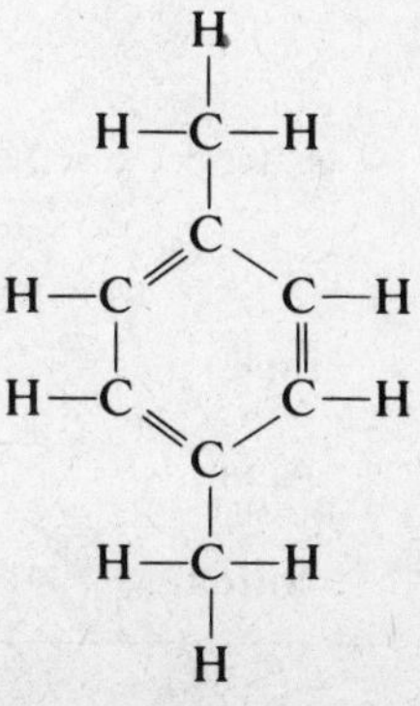   Xylene

Derivatives of Hydrocarbons. When one or more hydrogen atoms of a hydrocarbon molecule are replaced by other atoms or groups, the resulting compounds are said to be *derivatives* of the hydrocarbons. Thus, CH_3I and C_2H_5Cl are derived from methane and ethane by replacing a hydrogen atom in each molecule with a halogen atom. The first is methyl iodide; the second is ethyl chloride. The CH_3, C_2H_5, C_3H_7, etc., usually called *alkyl* radicals, are named from the parent hydrocarbon (CH_3, methyl; C_2H_5, ethyl; C_3H_7, propyl; etc.) and are frequently represented by R and R', etc. If a hydroxyl group replaces a hydrogen atom of a hydrocarbon molecule, we have a nonionizing compound called an *alcohol*. There are no stable organic compounds in which 2 (OH) groups are attached to one carbon atom. If two hydrogen atoms

Table 29-2

General Formulas of Hydrocarbon Derivatives

Type of Compound	General Formula		Functional Group
Alcohol	R OH	→	Hydroxyl Group
Aldehyde	R—C=O (H)	→	Aldehyde Group
Ketone	R—C—R (‖O)	→	Carbonyl Group
Organic Acid	R—C—OH (‖O)	→	Carboxyl Group
Ester	R—C—OR' (‖O)	→	Ester Group
Ether	R—O—R	→	Ether Group
Amine	R NH₂	→	Amino Group
Nitroalkyl	R NO₂	→	Nitro Group

are replaced by an oxygen atom, we have an *aldehyde*. Table 29-2 gives the general formulas of the more common derivatives of the hydrocarbons. The functional atom or group within the broken line inclosure is characteristic of that particular class of compounds.

Ring Compounds. The six carbon atoms in benzene (C_6H_6) are combined in such a way that a ring structure is obtained. (Note the structure of benzene in the structural formula above.) Compounds of the type having six carbon atoms per ring are frequently called *aromatic hydrocarbons* because, as a class, they possess distinctive odors. They also have single and double bonds alternating.

Naphthalene ($C_{10}H_8$) has a condensed two-ring structure, and anthracene ($C_{14}H_{10}$) has a condensed three-ring structure.

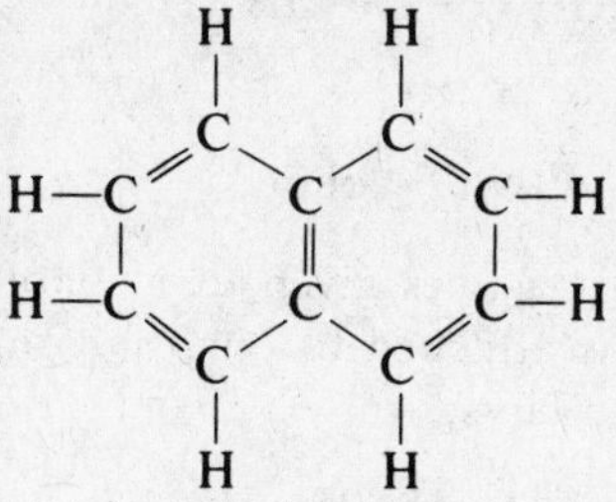

Naphthalene

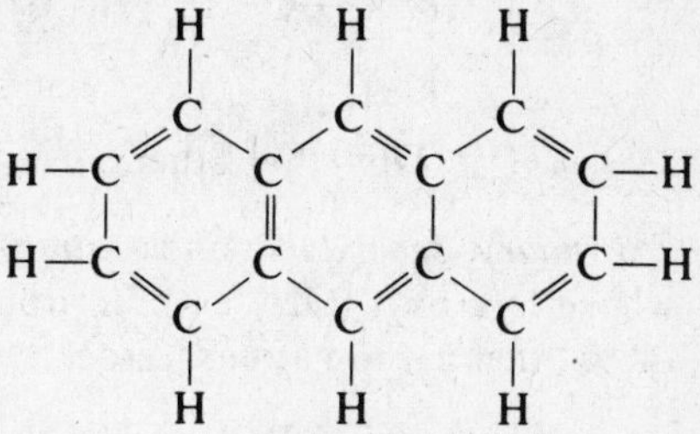

Anthracene

Compounds with a closed chain, such as cyclopentane, are called *alicyclic* compounds.

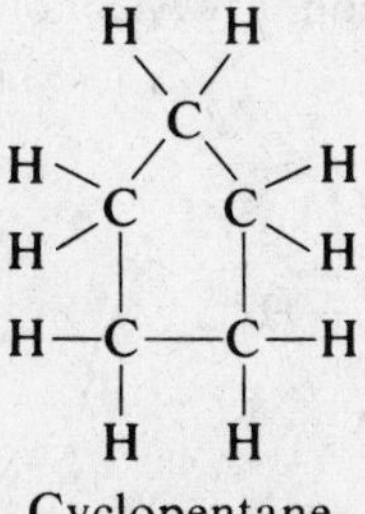

Cyclopentane

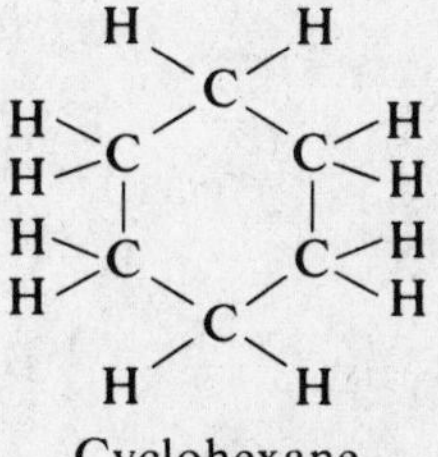

Cyclohexane

Isomerism. When two or more molecules have the same composition, but different structures, they are said to be *isomers*. For example, the carbon atoms in butane (C_4H_{10}) can be written in two ways:

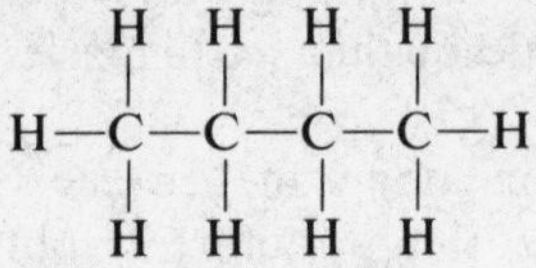

Normal Butane 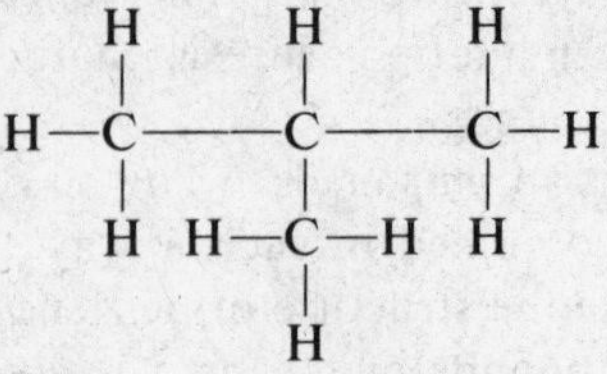   Isobutane

Derivatives of hydrocarbons may also be isomeric. Thus, a compound (C_2H_6O) can be written:

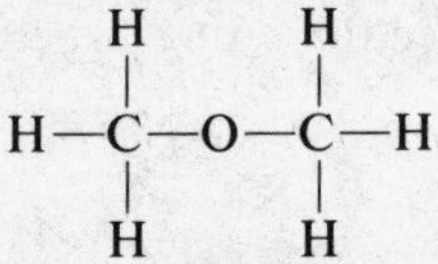

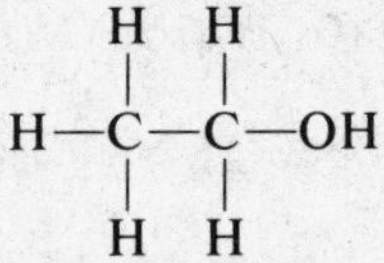

Dimethyl Ether   Ethyl Alcohol

Primary, *secondary*, and *tertiary* are names given to alcohols whose hydroxyl (OH) carbon atom is attached to one, two, or three other carbon atoms, respectively. Thus,

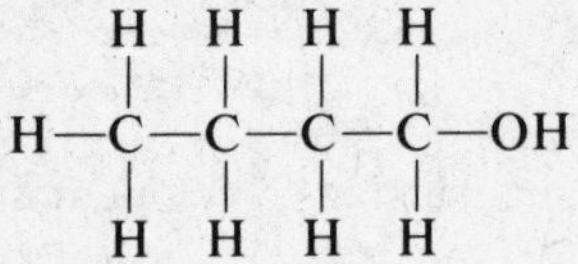

Primary Alcohol 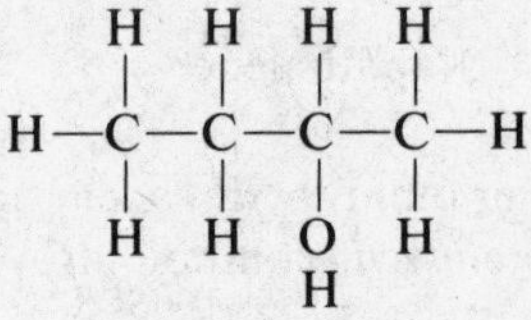   Secondary Alcohol

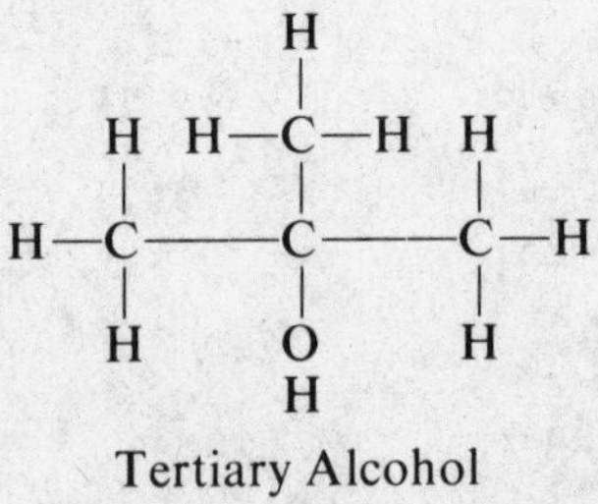

Tertiary Alcohol

ORTHO, META, AND PARA ISOMERISM. There are three-dimethyl benzenes. They have the following structures:

$$
\begin{array}{ccc}
\text{Ortho} & \text{Meta} & \text{Para} \\
\text{1-2 dimethylbenzene} & \text{1-3 dimethylbenzene} & \text{1-4 dimethylbenzene} \\
\text{(orthoxylene)} & \text{(metaxylene)} & \text{(paraxylene)}
\end{array}
$$

Note that the top carbon atom in the benzene ring is No. 1. The next to the right is No. 2 and so on. The differences in physical properties of the above isomers are given in Table 29-3.

Table 29-3
Physical Properties of Isomers

Substance	Molecular Weight	Melting Point	Boiling Point
Normal butane	58	−135°C	−0.6°C
ISO butane	58	−145°C	−10.2°C
Dimethyl ether	46	−140°C	−24.9°C
Ethyl alcohol	46	−114°C	+78.3°C
Normal butyl alcohol	74	−90°C	+117.7°C
Sec. butyl alcohol	74		+99.5°C
Tert. butyl alcohol	74	25°C	+82.5°C
Orthoxylene	106	−28°C	+144°C
Metaxylene	106	−54°C	+139°C
Paraxylene	106	+13°C	+138°C

Polyhydroxy Alcohols. Alcohols containing more than one hydroxyl group per molecule are said to be *polyhydroxy* alcohols. However, many organic compounds containing (OH) groups are not called alcohols. For example, a compound such as glucose, containing several (OH) groups, could be classed as a polyhydroxy alcohol. However, it is normally classed as a *carbohydrate* and called a *sugar*.

Alcohols containing 2 (OH) groups are called *glycols*.

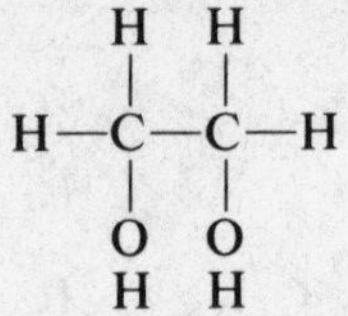

Ethylene Glycol 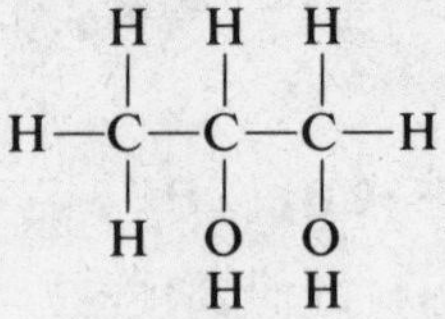   Propylene Glycol

Glycerol (glycerine) is a well-known trihydroxy alcohol. Its structural formula is:

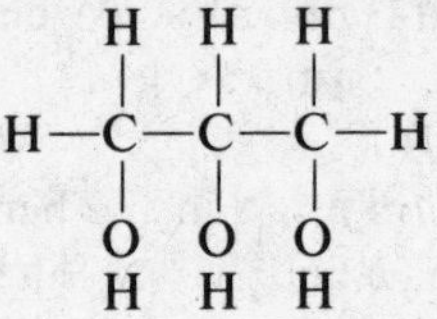

NAMES OF ALCOHOLS. The ending *ol* is used by chemists to indicate that a compound is an alcohol. Table 29-4 lists 4 of them.

Table 29-4

Alcohols: Chemical and Common Names

Formula of Alcohol	Compound Derived From	Chemical Name	Common Name
CH_3OH	Methane (CH_4)	Methan*ol*	Wood alcohol Methyl alcohol
C_2H_5OH	Ethane (C_2H_6)	Ethan*ol*	Ethyl alcohol Grain alcohol
$C_2H_4(OH)_2$	Ethylene $(H_2C{=}CH_2)$	Ethylene glyc*ol*	(Prestone and other trade names)
$C_3H_5(OH)_3$	Propane (C_3H_8)	Glycer*ol*	Glycerine

Naming Organic Compounds. The International Union of Pure and Applied Chemistry (IUPAC) has developed a system for naming organic compounds that simplifies the problem of naming the thousands of organic compounds. Table 29-1 contains 4 classes of hydrocarbons: the *alkanes* which have single carbon-to-carbon bonds, the *alkenes* which have one double bond carbon-to-carbon linkage per molecule, and the *alkynes* which have one triple bond carbon-to-carbon linkage per molecule. In

the benzene series are ring compounds of 6 carbon atoms in the
ring with 3 alternately spaced double bonds in the ring.

HYDROCARBON CHAIN COMPOUNDS.   Let us use octane as an
example. Octane is the eighth member of the alkane hydrocarbon
series. Its formula is C_8H_{18}. If this is written structurally we
have:

$$\begin{array}{c}
\text{H \quad H \quad H \quad H \quad H \quad H \quad H \quad H} \\
\text{|\quad |\quad |\quad |\quad |\quad |\quad |\quad |} \\
\text{H—C—C—C—C—C—C—C—C—H} \\
\text{|\quad |\quad |\quad |\quad |\quad |\quad |\quad |} \\
\text{H \quad H \quad H \quad H \quad H \quad H \quad H \quad H}
\end{array}$$

As a matter of convenience we often leave off the hydrogen atoms
and write octane as:

$$\begin{array}{c}
\text{|\quad |\quad |\quad |\quad |\quad |\quad |\quad |} \\
\text{—C—C—C—C—C—C—C—C—} \\
\text{|\quad |\quad |\quad |\quad |\quad |\quad |\quad |}
\end{array}$$

Now suppose one of the hydrogen atoms is removed from the
third carbon atom (counting from left to right) and in its place
we insert a methyl group (—CH_3). We now have:

$$\begin{array}{c}
\text{|\quad |\quad |\quad |\quad |\quad |\quad |\quad |} \\
\text{—C—C—C—C—C—C—C—C—} \\
\text{|\quad |\quad |\quad |\quad |\quad |\quad |\quad |} \\
\text{—C—} \\
\text{|}
\end{array}$$

Its name is *3-methyloctane*. What is the name of:

$$\begin{array}{c}
\text{C—C—C—C—C—C—C—C} \\
\text{|\qquad\quad |} \\
\text{—C—\quad —C—} \\
\text{|\qquad\quad |} \\
\text{—C—} \\
\text{|}
\end{array}$$

Since a methyl group (—CH_3) is attached to carbon atom 2 and
an ethyl group (—C_2H_5) is attached to carbon atom 4 the name is
2-methyl-4-ethyloctane.

Problem.   What is the name of:

$$\begin{array}{c}
\text{C—C—C—C—C—C} \\
\text{|\quad |} \\
\text{—C—C—} \\
\text{|\quad |} \\
\text{—C—} \\
\text{|}
\end{array}$$

Solution.   Use the longest continuous chain to obtain the basic name. In this case there are 6 carbon atoms in the chain so the basic name is hexane. Therefore the name is *2-methyl-3-ethylhexane*.

If the hydrocarbon has a double bond in the molecule, the name is changed to contain an *ene* in place of *ane* and the 2 before pentene indicates that carbon atom 2 contains the double bond.

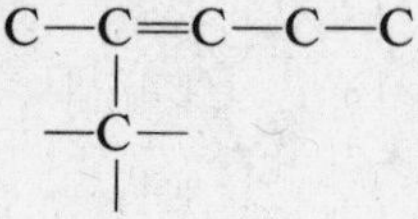

$$C-C=C-C-C$$
$$\quad | $$
$$-C-$$
$$\quad |$$

is *2-methyl-2-pentene*.

For a hydrocarbon with a triple bond such as

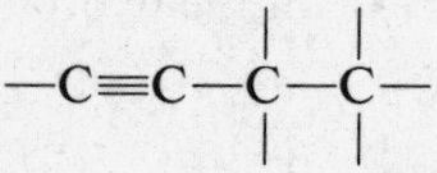

$$-C\equiv C-C-C-$$

the name is 1-butyne.

If a halogen atom, such as bromine, replaces a hydrogen atom in a hydrocarbon the *ine* ending of bromine is replaced by -o-. Thus:

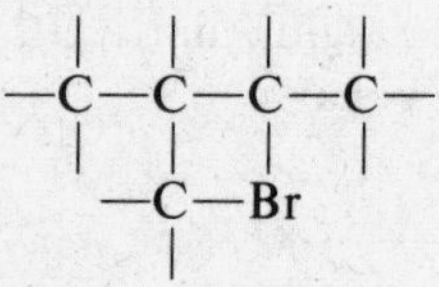

$$-C-C-C-C-$$
$$\quad -C-Br$$

is 2-methyl-3-brom*o*butane.

One can name an alcohol such as ethyl alcohol (C_2H_5OH) as a substituted hydrocarbon. Thus $-C-C-OH$ could be called monohydroxyethane. Or, ethylene glycol, $-C-C-$ could be
$$\quad O\;\;O$$
$$\quad H\;\;H$$
called dihydroxyethane.

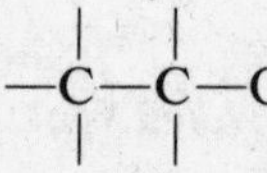

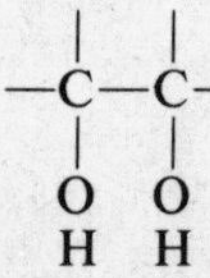

However, alcohols are usually named as primary, secondary, or tertiary alcohols or as polyhydroxy alcohols if more than one OH group is present in the molecule. (See discussion and illustrations above.)

Organic acids containing the carboxyl groups (—COOH) are often known by some common name such as *formic acid* (HCOOH) or *acetic acid* (CH_3COOH). However, they do have a formal name. Formic acid would be called *methanoic acid* and acetic acid would be called *ethanoic acid*.

The aromatic carboxylic acid, benzoic acid is:

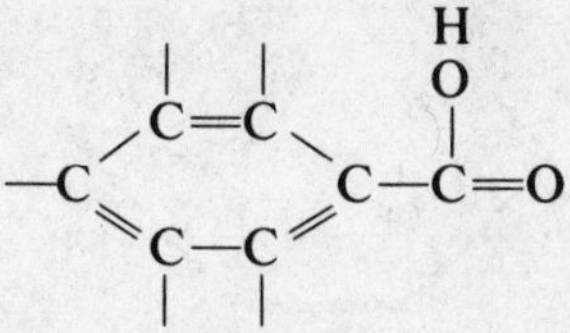

$$\begin{array}{c}
\text{H} \\
\text{O} \\
\end{array}$$

We can think of it as having been derived from benzene by replacing one of the hydrogen atoms by a carboxyl group, and so its name should be benzenoic acid. However the *ene* following the *z* is omitted and the name is *benzoic acid*.

Carbohydrates. Carbohydrates are compounds whose general formula is $C_x(H_2O)_y$. However, they are not hydrates. Many of the compounds in this group are *sugars* and have the suffix *ose* in their names. Thus, gluc*ose*, fruct*ose*, and sucr*ose* are common sugars. They can be thought of as the polyhydroxy derivatives of saturated aldehydes and ketones.

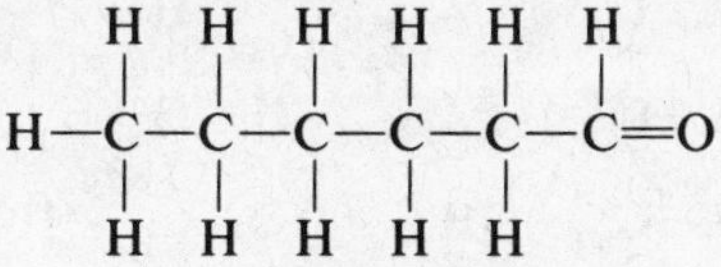

Aldehyde

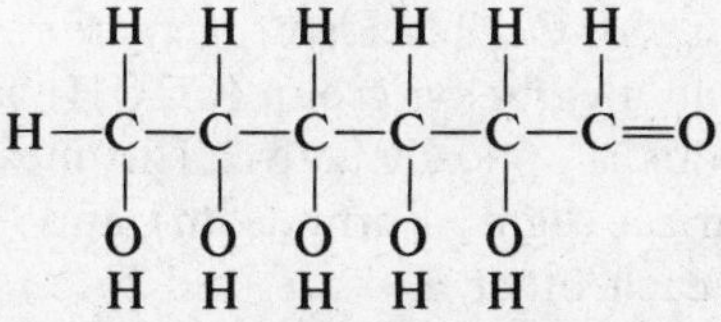

Glucose

Glucose and fructose are *monosaccharides* (six carbon atoms per molecule); sucrose and lactose are *disaccharides* (twelve carbon atoms per molecule). When sucrose is hydrolyzed in water (acid solution), it is changed into glucose and fructose (inversion of sugar).

$$C_{12}H_{22}O_{11} + H_2O \longrightarrow C_6H_{12}O_6 + C_6H_{12}O_6$$
      (Sucrose)   (Glucose)   (Fructose)

Although the formula for glucose given above is usually used, there is good experimental evidence to show that glucose has a

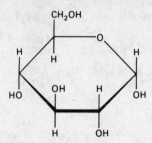

Fig. 29-3. A drawing to represent the spacial structure of D-glucose. Note that an oxygen atom is a member of the ring.

ring structure. Figure 29-3 is one recent drawing to show the ring structure of D-glucose. Note that an oxygen atom is a member of the ring.

Carbohydrates, like certain other compounds, contain *asymmetric carbon atoms*. This means there are four different groups attached to the same carbon atom. Thus, lactic acid is:

$$
\begin{array}{cc}
\text{H} & \text{H} \\
| & | \\
\text{H}-\text{C}-\text{H} & \text{H}-\text{C}-\text{H} \\
| & | \\
\text{H}-\text{C}_{(a)}-\text{OH} & \text{HO}-\text{C}_{(b)}-\text{H} \\
| & | \\
\text{O}{=}\text{COH} & \text{HO}-\text{C}{=}\text{O}
\end{array}
$$

The carbon atom $C_{(a)}$ or $C_{(b)}$ has a hydrogen atom, a (CH_3) group, an (OH) group, and a carboxyl group $(COOH)$ attached to it. A molecule such as this is optically active. This means it can rotate the plane of polarized light. Formula (*a*) and formula (*b*) are mirror images of each other and are said to be *optical isomers*. One rotates the plane of light to the left, the other to the right.

Proteins.   These important complex compounds are found in animal and vegetable matter. They are present in every living cell. Foods such as eggs, meat, peas, beans, and cheese are rich in proteins. *Albumin*, a protein, is found in eggs, blood, and milk. *Globulin*, another protein, is present in vegetables, and *gluten* or *glutelin* (protein material) is present in wheat flour.

Proteins are complex compounds containing carbon, hydrogen, oxygen, and nitrogen. Many contain sulfur, and a few contain phosphorus. Hemoglobin contains iron. In general, the proteins have high molecular weights. Hemoglobin, for example, is said to have a molecular weight of 68,000.

Proteins are long-chain molecules, probably made by the condensation of amino acids by splitting out water. For example, glycine (CH_2NH_2COOH) may condense as follows:

$$-N-C-C-\boxed{OH \quad H}-N-C-C-\boxed{OH \quad H}-N-C-C-OH \text{ etc.}$$

Sources of Organic Compounds. Many organic compounds have been made in the laboratory, but the greater portion of the supply comes from natural sources. Petroleum (crude oil) contains gasoline, kerosene, paraffin, naphtha, and other chemicals. These are sometimes called *petrochemicals*. By destructive distillation of soft coal, we obtain carbolic acid, creosote, coal tar, and coal gas. In the same manner, wood will yield wood alcohol, acetone, acetic acid, and charcoal. By fractional distillation of coal tar, we get benzene, toluene, naphthalene, and anthracene. From grains and fruits (by fermentation), we obtain ethyl alcohol, acetic acid, acetone, etc. By mechanical separation or by chemical means, we obtain starch, cellulose, and sugars. Fats and oils are obtained from animals, seeds, and nuts.

Preparation, Properties, and Uses of Organic Compounds. The most important organic compounds are listed and discussed below.

METHANE, CH_4. Millions of cubic feet of methane are obtained each year from gas wells. It is used as a fuel in place of coal or wood. On a small scale, methane can be made in the laboratory by heating sodium acetate and soda lime (NaOH + CaO).

$$CH_3COONa + NaOH \longrightarrow CH_4 + Na_2CO_3$$

ACETYLENE, C_2H_2. This compound is prepared by the action of water on calcium carbide.

$$CaC_2 + 2 H_2O \longrightarrow C_2H_2 + Ca(OH)_2$$

This colorless gas burns in oxygen and produces high temperatures. This explains its use in the acetylene torch. It is also used in organic synthesis.

BENZENE, C_6H_6. Benzene is prepared by the fractional distillation of coal tar. It is a colorless liquid of characteristic odor that boils at 80°C and freezes at 5°C. It reacts with nitric acid to form *nitrobenzene*. This compound is important because it is the starting substance in the manufacture of dyes.

$$C_6H_6 + HNO_3 \longrightarrow C_6H_5NO_2 + H_2O$$

CHLOROFORM ($CHCl_3$).　Chloroform is prepared by the action of iron powder on carbon tetrachloride.

$$8\,CCl_4 + 7\,Fe + 4\,H_2O \longrightarrow 8\,CHCl_3 + Fe_3O_4 + 4\,FeCl_2$$

This is a heavy, colorless liquid that boils at 61°C. It is used as a solvent and to a limited extent as an anesthetic.

METHANOL, CH_3OH (METHYL ALCOHOL, WOOD ALCOHOL). This compound is made by the destructive distillation of wood, and by a reaction between carbon monoxide and hydrogen in the presence of a catalyst.

$$2\,H_2 + CO \text{ (catalyst, ZnO)} \longrightarrow CH_3OH$$

Methanol is a colorless liquid with a characteristic odor, miscible with water in all proportions, very poisonous (small quantities cause blindness or death), and easily oxidized to formaldehyde. It reacts with acids to form esters.

$$CH_3OH + (O) \longrightarrow HCHO + H_2O$$
$$CH_3COOH + CH_3OH \longrightarrow CH_3COO \cdot CH_3 + H_2O$$
$$\text{(Ester)}$$

Methanol is used as a solvent, as a fuel in certain types of gas engines, in making formaldehyde, in denaturing ethyl alcohol, and as an antifreeze for automobile radiators. Large quantities are used in the manufacture of varnishes and shellacs. Millions of gallons of methanol are made in the United States each year.

ETHANOL, C_2H_5OH (ETHYL ALCOHOL, GRAIN ALCOHOL). Ethanol is prepared on a large scale by the fermentation of sugars, starch, or cellulose. This process is made possible by means of enzymes, which may be considered to be organic catalysts. Millions of gallons are made annually in the United States.

$$\text{Complex sugar} + \text{invertase (the catalyst)} \longrightarrow \text{simple sugar}$$
$$\text{Simple sugar} + \text{zymase (the catalyst)} \longrightarrow C_2H_5OH + CO_2$$

Ethanol is a colorless liquid with a characteristic odor and taste. It is miscible with water in all proportions, and boils and freezes at 78°C and −114°C, respectively.

Ethanol burns with a pale-blue flame. It reacts with sodium to give sodium ethoxide. It reacts with acids to form esters.

$$C_2H_5OH + 3 O_2 \longrightarrow 2 CO_2 + 3 H_2O$$
$$2 C_2H_5OH + 2 Na \longrightarrow 2 C_2H_5ONa + H_2$$

(Sodium ethoxide)

$$C_2H_5OH + CH_3COOH \longrightarrow CH_3COO \cdot C_2H_5 + H_2O$$

(Ethyl acetate)

Ethanol is used in organic synthesis, as a solvent, in beverages, as a fuel, in the preparation of ether and chloroform, in many patent medicines, and as an antifreeze in automobile radiators.

Absolute alcohol is prepared in the laboratory by removing the last trace of water from ordinary alcohol, which is about 95 per cent pure. This is accomplished by refluxing it (heating and collecting the condensed vapors in the same vessel) with lime, and then distilling in the usual way.

Denatured alcohol is ethyl alcohol that has been made unfit for use as a beverage by the addition of wood alcohol, benzene, carbolic acid, or other poisonous substances.

GLYCEROL, $C_3H_5(OH)_3$ (GLYCERINE). This is a trihydroxy alcohol obtained as a by-product in the manufacture of soap (see p. 314).

$$C_3H_5(C_{17}H_{35}COO)_3 + 3 NaOH \longrightarrow$$
(Fat)

$$3 C_{17}H_{35}COONa + C_3H_5(OH)_3$$
(Soap)   (Glycerine)

Glycerol is a thick, syrupy liquid which boils at 290°C with decomposition. It mixes with water in all proportions, is hygroscopic, and has a sweet taste.

Glycerol is used in making antifreeze mixtures, in the preparation of pharmaceuticals, and in the preparation of nitroglycerine and dynamite.

ACETONE, CH_3COCH_3. Acetone is the most important ketone. It is prepared by the following methods:

1. Heating calcium acetate.

$$(CH_3COO)_2Ca + heat \longrightarrow CH_3COCH_3 + CaCO_3$$

2. Fermentation of starch by certain bacteria.
3. Destructive distillation of wood.

Acetone is a colorless liquid which boils at 56°C, mixes with water in all proportions, and is an excellent solvent for acetylene, sulfur dioxide, and a wide variety of organic and inorganic compounds.

It is used as a solvent, in the preparation of chloroform, and in the manufacture of lacquers and explosives.

ACETIC ACID, CH_3COOH. Acetic acid is prepared by the fermentation of ethyl alcohol by a microorganism called *mycoderma aceti* ("mother of vinegar"), and by the destructive distillation of wood.

It is a colorless liquid with a sharp penetrating odor, freezing at 16°C and boiling at 117°C. It is miscible with water in all proportions, and has weak acid properties in water.

Acetic acid is used in making vinegar, in preparing white lead, in preparing esters, and in manufacturing dyes. Vinegar is made by allowing an alcohol solution to trickle over beechwood shavings which are inoculated with "mother of vinegar."

ETHYL ETHER, $(C_2H_5)_2O$. This is prepared by the action of sulfuric acid on ethyl alcohol. It is interesting to observe that sulfuric acid is used up and then reformed in this preparation.

$$C_2H_5OH + H_2SO_4 \longrightarrow C_2H_5HSO_4 + H_2O$$
$$C_2H_5HSO_4 + C_2H_5OH \longrightarrow C_2H_5-O-C_2H_5 + H_2SO_4$$

This compound is a light, colorless liquid with a characteristic odor that boils at 35°C and is flammable. It is a solvent and is also used as an anesthetic.

FATS AND OILS. These substances are the *glyceryl esters* of the fatty acids, such as stearic acid. The most common fats and oils are olein, palmitin, and stearin.

We can think of glyceryl stearate (stearin) as having been made by the reaction of 3 molecules of stearic acid with one molecule of glycerol.

$$3\,C_{17}H_{35}COOH + C_3H_5(OH)_3 \longrightarrow$$
(Stearic acid)   (Glycerol)

$$(C_{17}H_{35}COO)_3 \cdot C_3H_5 + 3\,H_2O$$
(Stearin)

Stearin and palmitin are white solids found in beef and mutton fat. Olive oil and cottonseed oil contain olein. These substances are used for food and in the manufacture of soap. Hydrogenated cottonseed oil is a substitute for lard. Margarine, a substitute for butter, is also prepared by the hydrogenation of suitable oils.

SOAPS. Soaps are the metallic salts of the fatty acids. They are prepared by boiling a fat, such as glyceryl stearate, with sodium hydroxide.

$$(C_{17}H_{35}COO)_3 \cdot C_3H_5 + 3\ NaOH \longrightarrow$$
$$3\ C_{17}H_{35}COONa + C_3H_5(OH)_3$$
$$\text{(Soap)}$$

Glycerine, $C_3H_5(OH)_3$, is always one product. It may be purified by distillation.

Cleansing Action of Soap. A grease film is removed by the addition of soap because the soap lowers the surface tension of the substance forming the film. This makes it easy for water to form an emulsion which can be washed away. Dirt particles are washed away because they are coated with a soap film.

Detergents. Soap substitutes and cleaning agents like sodium lauryl sulfate, $CH_3(CH_2)_{10}CH_2—O—SO_3Na$, are sold under trade names such as *Tide*, *Vel*, *Dreft*, *Dash*, and *All*. These will produce suds in hard water without first precipitating calcium and magnesium ions as ordinary soap does.

SUCROSE, $C_{12}H_{22}O_{11}$ (CANE SUGAR, BEET SUGAR). This *disaccharide* is the most important sugar. Millions of tons are produced annually. No other pure organic compound is produced on such a large scale. Sugar cane and sugar beets are the two most important sources of sucrose.

Raw sugar is obtained as follows: Sugar cane is cut into strips and soaked in water to remove the sugar. This solution is treated with lime to precipitate impurities (albuminoids), and then filtered. The filtrate is treated with sulfur dioxide to neutralize and bleach the syrup; and again, it is filtered. The solution is evaporated under reduced pressure until the raw sugar crystallizes out. Refined sugar is obtained from the raw sugar by the following method: The raw sugar is dissolved in water and then made alkaline with lime. The alkaline solution is heated to about 80°C and neutralized. The resulting solution is filtered through animal charcoal and then evaporated under reduced pressure until pure sugar crystallizes out. Sucrose is also prepared from the sugar beet.

Sucrose is a hard, white crystalline solid which melts at 160°C, is very soluble in water, changes to caramel at about 170°C, and is hydrolyzed by acids. This hydrolysis reaction is sometimes called *inversion of sugar*. Notice that one molecule of sucrose gives a molecule of glucose and a molecule of fructose.

$$C_{12}H_{22}O_{11} + H_2O + (acid) \longrightarrow C_6H_{12}O_6 + C_6H_{12}O_6$$
$$\text{(Glucose)} \quad \text{(Fructose)}$$

Sucrose is used as a food and also to preserve food (fruits).

STARCH, $(C_6H_{10}O_5)_n$. Starch is found in all plants except fungi. Grains and potatoes are rich in starch. It is prepared on a commercial scale from corn and potatoes. The material is softened by cold water and then ground fine. Upon standing, the starch separates out in practically pure form.

Starch is a white powder, insoluble in water and alcohol. It forms a colloidal solution, gives a blue color when treated with iodine (a test for starch), changes to dextrin at 204°C, and hydrolyzes in water to give sugar.

$$(C_6H_{10}O_5)_n + nH_2O \longrightarrow nC_6H_{12}O_6$$

Starch is used for food, for making glucose, and in laundering.

CELLULOSE, $(C_6H_{10}O_5)_x$. This substance is found in wood fiber, cotton, straw, hemp, and other similar substances. Cellulose is prepared by grinding a cellulose-containing material in water. That which does not dissolve is impure cellulose. The purity can be improved by grinding the material in an acid calcium sulfite solution. This dissolves out many gums and resins. Cellulose is a white solid, insoluble in ordinary reagents. It is resistant to most chemicals, but is acted upon by nitric acid to form cellulose nitrate (guncotton).

Cellulose is used for making paper and for the manufacture of artificial silk (rayon). Millions of pounds of rayon and related products are made in the United States each year.

NYLON. This is a synthetic fiber made by condensing adipic acid and hexamethylenediamine by the splitting out of water. Thus:

(Adipic acid)   (Hexamethylenediamine)

$$\underset{\text{(Portion of nylon molecule)}}{\text{HO}-\overset{\overset{\displaystyle O}{\|}}{\text{C}}-(\text{CH}_2)_4-\overset{\overset{\displaystyle O}{\|}}{\text{C}}-\overset{\overset{\displaystyle H}{|}}{\text{N}}-(\text{CH}_2)_6-\overset{\overset{\displaystyle H}{|}}{\text{N}}-\text{H}}$$

(Portion of nylon molecule)

This material is *thermoplastic*, so nylon thread is made by forcing the polymerized molten material (at 285°C) through tiny holes (spinnerets). Nylon is tough and elastic. It does not absorb water to the same degree as other fibrous material. It is a fairly good insulator and is not attacked by moths.

It is used for making hosiery, neckties, cord for rubber tires, tooth brushes, and many similar objects. Orlon, Dacron, and Teflon are other synthetic fibers that have important commercial uses.

PLASTICS. The term *plastic* refers to a substance which can be molded into any desired shape. More specifically, it refers to certain synthetic substances used in the manufacture of many common articles. Plastics may be grouped into two types: *thermosetting*, or those which flow when heated under pressure but "set" under continued heating, and *thermoplastic*, or those which flow under pressure when heated but do not "set" except at lower temperatures.

One of the spectacular triumphs of modern organic chemistry has been the synthesis of certain plastics by a process called *polymerization*. Here we are concerned with two types of polymers: (1) condensation polymers and (2) addition polymers. In the first type, bonding between monomers (small molecules) takes place by the splitting out of small molecules such as H_2O, HCl, NH_3, etc. Nylon is made by this process (p. 324). In the second type, bonding between monomers such as ethylene takes place with no splitting out of small molecules. In these polymerization reactions a suitable temperature, a relatively high pressure, and the presence of a catalyst such as an organic peroxide are necessary. Thus for ethylene (ethene) we have:

$$n\,[\text{H}_2\text{C}{=}\text{CH}_2] \xrightarrow[\text{ROOR (catalyst)}]{250°\text{C} + 2000 \text{ atm.}} \left[-\underset{\underset{\displaystyle H}{|}}{\overset{\overset{\displaystyle H}{|}}{\text{C}}} - \underset{\underset{\displaystyle H}{|}}{\overset{\overset{\displaystyle H}{|}}{\text{C}}} - \right]_n$$

Note. In this process the double bonds in the monomer are changed to single bonds in the polymer. Table 29-5, p. 326 contains examples of plastics made by the polymerization process.

Table 29-5

Polymerization Plastics

Monomer	Polymer	Uses
Ethylene $H_2C=CH_2$	Polyethylene $[-CH_2CH_2CH_2CH_2-]_n$	Electrical insulation, films, sheets, toys, plastic dishes
Propylene $H_2C=CH(CH_3)$	$[-CH_2CH_2CH(CH_3)-]_n$	Films, plastic bags, sheets
Vinyl Chloride $H_2C=CHCl$	Polyvinyl chloride $[-CH_2CHCH_2CH-]_n$ Cl Cl	Phonograph records, garden hose, floor coverings

Chloroprene
$CH_2=CHC=CH_2$
      Cl
	Polychloroprene
	$[-CH_2CH=CCH_2CH_2CH=CCH_2-]_n$
	      Cl Cl
		Synthetic rubber

Tetrafluorethylene $F_2C=CF_2$	Polytetrafluoroethylene (Teflon) $[-F_2C-CF_2-]_n$	Chemically inert items, coatings for cooking utensils

RUBBER. Natural rubber is an organic substance of high molecular weight obtained from the sap of the rubber tree. It consists of very long threadlike molecules arranged in random fashion and caused, probably, by the polymerization of isoprene (C_5H_8). Part of the rubber molecule is:

$$-CH_2-C=CH-CH_2\vdots CH_2-C=CH-CH_2-\text{etc.}$$
$$\qquad\quad CH_3 \qquad\qquad\qquad\quad CH_3$$

There are a number of substitutes for natural rubber. An interesting one is made from chloroprene (C_4H_5Cl). When this material polymerizes, a product is obtained which resembles natural rubber although many of its properties are quite different. For example, natural rubber deteriorates in contact with gasoline, oils, or grease. Synthetic rubber, on the other hand, resists the action of these substances. *Elastomers* is the general name for rubber substitutes.

Ordinary rubber is sticky, but its stickiness can be eliminated by *vulcanization*. When rubber and sulfur are heated together, the S_8 molecules open up and combine with the double bonds of the rubber molecule.

Vitamins. Vitamins are organic substances found in minute quantities in many of our foods. The chemical composition of several of them has now been accurately determined, and their function in maintaining health and vigor is a matter of common knowledge.

To discover the foods which contain vitamins and to determine the properties of the vitamins in those foods, experiments are conducted on rats, guinea pigs, and pigeons. These are divided into groups. Some of the groups are given diets containing vitamins while other groups are given vitamin-free foods. The effects produced are determined by keeping a record of the rate of growth and the state of health of the individuals in each group. See Table 29-6, pp. 328–329 for information concerning the vitamins.

The best-known vitamins are: A, B_1, B_2, B_6, B_{12}, niacin, C, D, E, and K. Several of the vitamins have rather complex structural formulas. Below are the structural formulas of 3 simpler ones.

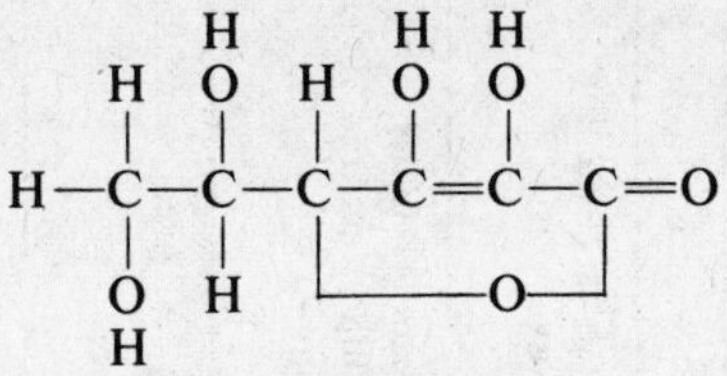

Vitamin C
(Ascorbic acid)

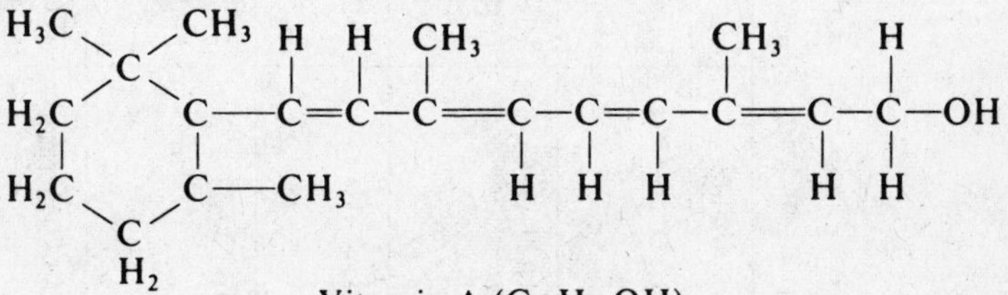

Vitamin A ($C_{20}H_{29}OH$)

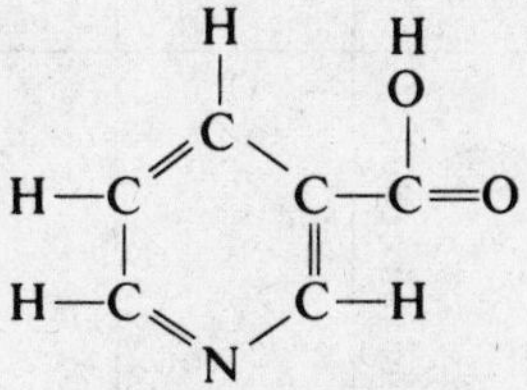

Nicotinic acid
(Niacin)

Hormones. The hormones, sometimes called chemical regulators, are chemical compounds secreted by the ductless glands of the body. They pass directly into the blood stream where they

Table 29-6

The Vitamins—Sources and Uses

Name, Properties of Vitamin	Formula	Source	Function in Body
A (anti-infective) Fat soluble Stable to cooking and drying	$C_{20}H_{29}OH$	Bread, vegetables, fruits, meat, milk, fish-liver oils.	Maintains healthy skin. Aids digestion. Prevents infection.
B_1 (Thiamin) Water soluble Destroyed by heat	$C_{12}H_{18}N_4OSCl$	Cereals, vegetables, fruits. (Thiamin is manufactured commercially.)	Aids in proper functioning of nervous system. Promotes growth. Helps resist infections.
B_2 (Riboflavin) (G) Water soluble Withstands mild heat	$C_{17}H_{20}N_4O_6$	Milk, eggs, yeast, green vegetables.	Aids in proper bodily development.
B_6 (Pyridoxine) Water soluble Withstands mild heat Unstable to light	$C_8H_{11}NO_3$	Yeast, egg yolk, cabbage, whole grains, liver.	Maintains normal health of the skin. May be important in protein metabolism.
B_{12} anti-anemia	$C_{63}H_{90}N_{14}O_{14}PCo$	Liver, muscle meat, egg yolk, seafood.	Helps control pernicious anemia.

Nicotinic Acid (Niacin) Water soluble Stable to heat	$C_6H_5NO_2$	Milk, dried figs, dates, green vegetables, rice, fish, lean meat.	Prevents pellagra.
C (Ascorbic Acid) Water soluble Destroyed by oxidation Withstands cooking in steam	$C_6H_8O_6$	Cereals, fruits, vegetables, raw milk. (Orange juice is rich in vitamin C.)	Prevents scurvy. Essential to normal bone and tooth development.
D_2, D_3 (sunshine vitamin) Fat soluble Stable to heat and oxidation	$D_2 : C_{28}H_{44}O$ $D_3 : C_{27}H_{44}O$	Fruits, vegetables, cereals, dairy products, meats; cod, halibut, and tuna fish-liver oils.	Necessary for bone and tooth development. Prevents rickets. Regulates the metabolism of calcium and phosphorus.
E (α Tocopherol) (fertility vitamin) Fat soluble Stable to cooking	$C_{29}H_{50}O_2$	Wheat germ oil, eggs, lettuce, milk, whole cereals, animal fats.	Necessary to normal reproduction in human beings and animals.
K (Menadione) Fat soluble Stable to heat Destroyed by light	$C_{31}H_{46}O_2$	Green vegetables, spinach.	Essential for clotting of blood.

play an important part in maintaining a proper balance in the functioning of the various organs of the body. There are three types: *amines* or *amino acids*, *peptides*, and *sterols*. There is very little known about the peptides. The following hormones are well known.

Thyroxine ($C_{15}H_{11}O_4NI_4$) is produced in the thyroid gland. It has also been prepared in the laboratory. A deficiency of this substance, often due to lack of iodine in the diet, causes goiter and, in extreme cases, the condition known as *cretinism*. It regulates the metabolic rate.

Adrenalin (epinephrine, $C_9H_{13}O_3N$) is secreted by the adrenal glands. Chemists have synthesized it in the laboratory. It regulates the blood pressure.

Insulin, a water-soluble protein, is the active principle found in the pancreas. Its absence causes the incomplete oxidation of sugar in the body. This causes the disease known as *diabetes*. Insulin, now injected subcutaneously, saves the lives of thousands of people suffering with this disorder.

Cortin is produced in the adrenal cortex and helps to regulate the water-salt balance of the body. *Cortisone* has been isolated from cortin by E. C. Kendall and his associates. It has been found to be especially effective in treating rheumatoid arthritis. ACTH (adrenocorticotropic hormone) is also an effective remedy for rheumatoid arthritis.

The sex hormones regulate sexual development. Modern research has enabled chemists and others to isolate certain hormones which play an important role in the development of the individual. The chief male hormones are androsterone ($C_{19}H_{30}O_2$) and testosterone ($C_{19}H_{28}O_2$). The female hormones are theelin (oestrone) ($C_{18}H_{22}O_2$), progesterone ($C_{21}H_{30}O_2$), and pregnandiol ($C_{21}H_{35}O_2$).

Drugs. A drug is a chemical substance (or substances) used to modify or correct a normal or abnormal body function of a human being or an animal. This is a simplified and, perhaps, an inadequate definition, but it is satisfactory for our purpose here.

ANTIBIOTICS. These are substances which kill or inhibit the activity of microorganisms. They are usually prepared from complexes produced by soil bacteria, fungi, or molds.

Penicillin. This substance is prepared from the mold called *penicillium chrysogenum*. Although nontoxic, it has remarkable curative powers in cases of pneumonia, infections, severe burns,

syphilis, meningitis, and many other disorders. Penicillin has been
synthesized in the laboratory. The formula is:

$$
\begin{array}{c}
\overset{\displaystyle CH_3 \quad H}{\underset{|}{}} \\
H_3C-C-C-C-OH \\
\end{array}
$$

At least four different modifications of this substance are
known. The R in the formula above is different for each form.
For example, in penicillin F the R is:

$$CH_2-CH=CH-CH_2-CH_3$$

Streptomycin.   This is prepared from *streptomyces griseus.* It is
effective against blood poisoning, typhoid fever, and pulmonary
tuberculosis.

Tyrothricin.   This antibiotic is effective in the treatment of
external ulcers and of nose and throat infections.

Chloromycetin.   This has been used for typhus and typhoid
fever.

Aureomycin.   This gold-colored crystalline substance was prepared from a mold called *streptomyces aureofaciens.* It has a low
toxicity for man and has proved very effective against certain
viruses. Typhus fever, virus pneumonia, atypical pneumonia, and
other bodily infections have yielded to its curative properties.

THE SULFA DRUGS.   These substances are synthesized in the
laboratory. They are useful in the treatment of pneumonia, gan-
grene, throat infections, and blood poisoning. The best-known
sulfa drugs are: sulfanilamide, sulfaguanidine, sulfadiazine, and
sulfathiazole. Sulfanilamide, the first one to be prepared, is synthesized as indicated by the unbalanced equations given below:

$$C_6H_5NH_2 + H_2SO_4 \longrightarrow C_6H_5NH_2SO_2(OH)$$
$$\text{(Aniline)} \qquad\qquad \text{(Sulfanilic acid)}$$
$$C_6H_5NH_2SO_2(OH) + NH_3 \longrightarrow C_6H_5NH_2(SO_2)NH_2$$
$$\text{(Sulfanilamide)}$$

ANESTHETICS. These are chemical substances which make an individual insensitive to pain. There are two types: general and local. General anesthetics include ether, chloroform, and nitrous oxide. Recently cyclopropane:

$$H_2C \underline{\hspace{2cm}} CH_2$$
$$CH_2$$

has come into favor because it leaves the patient with no after-sickness. Local anesthetics include cocaine, novocain, and nupercaine.

ANTISEPTICS AND GERMICIDES. The common ones are: sodium hypochlorite, hydrogen peroxide, boric acid, carbolic acid (phenol), iodine, and mercuric chloride. More recent antiseptics include merthiolate.

ALKALOIDS. These are for the most part poisonous substances which are used in small doses for a specific purpose. Some of them are habit-forming. Quinine is used in treating malaria. Caffeine is a stimulant. Atropine relaxes muscles. The molecule of an alkaloid contains at least one nitrogen atom. Some contain two nitrogen atoms. Nicotine ($C_{10}H_{14}N_2$) contains two nitrogen atoms and has the structural formula:

Nicotine

REVIEW QUESTIONS

1. What important experiment proved that organic compounds can be made in the laboratory?
2. What are the four important classes of hydrocarbons?
3. What is the general formula of a hydrocarbon belonging to the alkyne series?
4. Distinguish between an aromatic and an aliphatic hydrocarbon.
5. Distinguish between single, double, and triple bonds in organic molecules.
6. What happens when ammonium cyanate changes into urea?

7. Draw the structural formula of naphthalene. Draw the structural formula of a secondary alcohol.

8. What is meant by the term asymmetric carbon atom?

9. When glycine molecules condense to form a complex molecule (called a *protein*), water molecules split out. Show what happens.

10. How can one prepare (*a*) methane (CH_4)? (b) chloroform ($CHCl_3$)? (*c*) ethanol (C_2H_5OH)?

11. How is sucrose ($C_{12}H_{22}O_{11}$) prepared from sugar cane?

12. What are four methods of representing the structural features of the methane molecule?

13. What is the difference structurally between dimethyl ether and ethanol?

14. What is meant by ortho, meta, and para isomerism?

15. Write the formula of a dihydroxy alcohol.

16. Given $C_3H_5(OH)_3$. What is it, an alcohol or trihydroxy propane?

17. What are vitamins? Give the chemical name and formula of vitamin C.

18. What is the chemical name and formula of niacin?

19. Name three hormones. Give the empirical formula of adrenalin. What important function does it have in the body?

20. What is a drug? Name two important ones. Name a sulfa drug.

21. Cyclopropane is an anesthetic. What does that statement mean?

Chapter 30
Chemical Calculations

Certain calculations are not found in this chapter. For these look in the various chapters under the subject matter being considered. Thus:

Many college chemistry textbooks (and many chemistry teachers) do not emphasize the use of dimensions in solving mathematical problems. However, engineering subjects are usually taught with emphasis on solving problems dimensionally, as well as mathematically. Consequently, in this chapter, where practicable, we shall follow the same procedure. The following rules should be observed:

1. Quantities expressed in similar dimensions may be added or subtracted.

2. In equality expressions, both sides of the equation must be equal dimensionally, as well as numerically.

3. Unlike dimensions cannot be canceled or otherwise simplified.

A nonchemical problem will serve as an example to help us understand what is meant by solving a problem both mathematically and dimensionally.

Problem. A rectangular plot of land is 330 feet by 165 feet. (a) What is its area in square feet (ft)^2?

Solution of (a):

$$330 \text{ ft} \times 165 \text{ ft} = 54,450 \text{ (ft)}^2$$

(b) What is the area of this plot of land in square yards (yd)^2?
  Given: 1 yard = 3 feet.
  Solution of (b):

$$\left(\overset{110}{\cancel{330\,\text{ft}}} \times \frac{1\,\text{yd}}{\cancel{3\,\text{ft}}}\right)\left(\overset{55}{\cancel{165\,\text{ft}}} \times \frac{1\,\text{yd}}{\cancel{3\,\text{ft}}}\right) = 6050\,(\text{yd})^2$$

Note. In (b) the feet were canceled; in (a) they were squared: ft × ft = (ft)^2.

Temperature Measurements. The temperature of a body is measured by one of three scales, Centigrade (Celsius), Fahrenheit, or Absolute (Kelvin). The freezing point and boiling point of water on the three scales are:

Scale	Freezing Point	Boiling Point
Centigrade (Celsius)	0°C	100°C
Fahrenheit	32°F	212°F
Absolute (Kelvin)	273°A (K)	373°A (K)

In recent years scientists use the word *Kelvin (K)* to denote absolute degrees. This is a way of honoring *Lord Kelvin*, who did so much research in the area of low temperatures. Likewise the

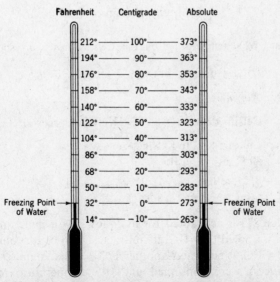

Fig. 30-1. The three temperature scales compared.

Centigrade scale is now called the *Celsius scale* to honor *Anders Celsius*, a Swedish astronomer who introduced a temperature scale where the freezing point of water is zero degrees (0°C) and the boiling point of water is 100°C. Keep in mind that °C still refers to Centigrade as well as Celsius degrees.

Temperature Conversions. The formulas for converting temperatures from one scale to another are as follows:

Celsius (Centigrade) to Fahrenheit.

$$\text{Fahrenheit degrees} = \frac{9}{5} \text{ Celsius degrees} + 32°$$

Problem. What is the Fahrenheit temperature that corresponds to 20°C?

Solution.

$$°F = \left(20°C \times \frac{9}{5}\right) + 32°$$

$$= \frac{180°C}{5} + 32°$$

$$= 36° + 32°$$

$$= 68°$$

Fahrenheit to Celsius

$$°C = \frac{5}{9} (\text{Fahrenheit degrees} - 32°)$$

Problem. What temperature on the Celsius scale corresponds to 80°F?

Solution. $°C = 5/9 (80°C - 32°C) = 5/9 (48°) = 26.7°C$

Celsius to Kelvin

$$\text{Celsius degrees} + 273 = \text{Kelvin degrees}$$

Problem. Convert 40°C to Kelvin degrees.
Solution. $40°C + 273 = 313°K$

Problem. Convert 265°K to Celsius degrees.
Solution. $265°K - 273° = -8°C$

The value 273 is accounted for as follows: If a given mass of gas at 0°C is cooled to −1°C, it contracts 1/273 of its volume. If cooled to −2°C, it contracts another 1/273 of its volume at 0°C. If this process were continued until the temperature reached −273°C, the volume of the gas would be zero. This low temperature is made the zero point on the absolute scale. However, we know that any gas changes to a liquid before −273°C is reached. See p. 73.

REVIEW QUESTIONS

1. Make a table showing the relationship between the freezing points and the boiling points of mercury on the three temperature scales. On the Celsius scale, mercury melts at −38.9°C and boils at 357°C.
2. Sulfur melts at 115°C. What is its melting point on the Fahrenheit scale? Show calculations. *Ans.* 239°F
3. In winter in some places in Siberia, the temperature is as low as −60°F. What is the temperature on the Celsius scale. *Ans.* −51.11°C
4. Metallic copper melts at 1083°C. What is the value on the Fahrenheit scale? *Ans.* 1981.4°F
5. The temperature of a certain room is 75°F. What is the temperature on the Celsius scale? The Kelvin scale? *Ans.* 23.9°C; 296.9 K

Density. The density of a substance is the number of units of mass of the substance in a unit volume.

Example 1. One liter of air at 0°C and 760 mm pressure weighs 1.293 g. Its density, therefore, is 1.293 g per liter.

Example 2. One cubic centimeter of aluminum weighs 2.699 g. Its density is, therefore, 2.699 g per cubic centimeter.

$$\text{Density} = \frac{\text{Mass (weight)}}{\text{Unit volume}} = \frac{2.699 \text{ g}}{1 \text{ cc}}$$

Problem. What is the density of a body that weighs 40 g and has a volume of 12 cubic centimeters?

$$\text{Density} = \frac{40 \text{ g}}{12 \text{ cc}} = 3.333 \text{ g/1 cc}$$

Specific Gravity. At known temperature the specific gravity is the mass of a known volume of a substance divided by the mass of an equal volume of water.

$$\text{Sp. gr.} = \frac{\text{Mass of known volume}}{\text{Mass of equal volume of water}}$$

Problem 1. Find the specific gravity of concentrated sulfuric acid at 0°C.

Given: at 0°C 1 cc of the acid = 1.84 g
at 0°C 1 cc of water = 1.0 g

Solution.

$$\text{Sp. gr.} = \frac{1.84\,g}{1.0\,g} = 1.84$$

Problem 2.   The specific gravity of a certain salt solution is 1.21. What is the weight of 125 cc?

Solution.

$$\frac{1.21\text{ g}}{1\,cc} \times 125\,cc = 151.25\text{ g}$$

Problem 3.   An 8% solution of table salt has a specific gravity of 1.11. If 120 ml of this solution were evaporated to dryness, what weight of salt would remain? Note that for all practical purposes 1 ml = 1 cc and 8% is equivalent to 0.08.

Solution.

$$\frac{1.11\text{ g}}{1\,ml} \times 120\,ml = 133.2\text{ g (total weight of solution)}$$

$$133.2\text{ g} \times 0.08 = 10.656\text{ g salt remaining}$$

REVIEW QUESTIONS

1. Describe a method of determining the specific gravity of a liquid.
2. Five ml of a liquid weighs 68 g. What is the density? *Ans.* 13.6 g per ml
3. At standard conditions, 1 ml of hydrogen weighs 0.000089 g. Calculate the absolute density of the gas. *Ans.* 0.089 g per liter
4. A bottle weighs 31 g. Filled with water it weighs 42 g. Filled with oil it weighs 40.8 g. Calculate the specific gravity of the oil. *Ans.* 0.89
5. The specific gravity of a 9 per cent salt solution is 1.14. How many grams of salt can be recovered from 115 ml of the solution? *Ans.* 11.799 g
6. In giving the density of a gas, why must the temperature and pressure be specified?
7. A certain volume of sulfuric acid (sp. gr. 1.84) weighs 36.8 g. (*a*) How many ml of water will have the same weight? (Assume 1 ml of water = 1 g.) (*b*) How many ml of the sulfuric acid weights 36.8 g? *Ans.* (*a*) 36.8 ml; (*b*) 20 ml

Formula Weights.   The formula weight of a compound is the sum of the weights of the atoms in one "molecule" of the substance. The simplest formula is usually used. Thus, the formula of calcium sulfate can be written $CaSO_4$. It can also be written $Ca_2S_2O_8$. The percentages of calcium, sulfur and oxygen are the same regardless of which formula is used. But $CaSO_4$ is simpler and, therefore, the one used. As a matter of fact calcium sulfate is a *macromolecule* and therefore we do not know its molecular weight. For most calculations making use of atomic weights, it is customary to round off the atomic weight values. Thus we use 23 for the atomic weight of sodium rather than 22.99. Also we use 56 for the atomic weight of iron, rather than 55.85, etc.

Problem.   Find the formula weight of calcium sulfate ($CaSO_4$).
Solution.

$$Ca + S + (4 \times O) = CaSO_4$$
$$40 + 32 + (4 \times 16) = 136$$

Problem.   Find the formula weight of sodium phosphate (Na_3PO_4).
Solution.

$$(3 \times Na) + P + (4 \times O) = Na_3PO_4$$
$$(3 \times 23) + 31 + (4 \times 16) = 164$$

Molecular Weights of Gases.   The following methods are useful in determining the molecular weights of gases.

VAPOR DENSITY METHOD.   The molecular weight of a gas is equal to twice its vapor density. This is apparent from the following:

$$\text{Vapor density} = \frac{\text{weight of 22.4 liters of the gas}}{\text{weight of 22.4 liters of hydrogen}}$$

Since 22.4 liters of hydrogen = 2 grams:

$$\text{Vapor density} = \frac{\text{weight of 22.4 liters}}{2}$$

or: $2 \times$ vapor density = molecular weight

Problem.   The vapor density of oxygen is 16. What is its molecular weight?
Solution. $2 \times 16 = 32$ (molecular weight)

METHOD BASED UPON AVOGADRO'S LAW.   The molecular weight (actually, one mole) of a gas is equal to 22.4 times the weight of one liter.

Problem.   What is the molecular weight of oxygen?
Solution.   One liter of oxygen weighs 1.429 g. Therefore:

$$\frac{1.429 \text{ g}}{\text{liter}} \times \frac{22.4 \text{ l}}{\text{mole}} = \frac{32 \text{ g}}{\text{mole}}$$

Molecular Weights of Nonvolatile Solutes (Nonelectrolytes). The molecular weight in grams (a mole weight) of a nonvolatile solute may be determined by measuring its effect upon the boiling point, freezing point, or vapor pressure of a solvent. This experimental value (omitting the grams term) is the molecular weight expressed in atomic weight units. Thus, if a mole of substance A is 46 g its molecular weight is 46. This means that a molecule of A weighs 46 AWU, if one atom of carbon-12 weighs 12 AWU.

BOILING POINT METHOD. One mole of a nonelectrolyte dissolved in 1000 g of water gives a solution that boils at 100.52°C. The boiling point constant, then, is 0.52°C.

Problem. 171 g of sugar in 1000 g of water boils at 100.26°C. What is the weight of a mole of sugar?

Solution. Let x equal the number of grams of sugar (1 mole) dissolved in 1000 g of water to give a solution that boils at 0.52°C above the boiling point of pure water (at 760 mm pressure).

$$\frac{x\,g}{0.52°C} = \frac{171\,g}{0.26°C} \quad \text{or} \quad x\,g = \frac{171\,g \times 0.52°C}{0.26°C} = 342\,g$$

Problem. Six grams of urea in 1000 g of water gives a solution that boils at 100.052°C. What is the weight of a mole of urea?

Solution.

$$\frac{x\,g}{0.52°C} = \frac{6\,g}{0.052°C} \quad \text{or} \quad x\,g = \frac{6\,g \times 0.52°C}{0.052°C} = 60\,g$$

FREEZING POINT METHOD. One gram-molecular weight (one mole) of a nonelectrolyte dissolved in 1000 g of water gives a solution that freezes at $-1.86°C$. Therefore, the freezing point constant is -1.86.

Problem. A solution of 30 g of urea in 1000 g of water freezes at $-0.93°C$. What is the gram-molecular weight of urea?

Solution.

$$\frac{x\,g}{-1.86°C} = \frac{30\,g}{-0.93°C} \quad \text{or} \quad x\,g = \frac{30\,g \times -1.86°C}{-0.93°C} = 60\,g$$

Problem. A solution of 17.1 g of sugar dissolved in 100 g of water freezes at $-0.93°C$. What is the weight of a mole of this sugar?

Solution. 17.1 g in 100 g is equivalent to 171 g in 1000 g.

$$\frac{x\,g}{-1.86°C} = \frac{171\,g}{-0.93°C} \quad \text{or} \quad x\,g = \frac{171\,g \times -1.86°C}{-0.93°C} = 342\,g$$

REVIEW QUESTIONS

1. Determine the molecular (or formula) weights of the following compounds from their formulas: $AgNO_3$, $BiCl_3$, $NaClO$, $C_{12}H_{22}O_{11}$.

2. The vapor density of nitrogen is 14; what is its molecular weight? *Ans.* 28

3. Five liters of nitrogen at standard conditions weighs 6.2535 g. What is its vapor density? What is its molecular weight? *Ans.* 14; 28

4. One liter of carbon dioxide weighs 1.96 g. What is its molecular weight? *Ans.* 44.

5. State the law governing the lowering of the freezing point by non-electrolytes.

6. State the law governing the raising of the boiling point by non-electrolytes.

7. Twenty-three grams of glycerine in 1000 g of water gives a solution that boils at 100.13°C. What is the molecular weight of glycerine? *Ans.* 92

8. Five grams of nonelectrolyte in 500 g of water freezes at −0.2°C. What is its molecular weight? *Ans.* 93

Percentage Composition. The percentage composition of a substance can be obtained as follows: (1) Find the molecular or formula weight of the substance. (2) Divide the atomic weight of each element (or its multiple) by the molecular weight. (3) Multiply each quotient by 100 to give the percentage.

Problem. Find the percentage composition of Na_2HPO_4.
Solution.

$$2\,Na + H + P + 4\,O = Na_2HPO_4$$
$$(2 \times 23) + 1 + 31 + (4 \times 16) = 142 \text{ (molecular or formula weight)}$$

$$\text{Per cent Na} = \frac{2 \times 23}{142} \times 100 = 32.4 \text{ per cent}$$

$$\text{Per cent H} = \frac{1}{142} \times 100 = 0.70 \text{ per cent}$$

$$\text{Per cent P} = \frac{31}{142} \times 100 = 21.83 \text{ per cent}$$

$$\text{Per cent O} = \frac{4 \times 16}{142} \times 100 = 45.07 \text{ per cent}$$

$$32.4\% + 0.70\% + 21.83\% + 45.07\% = 100\%$$

Derivation of Formula. The formula of a substance can be determined from its percentage composition by the following method: (1) Divide the percentage composition values by the atomic weights of the elements involved. (2) Determine the ratio between these quotients to obtain the relative number of atoms of each kind in the compound.

Problem.   What is the formula of a compound that contains 32.38 per cent sodium, 22.57 per cent sulfur, and 45.05 per cent oxygen? In the solution of this problem we shall round off atomic weights and report ratios to one decimal place. There should be no doubt what the correct formula is.

Solution.

$$\text{Sodium} = \frac{32.38}{23} = 1.408$$

$$\text{Sulfur} = \frac{22.57}{32} = 0.706$$

$$\text{Oxygen} = \frac{45.05}{16} = 2.816$$

Rounding off to one decimal place we have:

$$Na_{1.4}S_{0.7}O_{2.8}$$

Dividing each by 0.7:

$$Na_2S_1O_4$$

Therefore the formula is: Na_2SO_4.

Problem.   What is the formula of a compound that contains 43.64 per cent oxygen and 56.36 per cent phosphorus?

Solution.

$$\text{Phosphorus} = \frac{56.36}{31} = 1.82$$

$$\text{Oxygen} = \frac{43.64}{16} = 2.73$$

$$P_{1.82}O_{2.73} \quad \text{or} \quad P_1O_{1.5}$$

It appears from this ratio that 1 atom of phosphorus combines with 1.5 atoms of oxygen. Since atoms cannot be divided, it must be that 2 atoms of phosphorus combine with 3 atoms of oxygen. The formula is, then, P_2O_3.

REVIEW QUESTIONS

1. Find the percentage composition of the following compounds from their formulas: (*a*) C_2H_5OH, (*b*) C_6H_6, (*c*) $C_2H_4O_2$, (*d*) $C_6H_6O_2$, (*e*) $(NH_4)_2SO_4$. *Answer to* (*a*): C = 52.2%; H = 13%; O = 34.8%

2. A chemical compound contains calcium, carbon, and oxygen. Analysis shows the percentage composition to be: calcium, 40 per cent; carbon, 12 per cent; and oxygen, 48 per cent. Find the simplest formula. *Ans.* $CaCO_3$

3. Derive the simplest formula of the substance which gave, on analysis: hydrogen, 5.88 per cent and oxygen, 94.12 per cent. *Ans.* H_2O_2

4. The molecular weight of a substance is 90. It contains 2.22 per cent hydrogen, 26.67 per cent carbon, and 71.11 per cent oxygen. Find the simplest formula. *Ans.* $H_2C_2O_4$

5. Find the simplest formula of a substance that contains 16.08 per cent potassium, 40.15 per cent platinum, and 43.77 per cent chlorine. *Ans.* K_2PtCl_6

6. Find the simplest formula of an oxide of uranium that contains 83.3 per cent uranium and 16.7 per cent oxygen. *Ans.* UO_3

Weight-and-Weight Problems. There is a quantitative relationship between the weights of substances entering a reaction and the weights of the products formed. The following problems are based on this relationship. For many of the calculations, approximate atomic weights are used. Thus, 35.5, rather than 35.453 is used for the atomic weight of chlorine. Calculations made with the slide rule are satisfactory.

The method is outlined as follows: (1) Write the balanced equation for the reaction taking place. (2) Underline the formulas of the substances with which the problem is concerned. (3) Write the equation-weights (molecular weight times the number of molecules required to balance the equation) under those substances underlined. (4) Write the value of the weight of the known substance underneath its equation-weight. (5) Place an x under the substance whose weight is to be found. (6) Form an equation and solve for x.

Problem. How many grams of hydrogen are liberated when 50 g of sodium reacts with water? Let x = the grams of hydrogen liberated.
Solution.

$$\underset{46\,g}{2\,Na} + 2\,H_2O \longrightarrow 2\,NaOH + \underset{2\,g}{H_2}$$

$$50\,g \qquad\qquad\qquad\qquad\qquad x$$

$$\frac{46\,g}{2\,g} = \frac{50\,g}{x} \qquad x = 2.17\ \text{g hydrogen}$$

Problem. How many grams of calcium carbonate are necessary to make 100 g of calcium chloride? Let x = the grams of $CaCO_3$ required.
Solution.

$$\frac{CaCO_3}{100} + 2\,HCl \longrightarrow \frac{CaCl_2}{111} + H_2O + CO_2$$

$$x \qquad\qquad\qquad 100\,g$$

$$\frac{100\,g}{x} = \frac{111\,g}{100\,g} \qquad x = 90.09\,g$$

Weight-and-Volume Problems. According to Avogadro's hypothesis, one mole of a gas occupies 22.4 liters at standard conditions. This relationship enables one to convert weight to volume, or volume to weight. Two methods by which the calculation can be made follow.

Method 1. (1) Find the weight of the gas formed by the weight-and-weight method. (2) Convert this weight into volume as follows:

$$\frac{\text{Molecular weight (g)}}{22.4\ \text{liters}} = \frac{\text{weight of gas formed (g)}}{x\ \text{liters}}$$

Problem. How many liters of hydrogen chloride can be obtained from 100 g of sodium chloride by the action of concentrated sulfuric acid?
Solution.

$$\frac{NaCl}{58.5\,g} + H_2SO_4 \longrightarrow NaHSO_4 + \frac{HCl}{36.5\,g}$$

$$100\,g \qquad\qquad\qquad\qquad x$$

$$\frac{58.5\,g}{100\,g} = \frac{36.5\,g}{x\,g} \qquad x = 62.4\,g$$

Converting 62.4 g into liters:

$$\frac{36.5\,g}{22.4\,\ell} = \frac{62.4\,g}{x} \qquad x = 38.3\ \text{liters}$$

Method 2. Substitute 22.4 liters for each mole of gas formed.

Problem. How many liters of hydrogen may be obtained by the action of 48 g of magnesium on dilute hydrochloric acid?
Solution.

$$\frac{Mg}{24\,g} + 2\,HCl \longrightarrow MgCl_2 + \frac{H_2}{22.4\,\ell}$$

$$48\,g \qquad\qquad\qquad\qquad x$$

$$\frac{24 \text{ g}}{48 \text{ g}} = \frac{22.4 \, l}{x} \qquad x = 44.8 \text{ liters}$$

Volume-and-Volume Problems. These problems are based on Gay-Lussac's law and on Avogadro's law. Gay-Lussac's law states that the ratio of the volumes of combining gases and the volume of their products may be expressed by small whole numbers. Avogadro's law states that equal numbers of molecules of gases under the same conditions of temperature and pressure occupy equal volumes. A method based on these laws enables one to write the volume relationships from equations representing the reaction between gaseous substances and their products.

The method is as follows. (1) Write the balanced equation. (2) Form a proportion between the number of molecules reacting and their corresponding volumes. (3) Solve for x in the usual way.

Problem. How many liters of water vapor are formed when 20 liters of hydrogen react with an excess of oxygen?
Solution.

$$2 \text{ H}_2 + \text{O}_2 \longrightarrow 2 \text{ H}_2\text{O}$$

2 molecules $\longrightarrow$ 2 molecules

2 volumes $\longrightarrow$ 2 volumes

20 liters $\longrightarrow$ x liters

$$\frac{2 \text{ vols.}}{20 \text{ liters}} = \frac{2 \text{ vols.}}{x \text{ liters}} \qquad x = 20 \text{ liters}$$

Problem. Thirty liters of hydrogen are made to react with an excess of nitrogen. How many liters of ammonia are formed?
Solution.

$$3 \text{ H}_2 + \text{N}_2 \longrightarrow 2 \text{ NH}_3$$

3 volumes $\longrightarrow$ 2 volumes

30 liters $\longrightarrow$ x liters

$$\frac{3 \text{ vols.}}{30 \text{ liters}} = \frac{2 \text{ vols.}}{x} \qquad x = 20 \text{ liters}$$

Note. In these problems it is assumed that the temperature and pressure remain constant.

REVIEW QUESTIONS

1. How many grams of oxygen can be obtained from 245 g of potassium chlorate? *Ans.* 96 g

2. How many grams of sulfur are necessary to combine with 100 g of iron powder? *Ans.* 57.3 g

3. How many grams of ammonia can be prepared from 200 g of ammonium sulfate? *Ans.* 51.5 g

4. How many grams of sodium hydroxide are necessary to neutralize 40 g of hydrogen chloride? *Ans.* 43.8 g

5. How many liters of oxygen are liberated when 50 g of potassium nitrate is heated? *Ans.* 5.55 liters

6. How many liters of hydrogen chloride are liberated when 200 g of table salt is treated with concentrated sulfuric acid? *Ans.* 76.6 liters

7. Fifteen liters of hydrogen chloride (at standard conditions) are passed into a sodium hydroxide solution. How many grams of table salt will be formed? *Ans.* 39.2

8. How many liters of chlorine can be prepared from the decomposition of 10 liters of hydrogen chloride? *Ans.* 5 liters

9. A mixture of 5 liters of hydrogen and 20 liters of oxygen is made to combine by passing a spark through it.
   (*a*) How many liters of water vapor are produced?
   (*b*) What is the volume of the uncombined gas?
   Ans. (*a*) 5 liters water vapor; (*b*) 17.5 liters oxygen

Hydrogen Equivalent.   The hydrogen equivalent of a metal is the number of grams of a metal which will replace or combine with 1 g of hydrogen (or its equivalent).

A specific example will be considered to outline the method. (1) Add a small, accurately weighed piece of magnesium to some dilute hydrochloric acid. (2) Collect the hydrogen evolved (over water) in a tube that enables one to read the volume accurately (eudiometer). (3) Place the tube (now partially filled with gas) into a large jar or cylinder of water. (4) Let the gas come to the same temperature as the water over which it is confined. Record this temperature. (5) Adjust the tube so that the level of the water inside the tube is the same as the water level outside. Now read the volume of gas. (6) Tabulate the data.

Problem.   Thirty-five thousandths of a gram (0.035 g.) of magnesium was inserted into a eudiometer where it reacted with the dilute hydrochloric acid present. After the reaction was completed, the following data were recorded.

> Weight of magnesium = 0.035 g
> Temperature = 20°C
> Volume of hydrogen = 35.5 ml
> Barometer reading = 750 mm

Vapor pressure of water at 20°C = 17.54 mm

Pressure of hydrogen = 750 − 17.54 = 732.46 mm

Find the hydrogen equivalent of magnesium.

Solution. To find the volume occupied by the dry hydrogen at standard conditions:

$$\text{Volume of hydrogen} = 35.5 \text{ ml} \times \frac{273° \cancel{A}}{293° \cancel{A}} \times \frac{(750 - 17.54 \cancel{\text{mm}})}{760 \cancel{\text{mm}}}$$

$$= 31.85 \text{ ml}$$

To find the weight of this hydrogen:

$$31.85 \cancel{\text{ml}} \times \frac{0.0000899 \text{ g} \,[= \text{weight of 1 ml hydrogen}]}{\cancel{\text{ml}}} = 0.002863 \text{ g}$$

With this information, it is obvious that 0.035 g of magnesium liberated 0.002863 g of hydrogen. To find the weight of magnesium necessary to liberate 1 g of hydrogen, it is necessary to set up a proportion and solve for the unknown (x). Thus:

$$\frac{0.035 \text{ g Mg}}{0.002863 \text{ g H}} = \frac{x}{1 \text{ g H}}$$

x = 12.22 g magnesium to liberate 1 g hydrogen (hydrogen equivalent).

REVIEW QUESTIONS

1. Three grams of a certain metal liberated 0.333 g of hydrogen. What is the hydrogen equivalent? *Ans.* 9
2. Two and three-tenths grams (2.3 g) of sodium liberated 0.1 g of hydrogen from water. What is the hydrogen equivalent? *Ans.* 23
3. Calculate the hydrogen equivalent from the following data:
   Weight of metal = 0.033 g
   Volume of hydrogen liberated = 36 ml (measured over water)
   Temperature = 22°C
   Barometer reading = 738 mm
   Vapor pressure of water at 22°C = 19.83 mm
   Ans. 11.58 g
4. Assuming the pressure remains constant (760 mm) find the hydrogen equivalent of a metal if 0.327 g of it liberated 123 ml of hydrogen, measured over water, at 30°C. Be sure to make a correction for the vapor pressure of water at 30°C. *Ans.* 34.3 g

Chemical Equivalents. An equivalent of an element or compound is that weight in grams necessary to react (if reaction occurs) with 1 gram of hydrogen.

EQUIVALENT OF AN ACID. (1) That weight of acid which contains 1 g of proton (hydrogen ion or replaceable hydrogen). (2)

The molecular weight of the acid divided by the number of replaceable hydrogen atoms (protons) per molecule.

EQUIVALENT OF A BASE. (1) That weight of base which contains 17 g hydroxyl ion (replaceable OH). (2) The molecular weight of the base divided by the number of hydroxyl radicals. (ions) per molecule.

EQUIVALENT OF A SALT. (1) That weight of salt which contains one equivalent of its acidic or basic constituent. (2) The molecular weight of the salt divided by the valence characteristic of the metallic or acidic group. In the following table the atomic weights used to obtain molecular weights were rounded off.

Table 30-1
Table of Examples

Formula of Substance	Replaceable Atom or Group	Molecular Weight	Equivalent Weight
HCl	1 rep. hydrogen	36.46	36.46 g
H_2SO_4	2 rep. hydrogens	98	49 g
H_3PO_4	3 rep. hydrogens	98	32.66 g
NaOH	1 rep. hydroxyl	40	40 g
$Ca(OH)_2$	2 rep. hydroxyls	74	37 g
NaCl	1 rep. Na^+ or Cl^-	58.44	58.44 g
$Ca(NO_3)_2$	2 rep. NO_3^- or 1 rep. Ca^{++}	164	82 g

Problem. How many equivalents of hydrochloric acid are present in 12 g of the acid?

Solution. The equivalent weight of HCl is 36.46 g. Let x equal the number of equivalents. Thus:

$$\frac{x \text{ eq.}}{12 \text{ g}} = \frac{1 \text{ eq.}}{36.46 \text{ g}} \qquad x = 0.329 \text{ eq.}$$

Problem. Eleven grams of sodium were allowed to react with an excess of HCl. How many equivalents of the acid were used up?

Solution. The equivalent weight of sodium is 23 g. (See diagram, p. 13.) Therefore, 11 g of sodium is 11/23 of an equivalent. The reaction of sodium and HCl is:

$$2 \text{ Na} + 2 \text{ HCl} \longrightarrow 2 \text{ NaCl} + H_2$$

From this equation, we see that one equivalent of acid reacts with one equivalent of sodium. Therefore, 11/23 of an equivalent of acid is used up.

Problem. How many equivalents of silver chloride are formed by the action of 5.844 g of sodium chloride on an excess of silver nitrate?

Solution. The equation for the reaction is:

$$NaCl + AgNO_3 \longrightarrow AgCl + NaNO_3$$

From this equation we see that one equivalent of NaCl reacting with an excess of $AgNO_3$ forms one equivalent of AgCl. Since the equivalent weight of NaCl is 58.44 g, we see that 5.844 g of salt is 1/10 of an equivalent. Consequently, 1/10 of an equivalent of AgCl is formed.

Note. See Chapter 13 for equivalent weights of oxidizing and reducing agents.

Normal Solutions. A normal solution is one that contains one equivalent of the substance being considered per liter of solution.

PREPARATION OF A NORMAL ACID. Normal sulfuric acid is prepared by placing 49 g of pure sulfuric acid (one-half mole) in a volumetric flask and adding enough distilled water to make exactly one liter of solution.

PREPARATION OF A NORMAL BASE. Normal sodium hydroxide is prepared by adding 40 g of pure sodium hydroxide to a volumetric flask and then adding enough water to make one liter of solution. In actual practice, solutions are made up to the approximate normality desired, and then standardized with an acid (or base) of known normality. Solutions weaker or stronger than normal are designated as 0.2N, 3N, etc.

Note. In present practice, the term ml is used in place of cc. We now write 20 ml of acid, rather than 20 cc of acid.

Titration. The strength of an acid or base, i.e., the number of equivalents per liter (the normality of the solution) is conveniently determined by a process called *titration.* The volumes of acid and base used are accurately measured by means of *burets.* These are 50-ml or 100-ml glass tubes provided with glass stopcocks (or rubber tips) and graduated to fifths or tenths of a milliliter. One equivalent of an acid is neutralized by one equivalent of a base.

EXAMPLE OF TITRATION. The normality of a sodium hydroxide solution was determined by titration. The following steps were taken: (1) 20 ml of the base was placed in a small flask. (2) After adding a drop of suitable indicator, normal HCl was run into the flask, a few drops at a time, from a buret. It required 12 ml of the acid to neutralize 20 ml of base.

CALCULATION OF NORMALITY OF THE BASE.

$$\frac{12 \text{ ml}}{1000 \text{ ml}} \times 1 \text{ eq.} = 0.012 \text{ eq.}$$

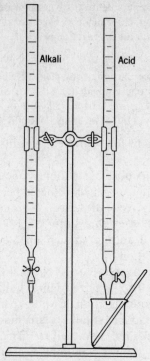

Fig. 30-2. Titrating burets.

This means 0.012 an equivalent of acid was used. Therefore, there was 0.012 equivalent of base in 20 ml. If 0.012 equivalent of base is present in 20 ml, then in 1000 ml there must be

$$\frac{0.012 \text{ eq.} \times 1000 \text{ ml}}{20 \text{ ml}} = 0.6 \text{ eq. of base per liter}$$

In other words, the base was 0.6 normal.

The normality of a solution is conveniently determined by the following relationship:

$$\text{Normality} \times \text{ml} = \text{normality} \times \text{ml}$$

Thus, in the above example we have:

$$1 \times 12 \text{ ml} = \text{normality} \times 20 \text{ ml}$$
$$\text{(acid)} \qquad\qquad\quad \text{(base)}$$

$$\text{Normality} = \frac{1 \times 12 \text{ ml}}{20 \text{ ml}} = 0.6$$

Normal Solutions of Oxidizing and Reducing Agents. For this discussion we shall consider the reaction between potassium permanganate ($KMnO_4$), the oxidizing agent, and iron(II) sulfate ($FeSO_4$), the reducing agent. In this reaction, which takes place in an acid solution, the $FeSO_4$ furnishes the electrons to reduce the permanganate. The ionic equations show what happens.

(a) $5\ Fe^{+2} \longrightarrow 5\ Fe^{+3} + 5\ e$

(b) $MnO_4^{-1} + 5\ e + 8\ H^{+1} \longrightarrow Mn^{+2} + 4\ H_2O$

Add (a),(b) $\overline{MnO_4^{-1} + 5\ Fe^{+2} + 8\ H^{+1} \longrightarrow 5\ Fe^{+3} + Mn^{+2} + 4\ H_2O}$

In reaction (a) 5 e are obtained to react with MnO_4^{-1}.

In reaction (b) we see that the MnO_4^{-1} is reduced to Mn^{+2} by the 5 electrons.

One formula weight (1 mole) of $FeSO_4$ is 151.85 g. This weight of iron(II) sulfate furnishes 1 mole of electrons (6.02×10^{23} electrons). The formula weight, 1 mole, of potassium permanganate ($KMnO_4$) is 158 g so $\dfrac{158\ g}{5}$ is 31.6 g, which is the equivalent weight of $KMnO_4$. Therefore a liter of N $KMnO_4$ is made by dissolving 31.6 g $KMnO_4$ in a liter of solution, and a liter of N $FeSO_4$ is made by dissolving 151.85 g of $FeSO_4$ in a liter of solution.

Problem. What is the normality of a potassium permanganate solution made by dissolving 20 g of the $KMnO_4$ in one liter of solution?

Solution. Let x = normality of $KMnO_4$ solution.

Set up a proportion:

$$\frac{\dfrac{31.6\ g}{1000\ ml}}{N} = \frac{\dfrac{20\ g}{1000\ ml}}{x}$$

Solve for x:

$$x = \frac{\dfrac{20\ g}{1000\ ml}}{\dfrac{31.6\ g}{1000\ ml}} \times N = \frac{20\ g \times 1000\ ml}{31.6\ g \times 1000\ ml} \times N$$

$$= \frac{20}{31.6} \times N = 0.633N$$

Problem 2. 50 ml of N $KMnO_4$ solution reacts with 40 ml of $FeSO_4$ solution. What is its normality?

Solution 1. Normality $\times$ ml = normality $\times$ ml

$$1 \times 50 = N \times 40$$

$$50 = 40N \qquad N = \frac{50}{40} = 1.25N$$

Solution 2. $\dfrac{50 \text{ ml}}{1000 \text{ ml}} \times 1 \text{ eq.} = 0.05 \text{ eq. of KMnO}_4$

$$\frac{0.05 \text{ eq.} \times 1000 \text{ ml}}{40 \text{ ml}} = \frac{50}{40} \text{ eq.} = 1.25 \text{ eq./liter or 1.25N}$$

REVIEW QUESTIONS

1. Give the equivalent weight of each of the following: Na_2SO_4, KOH, $H \cdot C_2H_3O_2$, $Ba(OH)_2$. *Ans.* 71 g, 56 g, 60 g, 85.67 g

2. How many equivalents of hydrochloric acid are necessary to neutralize (*a*) 5 g KOH; (*b*) 5 g $Ca(OH)_2$? *Ans.* (*a*) 0.0891 eq., (*b*) 0.135 eq.

3. How many equivalents are present in 100 ml of (*a*) normal HCl; (*b*) 1/5 normal $Ca(OH)_2$? *Ans.* (*a*) 0.1 eq., (*b*) 0.02 eq.

4. Suppose 50 ml of NaOH neutralizes 150 ml of normal HCl. What is the normality of the base? *Ans.* 3N

5. One gram of a solid acid was neutralized by 50 ml of 0.2N NaOH. What is the equivalent weight of the acid? *Ans.* 100 g

6. How many grams of oxalic acid ($H_2C_2O_4$) are necessary to neutralize 125 ml of 0.25N potassium hydroxide solution? (Oxalic acid has 2 g of replaceable hydrogen per mole.) *Ans.* 1.406 g

7. 10 g of $KMnO_4$ is present in a liter of solution. How many equivalents are present? *Ans.* 0.316. What is its normality?

8. How many grams of $FeSO_4$ can be oxidized in a dilute sulfuric acid solution by 24 ml of 0.25N potassium permanganate? *Ans.* 0.912

9. One gram of $KMnO_4$ crystals is dissolved in dilute sulfuric acid solution. How many ml of 0.1 N $FeSO_4$ solution are required to reduce the permanganate in the solution? *Ans.* 316.4 ml

Mole Fraction.   The mole fraction of a solution is obtained by dividing the number of moles of solute by the total moles of solute and solvent.

$$\frac{\text{Number of moles solute}}{\text{Moles of solute} + \text{moles of solvent}} = \text{Mole fraction}$$

Problem.   A solution of alcohol in water was made by dissolving 92 grams of ethyl alcohol in 180 g of water. What is the mole fraction of the alcohol?

Solution.   We find that 46 g is the weight of one mole of alcohol (C_2H_5OH) and 18 g is the weight of one mole of water.

$$\frac{\dfrac{92}{46}}{\dfrac{92}{46} + \dfrac{180}{18}} = \frac{2}{2 + 10} = \frac{2}{12} = \frac{1}{6} = \text{mole fraction}$$

REVIEW QUESTIONS

1. A solution of wood alcohol (CH_3OH) and glycerine ($C_3H_5(OH)_3$) was made by adding 64 g of the alcohol to 184 g of the glycerine. What is the mole fraction of the alcohol? *Ans.* 1/2

2. A carbon tetrachloride (CCl_4) chloroform ($CHCl_3$) solution was made by mixing 100 ml of each in a suitable container. Given: sp. gr. CCl_4 = 1.575; sp. gr. $CHCl_3$ = 1.489. What is the mole fraction of each? *Ans.* CCl_4, 0.452; $CHCl_3$, 0.548

Appendixes

I. IMPORTANT DEFINITIONS

Abrasive. A hard substance, with sharp edges and corners, used for grinding and polishing.

Absolute Temperature. The Centigrade temperature plus 273.

Absolute Zero. $-273°C = 0°$ A. At this temperature all molecular motion ceases.

Acid. (1) A substance which furnishes hydrogen ions in solution. (2) A substance which turns blue litmus red and contains hydrogen which can be replaced by a metal. An acid yields protons.

Acid Anhydride. The oxide of a nonmetal that reacts with water to give an acid.

Acid Salt. A salt containing replaceable hydrogen.

Actinide Series. The elements in the periodic table following actinium.

Activated Charcoal. Charcoal that has been treated to increase its adsorptive capacity.

Activated State (of an Atom). That state where the atom possesses more than the minimum of energy.

Alkane Series. Saturated hydrocarbons whose general formula is $C_n H_{2n+2}$. Also called *Methane Series* or *Paraffin Series.*

Adsorption. A process in which molecules of a gas or liquid condense, as a film, on the surface of a solid. The process may be largely physical in nature.

Adsorption, Activated. That form of adsorption in which sufficient heat energy must be supplied before the film forms. This process is probably chemical in nature.

Alchemists. The forerunners of the chemists. They were interested in making gold from the base metals.

Alcohol. (1) A compound, derived from a hydrocarbon, containing one or more hydroxyl groups. (2) An organic hydroxide.

Aldehyde. A compound, derived from a hydrocarbon, containing the HCO group.

Alkali. A strong water-soluble base.

Allotropy. The property shown by certain elements capable of existing in more than one form. This is due to the arrangement of the atoms or molecules.

Alloy. An alloy is made of two or more metals. It may be a solution, an intimate mixture, or a definite compound of the constituents.

Alpha Particles. Helium atoms that have lost two electrons. They are produced by radioactive disintegration.

Alum. The double sulfate of a monovalent and trivalent metal, containing a definite amount of water of hydration. ($KAl(SO_4)_2 \cdot 12 H_2O$)

Amalgam. An alloy containing mercury.

Ammonium Ion, NH_4^+. A cation produced by the ionization of an ammonium salt.

Ammonium Radical, NH₄. A group of atoms which plays the role of a metal in certain salts (e.g., NH_4Cl).

Amorphous. A substance without crystalline structure. The atoms or molecules are not arranged in a definite pattern.

Ampere. One coulomb of electricity per second. That current which deposits .001118 g silver per second.

Amphoteric Substance. A substance which has both acidic and basic properties.

Angstrom Unit (Å) $= 10^{-8}$ cm.

Anhydrous Substance. A substance free from water.

Anion. A negatively charged ion. It is attracted to the anode (+ electrode) during electrolysis.

Anode. (1) The positive electrode of an electrolytic cell. (2) The electrode where oxidation takes place.

Atmosphere. (1) The gaseous envelope surrounding the earth. (2) Unit of pressure. The average pressure of the atmosphere at sea level (76 cm of mercury).

Atom. (1) The smallest unit of an element that can take part in the formation of a compound. (2) A positive nucleus surrounded by electrons.

Atomic Number. (1) The net positive charge on the nucleus of an atom. (2) The ordinal number of an atom in the periodic table.

Atomic Weight. The weight of an atom referred to the carbon-12 atom as 12.0000.

Atomic Weight Unit (A.W.U.). This unit is equal to one-twelfth the weight of a carbon-12 atom. An atomic weight unit has the same meaning as an atomic mass unit (A.M.U.).

Avogadro's Hypothesis. Equal volumes of gases at the same temperature and pressure contain the same number of molecules.

Avogadro's Number. The number of molecules in a gram-molecule (mole), 6.023×10^{23}.

Avogram. This is another term that has the same meaning as atomic weight unit. We can say that an atom of sodium weighs 22.9898 avograms.

Base. (1) An oxide or hydroxide of a metal. (2) A substance which gives hydroxyl ions in solution. (3) A substance which neutralizes an acid to form a salt and water. (4) A substance that combines with a proton.

Base-Forming Element. An element that is easily oxidized.

Basic Anhydride. The oxide of a metal which reacts with water to form a base.

Basic Salt. A salt containing replaceable oxygen or hydroxyl groups.

Beta Particle (Ray). A negative electron given off by a radioactive substance.

Binary Compound. A compound containing two elements per molecule.

Boiling Point. The temperature at which the vapor pressure of a liquid reaches atmospheric pressure.

Bond. The force holding atoms in molecules.

Bond Energy. The energy required (in kcal/mole) to break a bond.

British Thermal Unit (B.T.U.). The quantity of heat necessary to raise the temperature of one pound of water one degree Fahrenheit.

Buffer. A suitable mixture of salt and acid (or salt and base) that regulates or stabilizes the pH of a solution.

Calcine. The product obtained by heating a mineral in an oxidizing atmosphere.

Calorie. A unit of heat. The heat required to raise the temperature of one gram of water 1°C.

Calorimeter. An apparatus for measuring the quantity of heat liberated or absorbed during a reaction.

Carbohydrate. A polyhydroxy derivative of a saturated aldehyde or ketone.

Catalysis. The change in the rate of a reaction by the presence of a substance which is unchanged at the end of the reaction.

Catalytic Agent. A substance which alters the rate of a chemical change and which remains unchanged at the end of the reaction: e.g., manganese dioxide in the preparation of oxygen from potassium chlorate.

Cathode. (1) The negative electrode of an electrolytic cell. (2) The electrode at which reduction takes place.

Cathode Ray. A stream of electrons shot out from a cathode.

Cation. (1) A positively charged ion. (2) The ion attracted to the cathode in electrolysis.

Celsius. Another name for Centigrade.

Centigrade, C. The temperature scale on which the freezing point of pure water is 0°C and the boiling point is 100°C.

Chain Reaction. One in which a reactant is produced by the reaction. *Example:* When a neutron strikes a uranium 235 atom, the atom splits into two large atoms and sets free neutrons. Some of these neutrons will react with other uranium 235 atoms to keep the reaction going.

Chemical Change. A change in which the composition of a substance is altered. The new substance or substances have new properties.

Chemical Equation. A qualitative and quantitative expression of a chemical change.

Chemistry. The science that investigates the composition of matter, the changes which take place in it, the amounts and kinds of energy necessary for these changes, and the laws which govern them.

Colligative Properties of Solutions. Properties which depend on the number of dissolved particles but not on the kind of particles. Thus the freezing point of a water solution will be the same whether we make the solution by adding 100 million molecules of ethyl alcohol to 1000 g of water or by adding 100 million molecules of acetone to 1000 g of water.

Colloids. Substances so finely divided that surface forces become an important factor in determining their properties.

Combining Weight. The number of grams of an element that will combine with or replace 7.9997 grams of oxygen or its equivalent.

Combustion. A chemical change producing heat and light.

Component. One of a minimum number of substances necessary to give the composition of a system.

Compound. Two or more elements chemically united in definite proportions by weight. It is homogeneous. The constituents may be separated only by chemical means.

Concentration. The amount of a substance (weight, moles, equivalents) per unit volume.

Cosmic Rays. Rays which come to the earth from somewhere in space, perhaps beyond the solar system.

Covalent Molecule. A molecule in which the bond between two atoms is a shared electron pair, such as H:Cl   H:H   Cl:Cl.

"Cracking" Petroleum. A process using heat and pressure in which hydrocarbons of high molecular weight are broken down ("cracked") into molecules which are in the gasoline range.

Critical Mass. The amount of fissionable material required for a self-propagating chain reaction.

Critical Pressure. The pressure of a system at its critical temperature.

Critical Temperature. The highest temperature at which a liquid and its vapor can coexist as separate phases.

Crystal. A solid in which atoms, ions, or molecules are arranged in a definite pattern.

Cupellation. A process of separating gold and silver from a base metal such as lead. An oxidizing flame converts the lead to the oxide (PbO), which is removed by a stream of air or is absorbed in the porous bottom of the reverberatory furnace. The silver-gold residue remains unchanged.

Curie. A unit of radioactivity. That quantity of radioactive substance which gives 3.7×10^{10} atomic disintegrations per second.

Cyclotron. A device that will increase the speed of "projectiles" such as alpha particles, deuterons, or protons until they reach velocities high enough to disrupt the nuclei of atoms.

Decomposition (of a Compound). The process of breaking down a substance into simpler substances.

Dehydrating Agent. A substance which extracts moisture from another substance, e.g., sulfuric acid.

Deliquescent Salt. A salt capable of absorbing moisture from the air and dissolving in it: e.g., fused calcium chloride.

Density. Mass per unit volume: e.g., grams per cubic centimeter.

Destructive Distillation. A process in which a substance (usually organic) is heated in the absence of air until decomposition takes place. The products of decomposition (volatile matter) are condensed and collected.

Deuterium. An isotope of hydrogen of mass 2.

Deuteron. The nucleus of the deuterium atom.

Developer. A reducing agent used in photography. It reacts more readily on that part of the film emulsion exposed to light.

Dialysis. A process in which a semipermeable membrane is used to separate colloidal particles from substances in true solution.

Dibasic Acid. An acid containing two replaceable hydrogen atoms per molecule.

Diffusion. A process in which unlike molecules intermingle (become mixed). Diffusion is rapid in gases, slower in liquids, and very slow in solids.

Dipole Moment. The product of the charge at the ends of the dipole times the effective distance separating the ends.

Distillate. A substance that has been vaporized and then condensed in a separate vessel.

Distillation. A process in which a liquid is vaporized and then condensed.

Double Salt. A salt in which two metal atoms are combined with one acid radical or one metal is combined with two acid radicals: e.g., $KAl(SO_4)_2 12 H_2O$.

Ductility. That property of a substance which permits its being drawn into wire.

Dyne. A unit of force. The force necessary to give a mass of one gram an acceleration of one centimeter per second per second.

Effervescence. The escape of gas from a liquid due to a decrease in pressure or an increase in temperature.

Efflorescent Substance. One which loses water of crystallization (water of hydration) on exposure to air: e.g., $Na_2CO_3 \cdot 10 H_2O$.

Elastomers. Chemical substances that serve as substitutes for rubber.

Electrochemical Series. A table in which elements are arranged in the order of their decreasing chemical activity.

Electrolytic Dissociation. The breaking up of molecules of electrolytes (acids, bases, salts) in solution to form positive and negative ions.

Electron. The unit of negative electricity. Its mass is $\frac{1}{1837}$ of the hydrogen atom.

Electron Microscope. A microscope in which a stream of electrons is used instead of light. Its resolving power is about 100 times that of the optical microscope.

Electron Volt. That quantity of energy which is equal to the kinetic energy of an electron accelerated by a potential difference of 1 volt.

Electronegative Element. An element which has a tendency to take up electrons.

Electroplating. A process by which a substance is deposited on an object which serves as one electrode in an electrolysis apparatus. For example, when a solution of copper sulfate is electrolyzed using a strip of platinum as the cathode, metallic copper is plated on the platinum.

Electropositive Element. An element which has a tendency to give away electrons.

Electrovalent Substance. One which is made up of ions. For example, the sodium chloride crystal consists of positive sodium ions surrounded by negative chloride ions. In turn the chloride ions are surrounded by sodium ions.

Emulsion. One liquid dispersed in another liquid: e.g., kerosene in water.

Energy. Work. The ability to do work.

Enzyme. A substance produced by living organisms which has catalytic properties. For example, yeast produces an enzyme which acts as a catalyst in converting sugar to alcohol.

Equilibrium (Chemical). A state in which a chemical reaction and the reverse reaction are taking place at the same rate. The concentrations (at equilibrium) of all substances remain constant.

Equilibrium Constant (K). The ratio (number) obtained by dividing the product of the active concentrations of the substances produced in a reaction, by the product of the active concentrations of the reactants, after equilibrium has been reached.

Equivalent Weight. *See* Combining Weight.

Erg. The work done by a force of one dyne per centimeter.

Ester. An 'ethereal salt' derived from an organic acid by replacing the carboxyl hydrogen atom by an alkyl radical.

Eutectic. A mixture of two or more substances with the lowest melting point.

Evaporation. The changing of a liquid to a vapor by heat.

Explosive Mixture. A mixture of gases in the proportion in which they combine. When ignited, combustion of the whole mass takes place within a very short interval of time.

Fahrenheit Temperature Scale. The temperature scale on which 32°F is the freezing point of water and 212°F is the boiling point.

Fermentation. A chemical reaction caused by living organisms or enzymes.

Filtrate. The liquid which passes through a filter.

Fission (Atomic). A process in which the neutron bombardment of a heavy atom nucleus, say, uranium 235, causes it to split into two fragments of about equal mass with the simultaneous liberation of several neutrons and a large amount of energy.

Fixation of Nitrogen. A process in which atmospheric nitrogen is converted into useful compounds.

Fixer (Photographic). A substance which stops the process of developing a plate (or film) by dissolving the unexposed silver halide.

Flotation (Ore). A process in which crushed ore is agitated in water containing a frother (pine oil) and a collector (potassium ethyl xanthate). The valuable mineral particles are attached to the froth and rise to the surface, from which they are removed.

Fluidity. The reciprocal of viscosity.

Fluorescence. The emission of light (not reflected light) by a substance under illumination.

Flux. A substance used to unite with impurities and form a low-melting mixture.

Formula. A chemical formula consists of one or a combination of symbols. If more than one atom of a given element is present in the molecule it is shown by a subscript. The formula shows the elements present and the number of atoms of each.

Fractional Distillation. A process of separating, by distillation, two or more liquids having different boiling points.

Functional Group. An atom or a group of atoms which replaces a hydrogen atom on a hydrocarbon molecule.

Fusion. The melting of a substance.

Fusion, Nuclear. A process in which the nuclei of light atoms combine (aided by high temperature) to form the nucleus of a heavier atom, with the simultaneous release of considerable energy.

Galvanizing. A process in which iron is coated with zinc. The iron is immersed in molten zinc.

Gamma Rays. Short-wave-length rays (X rays) given off by radioactive substances.

Gangue. The waste material, often siliceous matter, remaining after the valuable minerals have been removed from an ore.

Glass. A supercooled liquid composed of the silicates of sodium, calcium, etc.

Gram-Atom. The atomic weight in grams of an element.

Gram-Ion. The weight of an ion (obtained by adding the atomic weights of the atoms in the ion) expressed in grams.

Gram-Molecule. The molecular weight in grams of a substance; a mole.

Ground State (of an Atom). A stationary state in which the energy of the atom is a minimum.

Halogen. This word means salt-former and is given to the chlorine family.

Hard Water. Water which does not readily form a lather with soap. This is due to the presence of dissolved calcium and magnesium salts.

Heat of Formation. The number of calories of heat absorbed or liberated during the formation of a mole of a compound from the elements.

Heat of Fusion. The number of calories required to melt 1 g of substance at its melting point.

Heat of Neutralization. The number of calories liberated in the formation of 18 g of water from hydrogen and hydroxyl ions.

Heat of Vaporization. The number of calories required to convert 1 g of substance from liquid to vapor at its boiling point.

Heavy Water. Water containing deuterium atoms in place of ordinary hydrogen atoms.

Homologous Series. A series of compounds in which the composition may be represented by a general formula: e.g., Methane Series. $C_n H_{2n+2}$.

Hormones. Chemical compounds secreted by the ductless glands of the body. They play an important part in maintaining the proper functional balance between the various organs of the body.

Humidity. The amount of water vapor per unit of gas.

Humidity, Relative. The ratio of the actual amount of water vapor in the atmosphere to the amount necessary for saturation at the same temperature.

Hydrate. A compound containing water of hydration: e.g., $CuSO_4 \cdot 5 H_2O$.

Hydrocarbon. A compound containing only hydrogen and carbon.

Hydrogen Bond. A bond between hydrogen atoms and strongly electronegative atoms.

Hydrogenation. A process in which hydrogen is chemically added to certain unsaturated organic substances such as fats and oils.

Hydrolysis. A reaction between a salt and water in which an acid and base are formed, one or both of which is but slightly dissociated.

Hygroscopic Substance. A substance which takes up moisture but does not dissolve in it. *Example:* sodium nitrate crystals.

Indicator. A substance which changes color at a definite hydrogen ion (or other specific ion) concentration: e.g., litmus, starch, etc.

Inert Element. An element of the inert gas group of the periodic table. Elements in this group are chemically inactive.

Inversion (of Sugar). A process in which cane sugar (sucrose) is converted into glucose and fructose by hydrolysis.

Ion. A charged atom or group of atoms.

Ionization. The splitting up of a molecule into ions. The gain or loss of electrons by atoms.

Ionization Constant. The product of the concentration of the ions divided by the concentration of the un-ionized molecules of solute.

Ionization Energy. The smallest amount of energy required to remove an electron from the influence of the atom.

Ionization Potential. The energy necessary to remove an electron from a gaseous atom to form an ion. This energy is expressed in electron volts.

Isobars. Atoms of the same atomic weight but having different atomic numbers.

Isomers. Molecules having the same number and kinds of atoms but which are arranged differently within the molecule.

Isotopes. Atoms having the same atomic number but different atomic weights. Their chemical properties are identical.

Kindling Temperature. The lowest temperature at which a substance takes fire. This temperature varies with the physical state of the substance.

Lanthanide Series (Rare Earth Series). Those elements with atomic numbers 57–71 inclusive.

Latent Heat. The heat absorbed or liberated in changing a mole of substance from one state to another at a fixed temperature (e.g., converting 18 g of water to water vapor at 100°C).

Lattice, Crystal. The space arrangement of atomic nuclei in a crystal.

Malleability. The property of a substance which permits it to be rolled or hammered into sheets.

Mass. The property of a substance (body) that determines the acceleration it will acquire when acted upon by a given force.

Mass Action, Law of. The speed of a chemical change is proportional to the concentration of the reacting substances.

Mass Defect (of Atoms). All nuclei have smaller masses than the sum of the masses of protons and neutrons which make up those nuclei.

Matte. An intimate mixture of copper and iron sulfides (Cu_2S + FeS) produced in copper smelting.

Matter. That which occupies space and has mass.

Melting Point. The temperature at which the solid and liquid states of a substance are in equilibrium.

Metal. An element that is a good conductor of heat and electricity, reflects light and has luster, and forms hydroxides and oxides that are basic. Metal atoms contain one, two, three, or four electrons in their highest energy levels.

Metallurgy. The science of extracting metals from their ores.

Moderator. A substance such as pure graphite used to reduce the speed of high-energy neutrons.

Molal Solution. A solution which contains one mole of solute in 1000 grams of solvent.

Molar Solution. A solution which contains one mole of solute per liter.

Mole. The molecular weight of a substance expressed in grams.

Molecular Volume. The volume occupied by a mole of any gas at 0°C and 760 mm pressure, i.e., 22.4 liters.

Molecular Weight. The sum of the atomic weights of the atoms in a molecule.

Molecule. (1) The smallest physical unit of a substance possessing the properties of a mass of the substance. (2) The smallest unit of a substance capable of independent existence.

Monobasic Acid. An acid having one replaceable hydrogen atom per molecule.

Mordant. A substance used to fix a dye to cloth so it is not removed by washing.

Nascent. At the instant an element is liberated from a compound it is said to be in the nascent state. Nascent hydrogen is probably atomic hydrogen.

Neutralization. (1) The action of an acid and base to give a salt and water. (2) The union of hydrogen ions of an acid with the hydroxyl ions of a base to form water.

Neutron. A particle of unit mass but with no electric charge.

Nonelectrolyte. A substance (such as sugar) whose water solution does not conduct the electric current.

Nonmetal. An element that is a poor conductor of heat and electricity, is brittle, and has a rather low density level. Nonmetals contain five, six, or seven electrons in their highest energy levels.

Normal Salt. A salt containing neither replaceable hydrogen nor hydroxyl.

Normal Solution. A solution which contains one gram of replaceable hydrogen or its equivalent per liter.

Nuclear Reactor (formerly called an Atomic Pile). A structure in which controlled nuclear reactions take place. The reactor consists of (1) a fissionable element such as U235, (2) a moderator such as graphite to reduce the velocity of neutrons, and (3) a neutron absorber such as cadmium.

Nucleons. The fundamental constituents of atomic nuclei (protons and neutrons).

Nucleus of an Atom. The center of mass in the atom. It is made up of neutrons and protons.

Occlusion. The absorption of gases by solids.

Octet. The term applied to a group of eight electrons in the highest energy levels of atoms.

Ohm. The resistance to an electric current offered by a column of mercury (0°C) 14.4521 g in mass of a uniform cross-sectional area and 106.3 cm in length.

Orbital. Part of an atomic shell (a subshell) in an atom that is usually occupied by a pair of electrons.

Ore. A natural material suitable as a source of some substance (metal).

Organic Chemistry. That branch of chemistry which deals with the carbon compounds.

Osmotic Pressure. If a concentrated solution is separated from pure solvent by means of a semipermeable membrane, the solvent molecules pass through the membrane and dilute the solution. The force which causes this dilution is called *osmotic pressure* and the process is called *osmosis*.

Oxidation. (1) A process in which oxygen combines with another substance. (2) The loss of electrons by an atom or group of atoms.

Oxidation Number. The charge which an atom *appears* to have when its valence electron (or electrons) is assigned to the most electronegative atom in the molecule.

Oxide. A compound of oxygen with another element.

Oxidizing Agent. (1) A substance capable of giving oxygen to another substance without demanding an equivalent of another element in return. (2) A substance which contains an atom or group of atoms that gains electrons.

Paint Base. The particles suspended in the oil of a paint.

Paint Vehicle. A quick-drying oil that forms a flexible hornlike film. The paint base is suspended in this oil.

pH. The negative logarithm (base 10) of the hydrogen ion concentration of a solution. *Example:* If the hydrogen ion concentration (H^+) of a solution is 1×10^{-5}, then pH $= -\log 10^{-5} = 5$.

Phase. A homogeneous part of a system. For example, in the water system there are three phases: solid, liquid, and gaseous.

Phosphor. A material which emits light when struck by high-energy particles.

Photon. A unit of light (a particle of light).

Pile (Atomic). An apparatus designed to reduce the speed of the neutrons produced in atomic fission. A block of chemically pure graphite containing compartments in which are placed rods of metallic uranium (sealed in aluminum cans) is such a pile.

Plastic. A substance which, under pressure, assumes a definite form or shape that is retained after the pressure is released. Plastics are usually composed of an aggregate of long chainlike molecules.

pOH. The negative logarithm (base 10) of the hydroxyl ion concentration.

Polymorphism. The ability to exist in two or more crystalline forms.

Positron. A unit charge of positive electricity of approximately the same mass as the electron.

Precipitate. An insoluble solid formed by the chemical reactions between solutions: e.g., table salt reacts with a silver nitrate solution to give a white precipitate of silver chloride.

Properties. Characteristics by which a substance is identified: e.g., color, odor, state, taste, and solubility.

Protein. A complex nitrogen-containing compound which also contains carbon, hydrogen, oxygen, and, in some cases, sulfur and phosphorus.

Proton. Unit charge of positive electricity. The mass is approximately 1.

Quantum. A bundle of energy. A photon of light.

Radical. A group of atoms which act as a single unit in chemical changes.

Radioactivity. A partial disintegration of atoms. Alpha particles, electrons, or X rays are shot out from the nucleus.

Radiochemistry. That part of chemistry which makes use of radioactive atoms placed in molecules to study the mechanism of certain chemical reactions.

Redox. A word coined from two words, *red*uction and *ox*idation. It emphasizes the fact that reduction and oxidation reactions take place simultaneously.

Reducing Agent. (1) A substance which removes oxygen from another substance. (2) A substance which contains an atom that loses one or more electrons.

Reduction. Opposite of oxidation. (1) Removal of oxygen from a compound. (2) A process in which an atom gains one or more electrons.

Replacement Series. The arrangement of the metals in the order of their decreasing chemical activity.

Reverberatory Furnace. A furnace in which the charge is heated by the combustion of the fuel in the space above the charge. The heat, thus produced, is reflected (reverberates) from the ceiling onto the charge.

Roasting. Heating an ore in contact with air. Metallic sulfides are roasted to form oxides.

Salt. (1) A compound containing a metal or radical and an acid radical. (2) A compound which furnishes anions and cations when dissolved in water, but not hydrogen or hydroxyl ions.

Saponification. The hydrolysis of fats or oils by an alkali.

Saturated Solution. A solution in which the solute in solution is in equilibrium with undissolved solute.

Science. An organized body of facts which have been coordinated and generalized into a system.

Semiconductor. If an insulator, such as germanium, contains just a few atoms of an impurity, it becomes a semiconductor. There are two types: (1) the N type containing free electrons, and (2) the P type containing positive holes in the lattice.

Slag. A by-product of smelting. It is formed by the action of low-melting material (flux) on impurities in the ore. Slags contain calcium and aluminum silicates.

Smelting. A process in which a metal is obtained from its ore by heating in a special furnace with a suitable flux and a reducing agent.

Soap. The sodium or potassium salts of a fatty acid: e.g., $NaC_{18}H_{35}O_2$.

Solubility Product Constant. The product of the concentrations of the ions of slightly soluble salt at saturation.

Solute. The substance dissolved in a solvent.

Solution. A homogeneous body whose composition may be varied within certain limits. Solution may be liquid, solid, or gaseous.

Solvent. The constituent of a solution which is present in larger amount. The substance which does the dissolving.

Specific Gravity (Gases). The ratio of the weight of one liter of air to the weight of one liter of the gas.

Specific Gravity (Solid or Liquid). The ratio of the weight of a unit volume of a substance to the weight of the same volume of water.

Specific Heat. The heat required to raise the temperature of one gram of a substance one degree Centigrade.

Specific Volume. The volume of one gram of a substance.

Spectroscope. An instrument which separates light into its various wave lengths or colors. One form of this instrument utilizes a glass prism.

Spectrum. Light separated into its component parts with the aid of a prism or grating.

Speiss. Iron arsenide, a by-product of the lead blast furnace.

Spontaneous Combustion. Active burning of easily oxidizable substances due to the accumulation of the heat liberated during their oxidation.

Standard Conditions. 0°C and 1 atmosphere pressure (760 mm).

Stoichiometry. A term that refers to the weight relations represented in chemical formulas and equations.

Strong Acid. One which is completely ionized in water solutions.

Sublimation. A process in which a solid is vaporized and condensed to a solid without passing through the liquid state.

Substance. Matter that is homogeneous throughout.

Supersaturated Solution. One in which more solute is in solution than is present in a saturated solution of the same substances at the same temperature and pressure.

Surface Tension. The contractive force of a surface.

Synthesis. A process of preparing a compound from its constituent elements.

Temperature. That condition which determines whether heat will flow from one body to another.

Tempering. The process of changing the physical properties of a substance (steel) by heat treatments.

Thermochemistry. That branch of chemistry which deals with the heat changes accompanying chemical reactions.

Thermonuclear Reaction. A reaction in which atoms of low atomic numbers are combined (fused) to give atoms with heavier nuclei (higher atomic number) with the simultaneous liberation of large amounts of energy as heat and light.

Tincture. A solution the solvent of which is alcohol.

Transistor. A device in which N type and P type semiconductors are properly joined and included in an electric circuit that will allow an electric current to flow in one direction only.

Transmutation. The changing of one element to another through a radioactive process.

Transuranium Elements. Those elements whose atomic numbers are higher than that of uranium (92).

Tribasic Acid. An acid containing three replaceable hydrogen atoms per molecule.

Valence. A number which represents the number of atoms of hydrogen (or its equivalent) that will combine with or be replaced by the atom or radical in question.

Vapor Density. The ratio of the weight of a gas to the weight of an equal volume of hydrogen measured under the same conditions.

Vapor Pressure. The (partial) pressure exerted by a vapor.

Vapor Tension. The pressure which a vapor exerts on a liquid when the liquid and vapor are in equilibrium at a given temperature (the maximum vapor pressure for the given temperature).

Viscosity. (1) The resistance to flow of a liquid. (2) The internal friction of a liquid.

Vitamin. An organic chemical substance found in many different foods and essential to bodily health and vigor.

Volt. The electrical pressure required to make a current of one ampere through a resistance of one ohm.

Weak Acid. An acid which is but slightly ionized in water solutions.

Welding. A process of joining two metals together by heat and pressure.

X Rays. High frequency light waves which come from the bombardment of metals by electrons or from the radioactive disintegration of certain atoms.

II. A SHORT LIST OF SCIENTISTS WHO HAVE MADE NOTEWORTHY CONTRIBUTIONS TO CHEMISTRY

Anderson, Carl (1905–   ).   *American.*   Discovered the positron.

Arrhenius, Svante (1859–1927).   *Swedish.*   Developed the theory of electrolytic dissociation.

Aston, F. W. (1887–1945).   *English.*   Perfected the mass spectrographic method of detecting isotopes.

Becquerel, Henri (1852–1908).   *French.*   Showed that thorium and uranium salts would "expose" light-proof photographic paper. This was the discovery of radioactivity.

Berzelius, Jöns Jakob (1779–1848).   *Swedish.*   Introduced modern chemical symbols.

Bohr, Niels (1885–1962).   *Danish.*   Contributed much to our knowledge of atomic structure. He gave us the dynamic model of the atom.

Boyle, Robert (1627–1691).   *British.*   The first person to clearly define *element.* He also announced the gas law which bears his name.

Bragg, Sir William (1862–1942) and his son **Sir W. Lawrence Bragg** (1890–   ).   *English.*   Used X rays to determine the structure of crystals.

Brönsted, J. N. (1879–1947).   *Danish.*   Developed the generalized concept of acids and bases.

Bunsen, Robert (1811–1899).   *German.*   Early photochemist. Developed the Bunsen burner, the spectroscope, and other valuable devices.

Cavendish, Lord Henry (1731–1810).   *English.*   Discovered hydrogen and was the first to show that two volumes of hydrogen combine with one volume of oxygen.

Chadwick, James (1891–   ).   *English.*   Discovered the neutron in 1932.

Curie, Madame Marie (1867–1934).   *Polish.*   Discovered radium and contributed much to our knowledge of radioactivity.

Dalton, John (1766–1844).   *English.*   Developed the atomic theory and showed how to apply it.

Davy, Sir Humphrey (1778–1829).   *English.*   Brilliant experimentalist who discovered several elements including sodium, potassium, and magnesium.

Debye, Peter (1884–1966).   *Dutch.*   Contributed much to our knowledge of the properties of electrolytic solutions.

Einstein, Albert (1879–1955).   *German-American.*   Developed the theory of relativity and also showed the relationship between matter and energy ($E = MC^2$).

Eyring, Henry (1901–   ).   *American.*   Developed the absolute reaction-rate theory.

Faraday, Michael (1791–1867). *English.* The laws of electrolysis bear this man's name. He also was a pioneer in the liquefaction of gases.

Gibbs, J. Willard (1839–1903). *American.* Gave to the world the *phase rule.* It applies to all systems in equilibrium.

Graham, Thomas (1805–1869). *Scottish.* Discovered the law of diffusion which bears his name. He is also spoken of as the "father of colloid chemistry."

Haber, Frtiz (1868–1934). *German.* Invented the Haber Process of synthesizing ammonia from nitrogen and hydrogen.

Kekulé von Stradonitz, Friederich (1829–1896). *German.* Suggested the ring structure for benzene. He is sometimes spoken of as the "father of structural formulas of organic molecules."

Langmuir, Irving (1881–1957). *American.* Contributed to our knowledge of atomic structure and valence. He also invented the gas-filled electric light bulb.

Lavoisier, Antoine (1743–1794). *French.* He overthrew the phlogiston theory and is called the "father of modern chemistry."

Lawrence, E. O. (1901–1958). *American.* Invented the cyclotron.

Le Châtelier, Henry L. (1850–1936). *French.* Discovered the law, which bears his name, that predicts what happens when a system in equilibrium is disturbed by external forces.

Lewis, G. N. (1875–1946). *American.* Showed many relations between atomic structure and chemical properties.

Mendeleev, Dmitri (1834–1907). *Russian.* Announced the periodic law and made a successful periodic table.

Millikan, Robert A. (1868–1953). *American.* Measured the charge on the electron. He also contributed to our knowledge of cosmic rays.

Moseley, Henry (1887–1915). *English.* Announced the relationship between the atomic number of an element and its properties.

Pasteur, Louis (1822–1895). Showed the relation between fermentation and micro-organisms. He also pointed out the relation between disease and bacteria.

Pauling, Linus (1901–   ). *American.* Contributed much to our knowledge of the chemical bond.

Priestley, Joseph (1733–1804). *English.* Discovered oxygen and prepared a number of gases such as carbon monoxide and ammonia.

Richards, T. W. (1868–1928). *American.* First American chemist to receive the Nobel Prize—for his work on the precise determination of atomic weights.

Rutherford, Ernest (1871–1937). *English.* Showed that nuclei of atoms are very small and are positively charged. He also studied the nature of the rays given off by radioactive substances and helped develop the radioactive disintegration theory.

Seaborg, G. T. (1912–   ). *American.* Discovered, with the aid of the cyclotron, several of the "synthetic" elements such as americium, curium, berkelium, and californium.

Stas, J. S. (1813–1891). *Belgian.* Did some of the first very careful work on the atomic weights of the common elements.

Svedberg, The (1884–   ). *Swedish.* Famous colloid chemist. Developed the ultracentrifuge.

Thomson, J. J. (1856–1940). *English.* Discovered the electron and described some of its properties.

Urey, Harold C. (1893–   ). *American.* Discovered the isotopes of hydrogen. Also specialized in the separation of various isotopes.

III. THE METRIC SYSTEM

Length. 1 meter (1 m) = 10 decimeters = 100 centimeters (100 cm) = 1000 millimeters (1000 mm).

1 kilometer = 1000 meters (1000 m) = 0.6214 mile.
1 decimeter = 0.1 meter = 10 centimeters = 3.937 inches.
1 meter = 1.094 yards = 3.281 ft = 39.37 in.

Volume. 1 liter = 1000 milliliters (1000 ml) = a cube 10 cm × 10 cm × 10 cm.

1 liter = 0.03532 cu ft = 61.03 cu in = 1.057 quarts (U.S.) or 1.136 quarts (Brit.) = 34.1 fl oz (U.S.) = 35.3 oz (Brit.).
1 fluid ounce (U.S.) = 29.57 cc. 1 ounce (Brit.) = 28.4 cc.
1 cu ft = 28.32 liters.

Weight. 1 gram (g) = wt. of 1 cc of water at 4°C. 1 kilogram = 1000 g. 1 gram = 10 decigrams = 100 centigrams (100 cgm) = 1000 milligrams (1000 mgm).

1 kilogram = 2.205 lbs avoird. (U.S. and Brit.).
1000 kilograms = 2205 lbs = 1 metric ton.
1 lb avoird. = 453.6 g.
1 oz avoird. (U.S. and Brit.) = 28.35 g. 100 g = 3.5 oz.

IV. VAPOR PRESSURES OF WATER

Both the Fahrenheit (F), and Centigrade (C) temperatures are given.

Temperature		Pressure (mm of Hg)	Temperature		Pressure (mm of Hg)
F	C		F	C	
32°	0°	4.58	75.2°	24°	22.38
41	5	6.54	77.0	25	23.76
50	10	9.21	78.8	26	25.21
53.6	12	10.52	80.6	27	26.74
57.2	14	11.99	82.4	28	28.35
59.0	15	12.79	84.2	29	30.04
60.8	16	13.63	86.0	30	31.82
64.4	18	15.48	95.0	35	42.18
66.2	19	16.48	104.0	40	55.32
68.0	20	17.54	113.0	45	71.88
69.8	21	18.65	122.0	50	92.51
71.6	22	19.83	140.0	60	149.38
73.4	23	21.07	212.0	100	760.0

V. BORAX BEAD TESTS

A borax bead is made by fusing borax in a small loop of platinum wire. The bead is then dipped into some of the unknown solid and refused. After cooling, the following colors are obtained.

Element	Oxidizing Flame	Reducing Flame
Ni	Violet (hot), Brown (cold)	Gray
Mn	Violet	Colorless
Fe	Yellow	Green
Cu	Blue	Red (if concentrated)
Cr	Green	Green
Co	Blue	Blue

VI. FLAME TESTS

The substance to be tested is moistened with concentrated hydrochloric acid and placed, by means of a clean platinum wire, into the Bunsen flame.

Color Imparted to Flame	Substance Indicated
Fluffy Yellow	Sodium
Violet	Potassium
Deep Red	Strontium or Lithium
Brick Red	Calcium
Greenish-Yellow or Green	Copper, Barium, or Boric Acid
Pale Blue	Arsenic
Bright Blue	Copper Chloride

VII. SOLUBILITY RULES AND DATA

1. All nitrates, acetates, and chlorates are soluble in water. Due to hydrolysis, acetates of the heavy metals and salts of bismuth, tin, antimony, and copper need an excess of free acid to hold them in solution.
2. All chlorides, except $AgCl$, Hg_2Cl_2, and $PbCl_2$ are soluble in water.
3. All sulfates are soluble in water except $BaSO_4$, $SrSO_4$, $PbSO_4$, and $CaSO_4$. The last two are sparingly soluble.
4. All carbonates, phosphates, sulfites, chromates, silicates, borates, oxalates, arsenites, and arsenates, except those of the alkalies and certain acid phosphates, are insoluble or sparingly soluble in water.
5. All hydroxides are insoluble in water except $NaOH$, KOH, and NH_4OH. Barium and calcium hydroxides are sparingly soluble.
6. All sulfides, except those of the alkalies, are insoluble in water. Water decomposes those of magnesium, calcium, barium, and aluminum.

Formula of Salt	Color	Solubility g per 100g H_2O	Solubility in Other Solvents*	Formula of Salt	Color	Solutility g per 100g H_2O	Solubility in Other Solvents*
Aluminum				**Iron**			
$Al(C_2H_5O_2)_3$	White		S. HCl, H_2SO_4, Alk.	$FeOH(C_2H_3O_2)_2$	Brown	Insoluble	S. A.
$AlCl_3$	White	70 (15°C)	S. HCl, H_2SO_4, Alk.	$FeCl_3 \cdot 6H_2O$	Yellow	246 (0°C)	S. A.
$Al(NO_3)_3 \cdot 9H_2O$	White	Very Soluble	S. A.	$Fe(OH)_3$	Red-Brown	Insoluble	S. A.
$Al_2(SO_4)_3$	White	36.2 (20°C)	S. A.				
				$FeSO_4 \cdot 7H_2O$	Green	48 (20°C)	S. A.
Ammonium				**Manganese**			
$NH_4C_2H_3O_2$	White	148.1 (4°C)	S. A.	$Mn(C_2H_3O_2)_2 \cdot 4H_2O$	Pink-Red	3 (15°C)	S. A.
NH_4NO_3	White	183 (20°C)	S. A. :				
$(NH_4)_2SO_4$	White	75.4 (20°C)	S. A.	$MnCO_3$	Brown	.0065 (25°C)	S. A.
				$MnCl_2$	Pink	62.16 (10°C)	S. A.
Antimony				$Mn(NO_3)_2$	Pink-Red	166 (26°C)	S. HCl
$SbCl_3$	White	Decomposes	S. HCl	MnO_2	Black	Insoluble	S. HCl
Sb_2O_3	White	.0018 (15°C)	S. HCl, Alk.	$MnSO_4$	Pink-Red	66.3 (20°C)	S. A.
Sb_2S_3	Red	.00018 (18°C)	S. Alk.				

Compound	Color	Solubility	Acid/Alk.
Arsenic			
$AsCl_3$	Colorless liquid	Decomposes	S. HCl
As_2O_3	White	3.7 (20°C)	S. HCl
As_2S_3	Yellow	.000052	S. HNO_3, Alk.
Barium			
$Ba(C_2H_3O_2)_2H_2O$	White	69.2 (17.5°C)	S. HCl, HNO_3
$BaCO_3$	White	.002	S. HCl, HNO_3
$Ba(OH)_2 \cdot 8H_2O$	White	8.1 (20°C)	S. HNO_3, HCl
$Ba(NO_3)_2 \cdot H_2O$	White	79.5 (20°C)	S. A.
$BaSO_4$	White	.00024	Insol. in A.
Calcium			
$Ca(C_2H_3O_2)_2 \cdot 2H_2O$	White	34.7 (20°C)	S. HCl, HNO_3
$CaCO_3$	White	.0013	S. HCl, HNO_3
$CaCl_2$	White	74 (20°C)	S. A.
$Ca(OH)_2$	White	.17 (20°C)	S. HCl, HNO_3
$Ca(NO_3)_2 \cdot H_2O$	White	73 (18.5°C)	S. HCl, HNO_3
Cobalt			
Co_2S_3	Black	Insoluble	S. HNO_3
$CoCl_2$	Blue	43.3 (0°C)	S. A.
$Co(NO_3)_2 \cdot 6H_2O$	Red	99 (18°C)	S. A.
$CoSO_4$	Red	34.5 (20°C)	S. A.
Chromium			
$CrCl_3 \cdot 6H_2O$	Green	Very Soluble	S. A.
$Cr(OH)_3$	Green	Insoluble	S. in A., Alk.
CrO_3	Red	167 (20°C)	
$CrSO_4 \cdot 7H_2O$	Blue	12.35	S. A.

Compound	Color	Solubility	Acid
Nickel			
$Ni(C_2H_3O_2)_2$	Green	16.6 (15°C)	S. A.
$NiCl_2$	Green-Yellow	64 (20°C)	S. A.
$Ni(OH)_2$	Black	Insoluble	S. A.
$Ni(NO_3)_2$	Green	96.3 (20°C)	S. A.
$NiSO_4$	Yellow	39.7 (20°C)	S. A.
Potassium			
$KC_2H_3O_2$	White	256 (20°C)	
K_2CO_3	White	112 (20°C)	
KCl	White	34.7 (20°C)	
KOH	White	112 (20°C)	
KNO_3	White	31.2 (20°C)	
$K.HC_4H_4O_6$	White	.57 (20°C)	
K_2SO_4	White	10.9 (20°C)	
Sodium			
$NaC_2H_3O_2$	White	124 (20°C)	
$Na_2B_4O_7 \cdot 10H_2O$	White	7.9 (20°C)	
$NaCl$	White	36 (20°C)	
$NaNO_3$	White	87.5 (20°C)	
Na_2SO_4	White	19.5 (20°C)	
$Na_2HPO_4 \cdot 12H_2O$	White	11.8 (17°C)	
Tin			
$Sn(C_2H_3O_2)_2$	Yellow	Decomposes	S. HCl
$SnCl_2$	White	270 (15°C)	S. HCl
SnS	Brown	.000002 (18°C)	
SnS_2	Yellow	.00002 (18°C)	
$SnSO_4$	White	19 (19°C)	

*S. = Soluble   A. = Acids   Alk. = Alkalies

VIII. TESTS FOR THE COMMON CATIONS

For each test we shall assume that a solution of a soluble (nitrate) compound containing the cation is the starting substance. In the table an arrow pointing down ($\downarrow$) after a formula indicates a precipitate has formed. An arrow pointing up ($\uparrow$) indicates a gas has been liberated.

Cation	Reagent	Formula of Characteristic Substance	Color of Characteristic Substance	Remarks
Silver Ag^+	Cl^- S^{--}	$AgCl\downarrow$ $Ag_2S\downarrow$	White Black	(1) Insoluble in H_2O and dil. HNO_3. (2) Soluble in ammonia solution. Insoluble in H_2O and dil. acids except dil. HNO_3, in which it is readily soluble.
Mercurous Hg_2^{++}	Cl^-	$Hg_2Cl_2\downarrow$	White	(1) Insoluble in H_2O and dil. acids. (2) Reacts with ammonia to give $HgNH_2Cl$ (white) and Hg (black).
Lead Pb^{++}	Cl^- SO_4^{--}	$PbCl_2\downarrow$ $PbSO_4\downarrow$	White White	(1) Slightly soluble in cold water but readily soluble in warm or hot water. (1) Insoluble in H_2O, dil. HCl, dil. H_2SO_4. (2) Soluble in hot con. H_2SO_4. (3) Soluble in a solution of $(NH_4) \cdot C_2H_3O_2$.
	CrO_4^{--}	$PbCrO_4\downarrow$	Yellow	(1) Insoluble in H_2O, dil. $H \cdot C_2H_3O_2$, and NH_3. (2) Soluble in $NaOH$.
	H_2S	$PbS\downarrow$	Black	(1) Insoluble in H_2O and dil. HCl. (2) Soluble in hot dil. HNO_3.

Ion	Reagent	Product	Color	Solubility
Mercuric Hg^{++}	H_2S	$HgS\downarrow$	Black	(1) Insoluble in H_2O and dil. acids. (2) Soluble in aqua regia.
	$SnCl_2$	$Hg_2Cl_2\downarrow$ Hg	White to Black	(1) Insoluble in water and dil. acids.
Bismuth Bi^{++}	$Cl^- + H_2O$ OH^-	$BiOCl\downarrow$ $Bi(OH)_3\downarrow$	White White	(1) Soluble in dilute acids. (1) Insoluble in H_2O and dil. alkalies. (2) Soluble in dilute acids.
	H_2S	$Bi_2S_3\downarrow$	Black	(1) Insoluble in H_2O and dil. acids. (2) Soluble in hot con. HNO_3 and hot con. HCl.
	Na_2SnO_2 Sodium stannite	$Bi\downarrow$	Black	(1) Insoluble in water but soluble in HCl or HNO_3.
Cupric Cu^{++}	H_2S	$CuS\downarrow$	Black	(1) Insoluble in dil. acids except HNO_3. (2) The fresh precipitate dissolves in KCN.
	NH_3	$Cu(NH_3)_4{}^{++}$	Deep blue	(1) Deep blue color destroyed by acids. (2) KCN destroys deep blue color by forming colorless $Cu(CN)_3{}^{--}$ ion.
	$Fe(CN)_6{}^{----}$	$Cu_2Fe(CN)_6\downarrow$	Red-brown	(1) Insoluble in dilute acids. (2) Soluble in hot dil. NH_3.
Cadmium Cd^{++}	H_2S	$CdS\downarrow$	Yellow	(1) Insoluble in H_2O and dil. HCl. (2) Soluble in hot dil. HNO_3.

TESTS FOR THE COMMON CATIONS (Continued)

Cation	Reagent	Formula of Characteristic Substance	Color of Characteristic Substance	Remarks
Arsenous Acid H_3AsO_3 or $HAsO_2$	H_2S	$As_2S_3\downarrow$	Yellow	(1) Insoluble in con. HCl. (2) Soluble in alkali hydroxides and in con. HNO_3. (3) Soluble in yellow ammonium sulfide.
Arsenic Acid H_3AsO_4 or $H_2AsO_4^-$	H_2S Ag^+	$As_2S_5\downarrow$ $Ag_3AsO_4\downarrow$	Yellow Red-brown	Same as for As_2S_3: (1) Soluble in acids and ammonia. (2) Insoluble in Na_2CO_3 solution.
Antimony Chloride $SbCl_3$ in HCl	H_2O H_2S	$SbOCl\downarrow$ $Sb_2S_3\downarrow$	White Orange-red	Soluble in strong acids. Soluble in 12N HCl and in KOH.
Stannous Sn^{++}	OH^- H_2S	$Sn(OH)_2\downarrow$ $SnS\downarrow$	White Brown	Soluble in excess of NaOH or in mineral acids. (1) Soluble in con. HCl. (2) Soluble in yellow ammonium sulfide. (3) Insoluble in KOH.
Stannic Sn^{++++}	H_2S	$SnS_2\downarrow$	Light-yellow	Soluble in 6N HCl and in KOH.
Aluminum Al^{+++}	Ammonia solution	$Al(OH)_3\downarrow$	White	(1) Soluble in excess of NaOH. (2) Insoluble in excess of ammonium hydroxide solution. (3) Soluble in dilute acids.

Cation	Reagent	Precipitate/Ion	Color	Remarks
Chromic Cr^{+++}	Ammonia solution	$Cr(OH)_3\downarrow$	Green	(1) Soluble in excess of NaOH. (2) Soluble in acids. (3) Sodium peroxide solution converts $Cr(OH)_3$ or Na_3CrO_3 into yellow chromate ion.
Ferrous Fe^{++}	OH^-	$Fe(OH)_2\downarrow$	White	Oxidizes in air to $Fe(OH)_3$. Color changes from white to green to black and finally to red-brown.
	S^{--}	$FeS\downarrow$	Black	Soluble in acids with evolution of H_2S.
	$Fe(CN)_6^{----}$	$Fe_2Fe(CN)_6\downarrow$	Light blue	Insoluble in dil. HCl.
	$Fe(CN)_6^{---}$	$Fe_3[Fe(CN)_6]_2\downarrow$	Dark blue (Turnbull's blue)	
Ferric Fe^{+++}	OH^-	$Fe(OH)_3\downarrow$	Red-brown	Insoluble in excess of NaOH or NH_4OH.
	$Fe(CN)_6^{----}$	$Fe_4[Fe(CN)_6]_3$	Blue (Prussian blue)	Insoluble in dil. HCl.
	CNS^-	$[Fe(CNS)_6]^{---}$	Deep red solution	Very sensitive test.
Nickelous Ni^{++}	OH^-	$Ni(OH)_2\downarrow$	Light green	Soluble in acids and in ammonia.
	NH_3 excess	$[Ni(NH_3)_4]^{++}$	red-blue solution	
	H_2S Neutral or alkaline solution	$NiS\downarrow$	Black	
	Dimethylglyoxime $(C_4H_8N_2O_2)$	$(C_4H_7N_2O_2)_2Ni\downarrow$	Red	(1) Insoluble in dil. HCl. (2) Soluble in hot con. HNO_3 and in aqua regia. Very sensitive test.

TESTS FOR THE COMMON CATIONS (Continued)

Cation	Reagent	Formula of Characteristic Substance	Color of Characteristic Substance	Remarks
Cobaltous Co^{++}	OH^-	$Co(OH)_2\downarrow$	Pink	Soluble in presence of strong ammonium salts.
	S^{--}	$CoS\downarrow$	Black	Same as NiS.
	KNO_2	$K_2[Co(NO_2)_6]\downarrow$ $\cdot 3H_2O]\downarrow$	Yellow	Should be warmed and allowed to stand to give crystals chance to form.
Manganous Mn^{++}	OH^-	$Mn(OH)_2\downarrow$	White	The white precipitate soon turns dark due to oxidation. Finally $MnMnO_3$ is formed, which is black.
	S^{--}	$MnS\downarrow$	Pink	Turns brown on standing in air due to oxidation.
	$PbO_2 + HNO_3$	MnO_4^-	Purple	The purple MnO_4^- is easily changed to colorless Mn^{++} by reducing agents.
Zinc Zn^{++}	OH^-	$Zn(OH)_2$	White	Soluble in excess of NaOH and in acids.
	NH_3	$Zn(NH_3)_4^{++}$	Colorless	
	H_2S neutral or alkaline solution	ZnS	White	(1) Soluble in strong acids. (2) Insoluble in acetic acid.
Barium Ba^{++}	CO_3^{--}	$BaCO_3\downarrow$	White	Soluble in acids. CO_2 liberated.
	SO_4^{--}	$BaSO_4\downarrow$	White	Insoluble in acids.
	CrO_4^{--}	$BaCrO_4\downarrow$	Yellow	(1) Soluble in strong acids (except H_2SO_4). (2) Insoluble in dil. acids.

Strontium Sr^{++}	CO_3^{--} SO_4^{--}	$SrCO_3\downarrow$ $SrSO_4\downarrow$	White White	Same as barium carbonate. Insoluble in acids.
Calcium Ca^{++}	CO_3^{--} SO_4^{--}	$CaCO_3\downarrow$ $CaSO_4\downarrow$	White White	Soluble in acids. CO_2 liberated. Much more soluble than $BaSO_4$ or $SrSO_4$ in water and in dil. acids.
	$C_2O_4^{--}$	$CaC_2O_4\downarrow$	White	(1) Insoluble in water and acetic acid. (2) Soluble in strong acids.
Magnesium Mg^{++}	OH^-	$Mg(OH)_2\downarrow$	White	(1) Insoluble in excess of NaOH. (2) Insoluble in ammonia but dissolves if NH_4Cl is added.
	HPO_4^{--} in ammonia	$MgNH_4PO_4\downarrow$	White crystalline	(1) Precipitate forms slowly. (2) Soluble in acids.
Potassium K^+	$[Co(NO_2)_6]^{---}$	$K_3[Co(NO_2)_6]\downarrow$	Yellow	(1) Solution must not be alkaline. (2) Keep solution acid with acetic acid.
Sodium Na^+	Magnesium uranium acetate	$NaMg(UO_2)_3$ $(C_2H_3O_2)_9$ $\cdot 9H_2O\downarrow$ sodium magnesium uranyl acetate	Light Yellow	Precipitate forms slowly.
Ammonium NH_4^+	NaOH $PtCl_6^{--}$	$NH_3\downarrow$ $(NH_4)_2PtCl_6\downarrow$	Colorless Yellow	(1) Has characteristic odor. (2) The precipitate is decomposed by NaOH.

IX. TESTS FOR THE COMMON ANIONS

For each test we shall assume that a solution of a soluble (sodium) compound containing the anion is the starting substance. In the table an arrow pointing down (↓) after a formula indicates a precipitate has formed. An arrow pointing up (↑) indicates a gas has been liberated.

Anions	Reagent	Formula of Characteristic Substance	Color of Characteristic Substance	Solubility Information and Remarks
Chloride Cl⁻	Ag⁺	AgCl↓	White	(1) Insoluble in H_2O and HNO_3. (2) Soluble in ammonia solution.
Bromide Br⁻	(1) Ag⁺	AgBr↓	Yellowish white	(1) Insoluble in H_2O and dil. HNO_3. (2) Soluble in ammonia (slowly), KCN, and $Na_2S_2O_3$ solutions.
	(2) Cl_2 water	Br_2 in solution	Brown	Br_2 is readily soluble in CCl_4 to give a brown solution.
Iodide I⁻	(1) Ag⁺	AgI↓	Yellow	(1) Insoluble in HNO_3, H_2O, NH_3. (2) Soluble in KCN and $Na_2S_2O_3$ solutions.
	(2) Cl_2 water	I_2 in solution	Violet (brown)	I_2 is very soluble in CCl_4 to give a violet solution.

Ion	Reagent	Product	Observation	Remarks
Nitrate NO_3^-	(1) Saturated solution $FeSO_4 + H_2SO_4$ con.	$FeSO_4 \cdot NO$	Brown	(1) Mix the test solution with the $FeSO_4$ in test tube. Carefully pour the con. H_2SO_4 down the inside of the tilted tube. At junction of the two liquids a brown ring forms.
	(2) Copper turnings and con. H_2SO_4	$NO_2\uparrow$	Brown fumes	(2) Place clean copper turnings in 5 cc test solution. Add 5 cc con. H_2SO_4. Heat to boiling. Brown fumes (NO_2) given off. Solution becomes blue due to Cu^{++}.
Nitrite NO_2^-	(1) dil. H_2SO_4	$NO_2\uparrow$	Brown fumes	(1) Not necessary to heat.
	(2) $I^- + H \cdot C_2H_3O_2$	$NO + I_2$	Violet in CCl_4	(2) Iodine set free. Collected in CCl_4.
Acetate $C_2H_3O_2^-$	dil. H_2SO_4	$H \cdot C_2H_3O_2$	Colorless	Detected by vinegar odor.
Sulfide S^{--}	HCl(1:1)	$H_2S\uparrow$	Colorless	H_2S has a characteristic odor. A filter paper moistened with lead acetate solution is turned dark brown or black due to PbS.$\downarrow$
Oxalate $C_2O_4^{--}$	(1) Ca^{++}	$CaC_2O_4\downarrow$	White	(1) Insoluble in H_2O, acetic acid. Soluble in HCl, HNO_3.
	(2) dil. $KMnO_4 +$ dil. H_2SO_4:	CO_2	Colorless	(2) The $KMnO_4$ is decolorized.
Carbonate CO_3^{--}	HCl. Pass gas liberated into $Ba(OH)_2$	$BaCO_3\downarrow$	White	The $BaCO_3$ is soluble in an excess of carbonic acid (H_2CO_3).

IX. TESTS FOR THE COMMON ANIONS (Continued)

Phosphate PO_4^{---}	(1) HNO_3 + $(NH_4)_2MoO_4$ (2) Magnesia mixture ($MgCl_2$ + NH_4Cl) + NH_3	$(NH_4)_3PO_4 \cdot 12MoO_3\downarrow$ $MgNH_4PO_4\downarrow$	Yellow Small white crystals	(1) Insoluble in H_2O and HNO_3. (2) Soluble in H_3PO_4, NH_3, $NaOH$
Sulfate SO_4^{--}	Ba^{++}	$BaSO_4\downarrow$	White	Insoluble in H_2O and dilute acids.
Sulfite SO_3^{--}	(1) HCl (2) $Ba(OH)_2$	$SO_2\uparrow$ $BaSO_3\downarrow$	Colorless White	(1) Gas has odor of burning sulfur. (2) Turns moist blue litmus red. (1) Soluble in dil. HCl. (2) HCl + H_2O_2 oxidizes to SO_4^{--}.
Borate $B_4O_7^{--}$	C_2H_5OH + con. H_2SO_4	$(C_2H_5)_3BO_3$	Colorless vapor	Burn the alcohol solution. The $(C_2H_5)_3BO_3$ imparts a green color to flame.

X. TABLE OF COMMON CHEMICALS, THEIR CHEMICAL NAMES, FORMULAS, AND COMMON USES

Common Name	Chemical Name	Formula	Common Use
Aniline	Phenyl amine	$C_6H_5NH_2$	In making dyes.
Aqua fortis	Nitric acid	HNO_3	In chemical laboratory and in industrial chemistry.
Aqua regia	Nitric acid and hydrochloric acid	$HNO_3 + HCl$	Chemical analysis. Dissolves gold and platinum.
Aspirin	Acetyl-salicylic acid	$C_6H_4(OCOCH_3)COOH$	Medicine.
Bakelite	Phenol resin + formaldehyde	——	A plastic widely used in the manufacture of radios, toilet sets, buttons, etc.
Baking soda	Sodium bicarbonate	$NaHCO_3$	A leavening agent.
Barytes	Barium sulfate	$BaSO_4$	Manufacture of white paint.
Bleaching powder	Calcium hypochlorite	$CaOCl_2$	In bleaching.
Blue vitriol	Copper sulfate	$CuSO_4 \cdot 5H_2O$	In copper plating. As a fungicide.
Bone black	Carbon (animal charcoal)	C	In manufacturing sugar. Decolorizing agent.
Boric (boracic) acid	Boric acid	H_3BO_3	An antiseptic.
Borax	Sodium tetraborate	$Na_2B_4O_7 \cdot 10H_2O$	A water softener and household cleanser.
Brimstone	Sulfur	S	In vulcanizing rubber. In manufacturing matches and in insecticides, etc.
Calomel	Mercurous chloride	Hg_2Cl_2	Medicine.
Camphor (artificial)	Pinene hydrochloride	$C_{10}H_{17}Cl$	In the manufacture of plastics.

TABLE OF COMMON CHEMICALS (Continued)

Common Name	Chemical Name	Formula	Common Use
Carbolic acid	Phenol	C_6H_5OH	An antiseptic. In making synthetic resins.
Carborundum	Silicon carbide	SiC	Abrasive.
Caustic (or caustic soda)	Sodium hydroxide	NaOH	A strong base. In manufacture of laundry soaps, water softeners, etc.
Chili saltpeter	Sodium nitrate	$NaNO_3$	In making explosives. As a fertilizer.
Chloroform	Trichlormethane	$CHCl_3$	An anesthetic.
Chrome yellow	Lead chromate	$PbCrO_4$	A yellow pigment used in making paint.
Copperas (Green vitriol)	Ferrous sulfate	$FeSO_4 \cdot 7H_2O$	In making inks and dyes. A disinfectant.
Corrosive sublimate	Mercuric chloride	$HgCl_2$	A disinfectant.
Cream of Tartar	Potassium acid tartrate	$KHC_4H_4O_6$	In baking powder.
Dextrose	Glucose	$C_6H_{12}O_6 \cdot H_2O$	In manufacture of confections.
Emery powder	Aluminum oxide	Al_2O_3	An abrasive.
Epsom salts	Magnesium sulfate	$MgSO_4 \cdot 7H_2O$	A medicine.
Ether	Ethyl ether	$(C_2H_5)_2O$	An anesthetic.
Formalin	40% solution of formaldehyde in water	HCHO	A preservative used in the medical laboratory, etc.
Fuller's earth	Hydrated magnesium and aluminum silicates	—	A filtering and decolorizing medium.

Fusel oil	Mixed amyl alcohols	$C_5H_{11}OH$	Medicine.
Glauber's salt	Sodium sulfate	$Na_2SO_4 \cdot 10H_2O$	In making confections.
Glucose	Dextrose	$C_6H_{12}O_6$	In manufacturing explosives.
Glycerin	Glycerol	$C_3H_5(OH)_3$	In making inks and dyes.
Green vitriol	Ferrous sulfate	$FeSO_4 \cdot 7H_2O$	In making plaster of Paris.
Gypsum	Calcium sulfate	$CaSO_4 \cdot 2H_2O$	
Horn silver	Silver chloride	$AgCl$	Fixer in photography.
Hypo	Sodium thiosulfate	$Na_2S_2O_3 \cdot 5H_2O$	In the manufacture of porcelain.
Kaolin	Aluminum silicate	$Al_2O_3 \cdot 2SiO_2 \cdot 2H_2O$	To adsorb coloring matter from oils and liquids.
Kieselguhr	Silica	SiO_2	In making inks, carbon paper, etc.
Lampblack	Impure carbon	C	An anesthetic.
Laughing gas	Nitrous oxide	N_2O	In paints.
Lithopone	Zinc sulfide + barium sulfate	$ZnS + BaSO_4$	
Magnesia	Magnesium oxide	MgO	Medicine.
Marble	Calcium carbonate	$CaCO_3$	Building material. Monuments, etc.
Methanol	Methyl alcohol	CH_3OH	To denature grain alcohol. Anti-freezing solution. A solvent.
Microcosmic salt	Sodium ammonium hydrogen phosphate	$Na(NH_4)HPO_4 \cdot 4H_2O$	A test for magnesium ion.
Milk of magnesia	Magnesium hydroxide	$Mg(OH)_2$	A mild base. Medicine.

TABLE OF COMMON CHEMICALS (Continued)

Common Name	Chemical Name	Formula	Common Use
Muriatic acid	Hydrochloric acid	HCl	Used extensively in industry. A solvent.
Naphtha (solvent)	A coal-tar distillate	—	A solvent.
Paris green	Copper acetoarsenite	$Cu_5(C_2H_3O_2)_2 \cdot 3CuAs_2O_4$	An insecticide.
Petroleum ether	Benzine	A hydrocarbon mixture	A solvent.
Plaster of Paris	Calcium sulfate	$(CaSO_4)_2 \cdot H_2O$	For statuary, casts, etc.
Prussic acid	Hydrocyanic acid	HCN	A disinfectant.
Quicklime	Calcium oxide	CaO	In building mortar.
Quicksilver	Mercury	Hg	In thermometers, barometers. In treating certain ores.
Rochelle salts	Potassium sodium tartrate	$KNaC_4H_4O_6$	Medicine.
Spirit of hartshorn (Ammonia water)	Ammonia solution	NH_4OH	A solvent.
Silica	Silicon dioxide	SiO_2	In making glass. A building material.
Sugar of lead	Lead acetate	$Pb(C_2H_3O_2)_2 \cdot 3H_2O$	As a paint base.
Superphosphate	Calcium acid phosphate	$CaH_4(PO_4)_2$	A commercial fertilizer.
T.N.T.	Trinitrotoluene	$C_6H_2(CH_3)(NO_2)_3(1,2,4,6)$	An explosive.

Washing soda	Sodium carbonate	$Na_2CO_3 \cdot 10H_2O$	In soap powders.
Water glass	Sodium silicate	Na_2SiO_3	In manufacture of cements. In waterproofing porous materials. A preservative. An adhesive.
White lead	Basic lead carbonate	$Pb(OH)_2 \cdot 2PbCO_3$	A paint base.
Wood alcohol	Methyl alcohol	CH_3OH	A solvent. An antifreezing solution for automobile radiators. To denature grain alcohol.
Zinc white	Zinc oxide	ZnO	A paint base.

XI. THE ACTIVITY SERIES OF METALS

Definition. The activity series is a table of the metals arranged in the order of their decreasing chemical activity. This table is practically the same as the electrochemical series of the elements. However, nonmetals are not included in the table given here. The activity series represents the ease with which the metals give up one or more electrons to form positive ions. A stable substance is one that successfully resists attempts to decompose it. Heat, light, pressure, and electrical and chemical energy are applied in de-

THE ACTIVITY SERIES

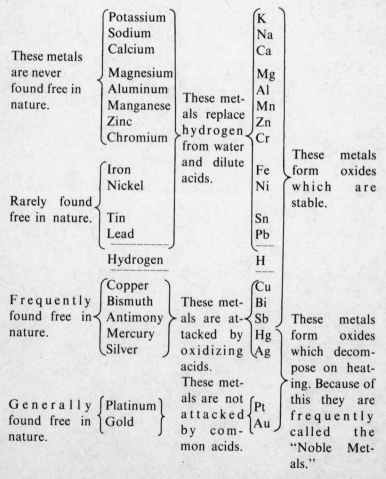

These metals are never found free in nature.	Potassium Sodium Calcium Magnesium Aluminum Manganese Zinc Chromium	These metals replace hydrogen from water and dilute acids.	K Na Ca Mg Al Mn Zn Cr
Rarely found free in nature.	Iron Nickel Tin Lead		Fe Ni Sn Pb
	Hydrogen		H
Frequently found free in nature.	Copper Bismuth Antimony Mercury Silver	These metals are attacked by oxidizing acids.	Cu Bi Sb Hg Ag
Generally found free in nature.	Platinum Gold	These metals are not attacked by common acids.	Pt Au

These metals form oxides which are stable.

These metals form oxides which decompose on heating. Because of this they are frequently called the "Noble Metals."

termining stability. A substance may be stable to one form of energy and unstable to another. For example, in moist air, metallic sodium is converted to sodium hydroxide. Calcium carbonate is decomposed when heated to a high temperature. Sodium is *active*; calcium carbonate is *unstable* to heat.

XII. ELECTRON POPULATIONS OF THE ATOMS OF THE ELEMENTS

Period	Atomic Number	Name of Element	Symbol of Element	n = 1 s	n = 2 s	n = 2 p	n = 3 s	n = 3 p	n = 3 d	n = 4 s	n = 4 p	n = 4 d	n = 4 f	n = 5 s	n = 5 p	n = 5 d	n = 5 f	n = 6 s	n = 6 p	n = 6 d	n = 6 f	n = 7 s
1	1	Hydrogen	H	1																		
	2	Helium	He	2																		
2	3	Lithium	Li	2	1																	
	4	Beryllium	Be	2	2																	
	5	Boron	B	2	2	1																
	6	Carbon	C	2	2	2																
	7	Nitrogen	N	2	2	3																
	8	Oxygen	O	2	2	4																
	9	Fluorine	F	2	2	5																
	10	Neon	Ne	2	2	6																
3	11	Sodium	Na	2	2	6	1															
	12	Magnesium	Mg	2	2	6	2															
	13	Aluminum	Al	2	2	6	2	1														
	14	Silicon	Si	2	2	6	2	2														
	15	Phosphorus	P	2	2	6	2	3														
	16	Sulfur	S	2	2	6	2	4														
	17	Chlorine	Cl	2	2	6	2	5														
	18	Argon	Ar	2	2	6	2	6														

				1s	2s	2p	3s	3p	3d	4s	4p	4d	5s
4	19	Potassium	K	2	2	6	2	6		1			
	20	Calcium	Ca	2	2	6	2	6		2			
	21	Scandium	Sc	2	2	6	2	6	1	2			
	22	Titanium	Ti	2	2	6	2	6	2	2			
	23	Vanadium	V	2	2	6	2	6	3	2			
	24	Chromium	Cr	2	2	6	2	6	5	1			
	25	Manganese	Mn	2	2	6	2	6	5	2			
	26	Iron	Fe	2	2	6	2	6	6	2			
	27	Cobalt	Co	2	2	6	2	6	7	2			
	28	Nickel	Ni	2	2	6	2	6	8	2			
	29	Copper	Cu	2	2	6	2	6	10	1			
	30	Zinc	Zn	2	2	6	2	6	10	2			
	31	Gallium	Ga	2	2	6	2	6	10	2	1		
	32	Germanium	Ge	2	2	6	2	6	10	2	2		
	33	Arsenic	As	2	2	6	2	6	10	2	3		
	34	Selenium	Se	2	2	6	2	6	10	2	4		
	35	Bromine	Br	2	2	6	2	6	10	2	5		
	36	Krypton	Kr	2	2	6	2	6	10	2	6		
5	37	Rubidium	Rb	2	2	6	2	6	10	2	6		1
	38	Strontium	Sr	2	2	6	2	6	10	2	6		2
	39	Yttrium	Y	2	2	6	2	6	10	2	6	1	2
	40	Zirconium	Zr	2	2	6	2	6	10	2	6	2	2
	41	Niobium	Nb	2	2	6	2	6	10	2	6	4	1
	42	Molybdenum	Mo	2	2	6	2	6	10	2	6	5	1

APPENDIX XII. (continued)

Orbital Distribution of Electrons in Various Energy Levels

Period	Atomic Number	Name of Element	Symbol of Element	n=1 s	n=2 s	n=2 p	n=3 s	n=3 p	n=3 d	n=4 s	n=4 p	n=4 d	n=4 f	n=5 s	n=5 p	n=5 d	n=5 f	n=6 s	n=6 p	n=6 d	n=6 f	n=7 s
5	43	Technetium	Tc	2	2	6	2	6	10	2	6	6		1								
	44	Ruthenium	Ru	2	2	6	2	6	10	2	6	7		1								
	45	Rhodium	Rh	2	2	6	2	6	10	2	6	8		1								
	46	Palladium	Pd	2	2	6	2	6	10	2	6	10										
	47	Silver	Ag	2	2	6	2	6	10	2	6	10		1								
	48	Cadmium	Cd	2	2	6	2	6	10	2	6	10		2								
	49	Indium	In	2	2	6	2	6	10	2	6	10		2	1							
	50	Tin	Sn	2	2	6	2	6	10	2	6	10		2	2							
	51	Antimony	Sb	2	2	6	2	6	10	2	6	10		2	3							
	52	Tellurium	Te	2	2	6	2	6	10	2	6	10		2	4							
	53	Iodine	I	2	2	6	2	6	10	2	6	10		2	5							
	54	Xenon	Xe	2	2	6	2	6	10	2	6	10		2	6							
6	55	Cesium	Cs	2	2	6	2	6	10	2	6	10		2	6			1				
	56	Barium	Ba	2	2	6	2	6	10	2	6	10		2	6			2				
	57	Lanthanum	La	2	2	6	2	6	10	2	6	10		2	6	1		2				
	58	Cerium	Ce	2	2	6	2	6	10	2	6	10	1	2	6	1		2				
	59	Praseodymium	Pr	2	2	6	2	6	10	2	6	10	3	2	6			2				
	60	Neodymium	Nd	2	2	6	2	6	10	2	6	10	4	2	6			2				
	61	Promethium	Pm	2	2	6	2	6	10	2	6	10	5	2	6			2				

	Name	Symbol	1s	2s	2p	3s	3p	3d	4s	4p	4d	4f	5s	5p	5d	6s	6p
62	Samarium	Sm	2	2	6	2	6	10	2	6	10	6	2	6		2	
63	Europium	Eu	2	2	6	2	6	10	2	6	10	7	2	6		2	
64	Gadolinium	Gd	2	2	6	2	6	10	2	6	10	7	2	6	1	2	
65	Terbium	Tb	2	2	6	2	6	10	2	6	10	9	2	6		2	
66	Dysprosium	Dy	2	2	6	2	6	10	2	6	10	10	2	6		2	
67	Holmium	Ho	2	2	6	2	6	10	2	6	10	11	2	6		2	
68	Erbium	Er	2	2	6	2	6	10	2	6	10	12	2	6		2	
69	Thulium	Tm	2	2	6	2	6	10	2	6	10	13	2	6		2	
70	Ytterbium	Yb	2	2	6	2	6	10	2	6	10	14	2	6		2	
71	Lutetium	Lu	2	2	6	2	6	10	2	6	10	14	2	6	1	2	
72	Hafnium	Hf	2	2	6	2	6	10	2	6	10	14	2	6	2	2	
73	Tantalum	Ta	2	2	6	2	6	10	2	6	10	14	2	6	3	2	
74	Tungsten	W	2	2	6	2	6	10	2	6	10	14	2	6	4	2	
75	Rhenium	Re	2	2	6	2	6	10	2	6	10	14	2	6	5	2	
76	Osmium	Os	2	2	6	2	6	10	2	6	10	14	2	6	6	2	
77	Iridium	Ir	2	2	6	2	6	10	2	6	10	14	2	6	7	2	
78	Platinum	Pt	2	2	6	2	6	10	2	6	10	14	2	6	9	1	
79	Gold	Au	2	2	6	2	6	10	2	6	10	14	2	6	10	1	
80	Mercury	Hg	2	2	6	2	6	10	2	6	10	14	2	6	10	2	
81	Thallium	Tl	2	2	6	2	6	10	2	6	10	14	2	6	10	2	1
82	Lead	Pb	2	2	6	2	6	10	2	6	10	14	2	6	10	2	2
83	Bismuth	Bi	2	2	6	2	6	10	2	6	10	14	2	6	10	2	3
84	Polonium	Po	2	2	6	2	6	10	2	6	10	14	2	6	10	2	4
85	Astatine	At	2	2	6	2	6	10	2	6	10	14	2	6	10	2	5
86	Radon	Rn	2	2	6	2	6	10	2	6	10	14	2	6	10	2	6

6

Period	Atomic Number	Name of Element	Symbol of Element	n=1 s	n=2 s	n=2 p	n=3 s	n=3 p	n=3 d	n=4 s	n=4 p	n=4 d	n=4 f	n=5 s	n=5 p	n=5 d	n=5 f	n=6 s	n=6 p	n=6 d	n=6 f	n=7 s
	87	Francium	Fr	2	2	6	2	6	10	2	6	10	14	2	6	10		2	6			1
	88	Radium	Ra	2	2	6	2	6	10	2	6	10	14	2	6	10		2	6			2
	89	Actinium	Ac	2	2	6	2	6	10	2	6	10	14	2	6	10		2	6	1		2
	90	Thorium	Th	2	2	6	2	6	10	2	6	10	14	2	6	10		2	6	2		2
	91	Protactinium	Pa	2	2	6	2	6	10	2	6	10	14	2	6	10	2	2	6	1		2
	92	Uranium	U	2	2	6	2	6	10	2	6	10	14	2	6	10	3	2	6	1		2
	93	Neptunium	Np	2	2	6	2	6	10	2	6	10	14	2	6	10	4	2	6	1		2
7	94	Plutonium	Pu	2	2	6	2	6	10	2	6	10	14	2	6	10	6	2	6			2
	95	Americium	Am	2	2	6	2	6	10	2	6	10	14	2	6	10	7	2	6			2
	96	Curium	Cm	2	2	6	2	6	10	2	6	10	14	2	6	10	7	2	6	1		2
	97	Berkelium	Bk	2	2	6	2	6	10	2	6	10	14	2	6	10	9	2	6			2
	98	Californium	Cf	2	2	6	2	6	10	2	6	10	14	2	6	10	10	2	6			2
	99	Einsteinium	Es	2	2	6	2	6	10	2	6	10	14	2	6	10	11	2	6			2
	100	Fermium	Fm	2	2	6	2	6	10	2	6	10	14	2	6	10	12	2	6			2
	101	Mendelevium	Md	2	2	6	2	6	10	2	6	10	14	2	6	10	13	2	6			2
	102	Nobelium	No	2	2	6	2	6	10	2	6	10	14	2	6	10	14	2	6			2
	103	Lawrencium	Lw	2	2	6	2	6	10	2	6	10	14	2	6	10	14	2	6	1		2

Orbital Distribution of Electrons in Various Energy Levels

XIII. GENERAL REVIEW QUESTIONS AND ANSWERS
Part 1

The following are multiple-choice test questions. Select, from the alternative answers given, the correct one, or the best one, and place the number of that answer on the line at the left of the question. Select only ONE answer per question. Omissions and questions marked with more than one alternative should be counted as errors.

Example. __3__ Chemistry is one of the (1) social sciences; (2) biological sciences; (3) physical sciences; (4) metaphysical sciences.

_____ 1. One of the following is *not* a physical property of matter: (1) color; (2) solubility; (3) stability toward heat; (4) density.

_____ 2. One of the following is *not* a type of chemical change: (1) combination; (2) decomposition; (3) displacement; (4) fusion.

_____ 3. One physical property of a nonmetal is: (1) ductility; (2) high density; (3) low tensile strength; (4) good conductor of heat.

_____ 4. (1) Over 50%; (2) about 50%; (3) less than 45%, of the weight of the compounds comprising the earth's crust is oxygen.

_____ 5. A chemical change occurs when:
(1) snow melts; (2) sugar dissolves in water; (3) steam condenses; (4) iron rusts.

_____ 6. Hydrogen is contained in:
(1) all acids; (2) all bases; (3) all salts.

_____ 7. The molecular theory assumes that:
(1) in chemical changes molecules combine, separate, or change places; (2) the mass of a molecule is the sum of the masses of the atoms composing it; (3) the word molecule always means a compound.

_____ 8. One of the following is *not* a common commercial method of producing oxygen:
(1) potassium chlorate plus heat; (2) electrolysis of water; (3) fractional distillation of liquid air.

_____ 9. In one of the following the metal is bivalent:
(1) KNO_3; (2) $NaCl$; (3) $Cu(NO_3)_2$; (4) $Al(NO_3)_3$.

_____ 10. The Phlogiston Theory was proved to be false by: (1) Boyle; (2) Priestley; (3) Lavoisier; (4) Graham.

_____ 11. One of the following is *not* a usual laboratory method of preparing hydrogen:
(1) electrolysis of water; (2) action of steam on hot coke; (3) sodium plus water; (4) zinc plus HCl.

_____ 12. $2 NaCl + H_2SO_4$ (heat) →
(1) $Na_2SO_4 + HCl$; (2) $NaSO_4 + 2 HCl$;
(3) $Na_2SO_4 + 2 HCl$; (4) $Na_2(SO_4)_2 + HCl$

_____ 13. Isotopes are:

(1) atoms of unequal weight but of the same nuclear charge; (2) atoms which have the same number of protons but an unequal number of neutrons; (3) neither of these alternatives; (4) both of these alternatives.

_____ 14. Rutherford's experiments showed that all atoms: (1) are uniform in structure; (2) are made exclusively of electrons; (3) have very small positively charged nuclei.

_____ 15. The density of deuterium oxide (heavy water) is greater than that of ordinary water by about:

(1) 10%; (2) 15%; (3) 20%; (4) 25%.

_____ 16. If a small crystal of the solute added to a solution results in the formation of additional crystals of the solute, the solution is said to be:

(1) normal; (2) concentrated; (3) saturated; (4) supersaturated.

_____ 17. Atoms are made of: (1) electrons; (2) neutrons, electrons, and protons; (3) protons and neutrons; (4) electrons and positrons.

_____ 18. Concentrated sulfuric acid is:

(1) an oxidizing agent; (2) a dehydrating agent; (3) neither of these alternatives; (4) both of these alternatives.

_____ 19. In the reaction $3 P + 5 HNO_3 + 2 H_2O \rightarrow 5 NO + 3 H_3PO_4$, the oxidizing agent is:

(1) nitric acid; (2) phosphorus; (3) neither of these alternatives.

_____ 20. One of these metals is never found free in nature:

(1) gold; (2) silver; (3) copper; (4) zinc.

_____ 21. An 8% solution of table salt has a specific gravity of 1.11. If 120 ml of this solution is evaporated to dryness, the weight of the remaining salt will be:

(1) 10.65 g; (2) 11.0 g; (3) 11.35 g; (4) 11.65 g.

_____ 22. Neutrons are: (1) protons and electrons; (2) positrons and electrons; (3) particles with zero electric charge.

_____ 23. $MgCO_3 + 2 HCl \rightarrow$

(1) $MgCl_2 + H_2O + CO_2$; (2) $MgCO_2 + 2 H_2O + CO_2$; (3) $MgCl_2 + H_2O + 2 CO_2$; (4) $MgCl_3 + HCO_3 + CO_2$.

_____ 24. One of these metals is not attacked by the common acids:

(1) copper; (2) platinum; (3) silver; (4) mercury.

_____ 25. One of the following is an incorrect statement of a chemical property of chlorine:

(1) it does not combine directly with nonmetals; (2) active metals burn in chlorine; (3) it replaces bromine and iodine

from their salt solutions; (4) it combines with hydrogen in the presence of sunlight with explosive violence.

_____ 26. In the formula Na_2HPO_4, the percentage of oxygen is approximately:
(1) 40; (2) 45; (3) 50; (4) 55.

_____ 27. An acid is:
(1) a substance which gives up protons; (2) a molecule or ion which will combine with a proton; (3) neither of these alternatives; (4) both of these alternatives.

_____ 28. The concentration of an acid or a base, i.e., the number of equivalents per liter, is conveniently determined by:
(1) filtration; (2) electrolysis; (3) titration; (4) fusion.

_____ 29. $Cu + 2 H_2SO_4$ (heat) $\rightarrow$
(1) $Cu_2SO_4 + 2 H_2O + SO_2$; (2) $CuSO_2 + H_2O + 2 SO_2$;
(3) $CuSO_4 + 2 H_2O + SO_2$; (4) $CuSO_4 + H_2O + 2 SO_2$.

_____ 30. (1) Graham's Law, (2) Boyle's Law, (3) Charles's Law, (4) Gay-Lussac's Law, states that the rate of diffusion of two gases is inversely proportional to the square root of their densities.

_____ 31. The electrons in an atom: (1) are, relatively, very close to the nucleus; (2) rotate in circular orbits outside the nucleus; (3) are located in regions in the atom called orbitals.

_____ 32. A curdy, white precipitate soluble in ammonia solution is formed if silver nitrate is added to:
(1) nitric acid; (2) hydrochloric acid; (2) acetic acid; (4) sulfuric acid.

_____ 33. (1) The oxygen, (2) the nitrogen, (3) the water vapor, (4) the carbon dioxide, (5) the inert gases, will be removed from the ordinary air passed through a tube containing phosphorus pentoxide.

_____ 34. (1) Argon, (2) helium, (3) krypton, (4) neon, (5) xenon, is present in the largest amount in ordinary air.

_____ 35. The most active halogen is:
(1) astatine; (2) bromine; (3) chlorine; (4) fluorine; (5) iodine.

_____ 36. The law which states that the properties of elements are periodic functions of their atomic numbers was announced by:
(1) Mendeleev; (2) Moseley; (3) Newlands; (4) Dobereiner.

_____ 37. (1) Antimony, (2) arsenic, (3) bismuth, (4) phosphorus, is not amphoteric.

_____ 38. An atomic orbital: (1) may contain up to 5 electrons; (2) may contain not more than 2 electrons and these must be of opposite spin; (3) contains an orderly arrangement of protons and electrons.

_____ 39. One of the following is *not* a property typical of the lyophobic colloids: (1) low concentration; (2) a sol that is not easily reversible; (3) low molecular weight; (4) high viscosity.

_____ 40. One of the following is an incorrect statement of a general chemical property of the alkali metals:
(1) they are active electropositive metals; (2) they are bivalent; (3) they decompose water vigorously; (4) they are good reducing agents.

_____ 41. The shape of the orbital of the hydrogen atom is: (1) cone shaped; (2) spherical; (3) elliptical; (4) cubical.

_____ 42. The molecular weight of $(NH_4)_2CO_3$ is:
(1) 90; (2) 92; (3) 94; (4) 96.

_____ 43. Metallic copper is made in:
(1) a reverberatory furnace; (2) a Dwight-Lloyd machine; (3) a copper converter; (4) a blast furnace.

_____ 44. Magnesium is obtained from:
(1) sea water; (2) bauxite; (3) calcite; (4) dolomite.

_____ 45. Low-grade copper sulfide ores are concentrated by:
(1) smelting; (2) milling and flotation; (3) fluxing; (4) leaching.

_____ 46. Bessemer converters are used to prepare:
(1) lead; (2) steel; (3) cobalt; (4) wrought iron.

_____ 47. Gray cast iron contains:
(1) martensite; (2) graphite; (3) cementite; (4) pearlite.

_____ 48. Silicones are compounds of:
(1) silicon and oxygen; (2) silicon and hydrogen; (3) sulfur and silicon; (4) carbon, silicon, oxygen, and hydrogen.

_____ 49. A lead storage battery is made up of:
(1) lead and lead dioxide; (2) sulfuric acid and lead; (3) graphite, lead, and sulfuric acid; (4) lead dioxide, lead, and sulfuric acid.

_____ 50. Cortisone is:
(1) a hormone; (2) a drug; (3) a vitamin; (4) an anesthetic.

_____ 51. Most metals crystallize in the: (1) hexagonal system; (2) cubic system; (3) tetragonal system.

_____ 52. Metallic crystals are *not* made up of: (1) atoms; (2) ions; (3) molecules.

_____ 53. Most crystals are imperfect. Imperfections may be due to: (1) the temperature at which the crystals were formed; (2) the ratio of positive to negative ions present in the crystals; (3) unoccupied lattice points or atoms not in their proper places in the space lattice; (4) the presence of foreign atoms in the crystals.

_____ 54. Crystals of a solid solution: (1) have a definite melting temperature; (2) are all the same size; (3) have no surface imperfections; (4) have atoms of more than one element present in the solid solution crystals.

_____ 55. The various lanthanides are difficult to separate (chemically) from each other because: (1) their electronic structures are very similar; (2) they have no valence electrons; (3) they contain no *f* orbitals.

_____ 56. The modern periodic table of the elements is based on their: (1) atomic weights; (2) chemical properties; (3) atomic numbers; (4) physical properties.

_____ 57. In making periodic tables there was difficulty in placing: (1) the nonmetallic elements; (2) the lanthanide elements; (3) the inert elements.

_____ 58. In nuclear chemical reactions: (1) the nuclei of atoms are not changed chemically; (2) no heat is liberated or absorbed; (3) the original nuclei are changed into new ones.

_____ 59. In the radioactive disintegrations of elements the process depends on: (1) nuclear energy; (2) temperature; (3) pressure; (4) ultraviolet light.

_____ 60. Ordinary elements can be made radioactive by: (1) exposing them to solar energy; (2) bombarding their nuclei with alpha particles, protons, or neutrons; (3) passing high voltage currents through them.

Part 2

Place a T before each statement that is true, an F before each statement that is false.

_____ 61. Oxidation is a process in which an atom or ion loses one or more electrons.

_____ 62. In any reaction in which one substance is oxidized, some other substance is reduced.

_____ 63. Since bismuth has metallic properties, there are several bismuth acids.

_____ 64. According to Avogadro's Hypothesis, a liter of hydrogen and a liter of chlorine, under the same conditions, contain the same number of molecules.

_____ 65. The electron is the unit of positive electricity.

_____ 66. An atom which gives up electrons has positive valence or a positive oxidation number.

_____ 67. The nucleus of an atom is charged negatively because it contains an excess of protons.

_____ 68. There are three types of solutions—solid, liquid, and gaseous.

_____ 69. Red phosphorus has a higher kindling temperature than white phosphorus.

_____ 70. When two or more molecules have the same composition but different structures they are called *isotopes*.

_____ 71. When two or more molecules have the same composition but different structures they are called *isomers*.

_____ 72. Aluminum is the most abundant metal in the earth's crust.

_____ 73. Natural gas is largely methane (CH_4).

_____ 74. Wood alcohol is made from CO and H_2.

_____ 75. Vinegar contains about 5% ethyl alcohol (C_2H_5OH).

_____ 76. The word *atom* refers to elements.

_____ 77. It is incorrect to speak of an atom of water.

_____ 78. All the atoms of oxygen are not precisely the same physically.

_____ 79. The weight (in grams) of any element that combines with 8 g of oxygen is its equivalent weight.

_____ 80. Lavoisier originated the atomic theory.

_____ 81. Atoms are complex in structure because electrons can be removed from them.

_____ 82. The nucleus of an atom contains one or more protons.

_____ 83. A neutron has approximately the same mass as a proton.

_____ 84. In atomic structure nomenclature, the term *orbital* means the same as *shell*.

_____ 85. Sodium imparts a yellow color to a flame. This is due to the electrons of sodium atoms moving from one orbit to another under the influence of heat.

_____ 86. Vitamins are present in many foods.

_____ 87. Ascorbic acid is the chemical name for vitamin C.

_____ 88. Lyophobic colloids have a strong affinity for water.

_____ 89. The word *mineral* has the same meaning as the word *ore*.

_____ 90. From an industrial point of view, aluminum is the most important metal in Group IIIA of the periodic table.

_____ 91. Gold was first discovered early in the twentieth century.

_____ 92. Certain silver salts are used in photography.

_____ 93. For commercial use as the metal, copper should be at least 99.9% pure.

_____ 94. Mercury is the only metal that is liquid at 0°C.

_____ 95. Low-grade lead sulfide ores are concentrated by flotation.

_____ 96. "The current strength in a circuit is directly proportional to the electromotive force and inversely proportional to the resistance" is a statement of Faraday's second law.

_____ 97. The heat of formation of a compound is the heat (measured in calories) absorbed or liberated when one mole of the compound is formed.

_____ 98. A calorimeter is an apparatus for producing heat.

Part 3

For each of the following exercises write the letter of the most appropriate item from column (2) in the blank at the left of the corresponding item in column (1).

	(1)		(2)
99. Electronic Arrangement			
_____ 1.	(2,7)	A.	Silicon
_____ 2.	(2,8,6)	B.	Fluorine
_____ 3.	(2,8,4)	C.	Sulfur
_____ 4.	(2,8,8,1)	D.	Aluminum
_____ 5.	(2,8,8,2)	E.	Zinc
_____ 6.	(2,8,3)	F.	Potassium
_____ 7.	(2,8,18,1)	G.	Antimony
		H.	Calcium
		I.	Boron
		J.	Copper

	(1)		(2)
100. _____ 1.	Sodium silicate	A.	HNO_3
_____ 2.	Lead acetate	B.	$(C_2H_5)_2O$
_____ 3.	Phenol	C.	Na_2SiO_3
_____ 4.	Calcium hypochlorite	D.	$C_6H_{12}O_6$
_____ 5.	Glucose	E.	C_6H_5OH
		F.	$CaSO_4 \cdot 2\,H_2O$
		G.	$Pb(C_2H_3O_2)_2$
		H.	$Ca(ClO)_2$

	(1)		(2)
101.			
_____ 1.	CaO	A.	Acid anhydride
_____ 2.	HCN	B.	Very weak acid
_____ 3.	HNO_3	C.	Strong acid
_____ 4.	C_2H_6	D.	Dibasic acid
_____ 5.	H_2SO_4	E.	Hydrocarbon
_____ 6.	P_2O_5	F.	Basic anhydride
		G.	Monoacid base
		H.	Salt

	(1)		(2)
102.			
_____ 1.	Radioactive metal	A.	Positive electron
_____ 2.	Fissionable substance	B.	Californium
_____ 3.	Cyclotron-made metal	C.	Curium
_____ 4.	Element 98	D.	Thorium
_____ 5.	Substance with half-life of 1590 years	E.	Uranium 235
		F.	Alpha particle
_____ 6.	Positron	G.	Deuteron
		H.	Radium

103. (1) Exercise in naming compounds　　　　　　　(2)

　　　_____ 1. *ide* ending　　　　　　A.　H_2SO_3

　　　_____ 2. *ite* ending　　　　　　B.　$KClO_4$

　　　_____ 3. per.....ate　　　　　　C.　CH_3OH

　　　_____ 4. ous　　　　　　　　　D.　$SrCl_2$

　　　_____ 5. ic　　　　　　　　　　E.　H_3PO_4

　　　_____ 6. hypo....ite　　　　　　F.　$KBrO$

　　　_____ 7. per.....ide　　　　　　G.　KNO_3

　　　　　　　　　　　　　　　　　H.　$NaNO_2$

　　　　　　　　　　　　　　　　　I.　H_2O_2

　　　　　　　　　　　　　　　　　J.　$HCHO$

104. (1) Exercise in classification of compounds　　　　(2)

　　　_____ 1. monobasic acid　　　　A.　$Ca(OH)_2$

　　　_____ 2. primary alcohol　　　　B.　$CH_3—CO—C_2H_5$

　　　_____ 3. acid salt　　　　　　　C.　$HC_2H_3O_2$

　　　_____ 4. diacid base　　　　　　D.　C_2H_5OH

　　　_____ 5. secondary alcohol　　　E.　$KHSO_4$

　　　_____ 6. ketone　　　　　　　　F.　$Na_2C_2O_4$

　　　　　　　　　　　　　　　　　G.　$CH_3CHOHCH_3$

　　　　　　　　　　　　　　　　　H.　$Bi(OH)Cl_2$

Part 4

　　Many students have considerable experience in answering questions like those in parts 1, 2, and 3 of this list. On the other hand they seldom have to answer discussion (or essay) type questions. Part 4 questions are of this important type and should be carefully answered. The object is to write out the chemically correct answers in acceptable English. The answers to questions in part 4 are not given in this book. Have your chemistry friend or teacher read your written answers.

105. How does the kinetic theory account for the properties of gases? This should be answered in terms of (1) expansibility, (2) compressibility, (3) diffusibility, (4) liquefiability.

106. Describe a laboratory method of showing that salt water can be purified by distillation.

107. Suggest a method to prove that copper oxide contains oxygen. (Hint! use hydrogen to reduce copper oxide.)

108. Draw an apparatus suitable for preparing hydrogen from steam by passing it over iron shavings (steel wool).

109. Show by suitable chemical equations what happens when: (1) chlorine is passed into cold potassium hydroxide solution; (2) chlorine is passed into hot potassium hydroxide solution.

110. What are the conditions necessary for converting $KClO_3$ into $KClO_4$? Write an equation showing what happens chemically.

111. (a) Describe a simple method of preparing impure nitrogen from the atmosphere. (b) Write an equation for the preparation of essentially pure nitrogen from a nitrogen compound.

112. (a) Describe a common laboratory method of preparing ammonia. (b) Write equations showing how ammonia is prepared from the elements.

113. Explain why an all-glass apparatus is required for making nitric acid from sodium nitrate and concentrated sulfuric acid.

114. (a) Describe the contact method of making sulfuric acid. (b) Write the chemical equations for this process.

115. What is meant by the term "critical temperature" of a gas?

116. Which of the following solutions has the lower freezing point: (1) one prepared by dissolving a formula weight of table salt in 1000 g of water, or (2) one containing a formula weight of sucrose (cane sugar) dissolved in 1000 g of water? Explain in terms of the electrolytic dissociation theory.

117. Define each of the following terms: (a) ion; (b) cation; (c) ionic reaction; (d) pH (of a solution); (e) indicator.

118. Distinguish between a reversible and irreversible reaction. Why do we think of the reaction between magnesium and oxygen as one that goes to completion?

119. List the factors that affect the rates of chemical reactions.

120. By means of a specific example explain the function of a catalyst.

121. What is meant by the term pH? What advantages are obtained by expressing the hydrogen ion concentration of a solution in terms of pH?

122. The solubility product constant $(K_{s.p.})$ is a special equilibrium constant. Explain.

123. Write an acceptable definition for a chemical equation.

124. Describe an experimental method of showing that acetic acid is weaker than hydrochloric acid.

125. The chemical properties of the lanthanide series of elements (atomic numbers 58–71 inc.) are very similar. Explain why this is so from the standpoint of their atomic structures.

126. Dimethyl ether and ethyl alcohol have the same composition but different properties. (a) What name is given to such substances? (b) Write the structural formula of each.

127. (a) Write the structural formula of: 2-4-dimethylhexane. (b) What is the name of the substance whose formula is:

$$-\overset{|}{C}-\overset{|}{\underset{|}{C}}-\overset{|}{C}-\overset{|}{C}=\overset{|}{C}-$$
$$\overset{|}{\underset{|}{C}}-$$

128. In terms of atomic structure why is chlorine a nonmetal and potassium a metal?
129. What is the relationship between elements and their position in a modern periodic table?
130. Explain why certain silver salts are used in photography.

ANSWERS TO GENERAL REVIEW QUESTIONS— PARTS 1, 2, AND 3

PART 1

1. (3)	16. (4)	31. (3)	46. (2)
2. (4)	17. (2)	32. (2)	47. (2)
3. (3)	18. (4)	33. (3)	48. (4)
4. (2)	19. (1)	34. (1)	49. (4)
5. (4)	20. (4)	35. (4)	50. (1)
6. (1)	21. (1)	36. (2)	51. (2)
7. (2)	22. (3)	37. (4)	52. (3)
8. (1)	23. (1)	38. (2)	53. (3)
9. (3)	24. (2)	39. (4)	54. (4)
10. (3)	25. (1)	40. (2)	55. (1)
11. (2)	26. (2)	41. (2)	56. (3)
12. (3)	27. (1)	42. (4)	57. (2)
13. (2)	28. (3)	43. (3)	58. (3)
14. (3)	29. (3)	44. (1)	59. (1)
15. (1)	30. (1)	45. (2)	60. (2)

PART 2

61. T	71. T	81. T	91. F
62. T	72. T	82. T	92. T
63. F	73. T	83. T	93. T
64. T	74. T	84. F	94. T
65. F	75. F	85. T	95. T
66. T	76. T	86. T	96. F
67. F	77. T	87. T	97. T
68. T	78. T	88. F	98. F
69. T	79. T	89. F	
70. F	80. F	90. T	

PART 3

99.		100.		101.	
B	1.	C	1.	F	1.
C	2.	G	2.	B	2.

A	3.
F	4.
H	5.
D	6.
J	7.

E	3.
H	4.
D	5.

C	3.
E	4.
D	5.
A	6.

102.

D	1.
E	2.
C	3.
B	4.
H	5.
A	6.

103.

D	1.
H	2.
B	3.
A	4.
E	5.
F	6.
I	7.

104.

C	1.
D	2.
E	3.
A	4.
G	5.
B	6.

XIV. FOUR-PLACE LOGARITHMS
AND ANTILOGARITHMS

FOUR-PLACE LOGARITHMS[1]

N	0	1	2	3	4	5	6	7	8	9	1	2	3	4	5
													Proportional Parts		
10	0000	0043	0086	0128	0170	0212	0253	0294	0334	0374	4	8	12	17	21
11	0414	0453	0492	0531	0569	0607	0645	0682	0719	0755	4	8	11	15	19
12	0792	0828	0864	0899	0934	0969	1004	1038	1072	1106	3	7	10	14	17
13	1139	1173	1206	1239	1271	1303	1335	1367	1399	1430	3	6	10	13	16
14	1461	1492	1523	1553	1584	1614	1644	1673	1703	1732	3	6	9	12	15
15	1761	1790	1818	1847	1875	1903	1931	1959	1987	2014	3	6	8	11	14
16	2041	2068	2095	2122	2148	2175	2201	2227	2253	2279	3	5	8	11	13
17	2304	2330	2355	2380	2405	2430	2455	2480	2504	2529	2	5	7	10	12
18	2553	2577	2601	2625	2648	2672	2695	2718	2742	2765	2	5	7	9	12
19	2788	2810	2833	2856	2878	2900	2923	2945	2967	2989	2	4	7	9	11
20	3010	3032	3054	3075	3096	3118	3139	3160	3181	3201	2	4	6	8	11
21	3222	3243	3263	3284	3304	3324	3345	3365	3385	3404	2	4	6	8	10
22	3424	3444	3464	3483	3502	3522	3541	3560	3579	3598	2	4	6	8	10
23	3617	3636	3655	3674	3692	3711	3729	3747	3766	3784	2	4	5	7	9
24	3802	3820	3838	3856	3874	3892	3909	3927	3945	3962	2	4	5	7	9
25	3979	3997	4014	4031	4048	4065	4082	4099	4116	4133	2	3	5	7	9
26	4150	4166	4183	4200	4216	4232	4249	4265	4281	4298	2	3	5	7	8
27	4314	4330	4346	4362	4378	4393	4409	4425	4440	4456	2	3	5	6	8
28	4472	4487	4502	4518	4533	4548	4564	4579	4594	4609	2	3	5	6	8
29	4624	4639	4654	4669	4683	4698	4713	4728	4742	4757	1	3	4	6	7
30	4771	4786	4800	4814	4829	4843	4857	4871	4886	4900	1	3	4	6	7
31	4914	4928	4942	4955	4969	4983	4997	5011	5024	5038	1	3	4	6	7
32	5051	5065	5079	5092	5105	5119	5132	5145	5159	5172	1	3	4	5	7
33	5185	5198	5211	5224	5237	5250	5263	5276	5289	5302	1	3	4	5	6
34	5315	5328	5340	5353	5366	5378	5391	5403	5416	5428	1	3	4	5	6
35	5441	5453	5465	5478	5490	5502	5514	5527	5539	5551	1	2	4	5	6
36	5563	5575	5587	5599	5611	5623	5635	5647	5658	5670	1	2	4	5	6
37	5682	5694	5705	5717	5729	5740	5752	5763	5775	5786	1	2	3	5	6
38	5798	5809	5821	5832	5843	5855	5866	5877	5888	5899	1	2	3	5	6
39	5911	5922	5933	5944	5955	5966	5977	5988	5999	6010	1	2	3	4	6
40	6021	6031	6042	6053	6064	6075	6085	6096	6107	6117	1	2	3	4	5
41	6128	6138	6149	6160	6170	6180	6191	6201	6212	6222	1	2	3	4	5
42	6232	6243	6253	6263	6274	6284	6294	6304	6314	6325	1	2	3	4	5
43	6335	6345	6355	6365	6375	6385	6395	6405	6415	6425	1	2	3	4	5
44	6435	6444	6454	6464	6474	6484	6493	6503	6513	6522	1	2	3	4	5
45	6532	6542	6551	6561	6571	6580	6590	6599	6609	6618	1	2	3	4	5
46	6628	6637	6646	6656	6665	6675	6684	6693	6702	6712	1	2	3	4	5
47	6721	6730	6739	6749	6758	6767	6776	6785	6794	6803	1	2	3	4	5
48	6812	6821	6830	6839	6848	6857	6866	6875	6884	6893	1	2	3	4	4
49	6902	6911	6920	6928	6937	6946	6955	6964	6972	6981	1	2	3	4	4
50	6990	6998	7007	7016	7024	7033	7042	7050	7059	7067	1	2	3	3	4
51	7076	7084	7093	7101	7110	7118	7126	7135	7143	7152	1	2	3	3	4
52	7160	7168	7177	7185	7193	7202	7210	7218	7226	7235	1	2	2	3	4
53	7243	7251	7259	7267	7275	7284	7292	7300	7308	7316	1	2	2	3	4
54	7324	7332	7340	7348	7356	7364	7372	7380	7388	7396	1	2	2	3	4
N	0	1	2	3	4	5	6	7	8	9	1	2	3	4	5

[1] *An enlarged reproduction of a page from the* Handbook of Chemistry and Physics, Edited by Charles D. Hodgman, *by permission of* Chemical Rubber Publishing Company.

FOUR-PLACE LOGARITHMS[1]
(continued)

N	0	1	2	3	4	5	6	7	8	9	Proportional Parts				
											1	2	3	4	5
55	7404	7412	7419	7427	7435	7443	7451	7459	7466	7474	1	2	2	3	4
56	7482	7490	7497	7505	7513	7520	7528	7536	7543	7551	1	2	2	3	4
57	7559	7566	7574	7582	7589	7597	7604	7612	7619	7627	1	2	2	3	4
58	7634	7642	7649	7657	7664	7672	7679	7686	7694	7701	1	1	2	3	4
59	7709	7716	7723	7731	7738	7745	7752	7760	7767	7774	1	1	2	3	4
60	7782	7789	7796	7803	7810	7818	7825	7832	7839	7846	1	1	2	3	4
61	7853	7860	7868	7875	7882	7889	7896	7903	7910	7917	1	1	2	3	4
62	7924	7931	7938	7945	7952	7959	7966	7973	7980	7987	1	1	2	3	3
63	7993	8000	8007	8014	8021	8028	8035	8041	8048	8055	1	1	2	3	3
64	8062	8069	8075	8082	8089	8096	8102	8109	8116	8122	1	1	2	3	3
65	8129	8136	8142	8149	8156	8162	8169	8176	8182	8189	1	1	2	3	3
66	8195	8202	8209	8215	8222	8228	8235	8241	8248	8254	1	1	2	3	3
67	8261	8267	8274	8280	8287	8293	8299	8306	8312	8319	1	1	2	3	3
68	8325	8331	8338	8344	8351	8357	8363	8370	8376	8382	1	1	2	3	3
69	8388	8395	8401	8407	8414	8420	8426	8432	8439	8445	1	1	2	3	3
70	8451	8457	8463	8470	8476	8482	8488	8494	8500	8506	1	1	2	2	3
71	8513	8519	8525	8531	8537	8543	8549	8555	8561	8567	1	1	2	2	3
72	8573	8579	8585	8591	8597	8603	8609	8615	8621	8627	1	1	2	2	3
73	8633	8639	8645	8651	8657	8663	8669	8675	8681	8686	1	1	2	2	3
74	8692	8698	8704	8710	8716	8722	8727	8733	8739	8745	1	1	2	2	3
75	8751	8756	8762	8768	8774	8779	8785	8791	8797	8802	1	1	2	2	3
76	8808	8814	8820	8825	8831	8837	8842	8848	8854	8859	1	1	2	2	3
77	8865	8871	8876	8882	8887	8893	8899	8904	8910	8915	1	1	2	2	3
78	8921	8927	8932	8938	8943	8949	8954	8960	8965	8971	1	1	2	2	3
79	8976	8982	8987	8993	8998	9004	9009	9015	9020	9025	1	1	2	2	3
80	9031	9036	9042	9047	9053	9058	9063	9069	9074	9079	1	1	2	2	3
81	9085	9090	9096	9101	9106	9112	9117	9122	9128	9133	1	1	2	2	3
82	9138	9143	9149	9154	9159	9165	9170	9175	9180	9186	1	1	2	2	3
83	9191	9196	9201	9206	9212	9217	9222	9227	9232	9238	1	1	2	2	3
84	9243	9248	9253	9258	9263	9269	9274	9279	9284	9289	1	1	2	2	3
85	9294	9299	9304	9309	9315	9320	9325	9330	9335	9340	1	1	2	2	3
86	9345	9350	9355	9360	9365	9370	9375	9380	9385	9390	1	1	2	2	3
87	9395	9400	9405	9410	9415	9420	9425	9430	9435	9440	0	1	1	2	2
88	9445	9450	9455	9460	9465	9469	9474	9479	9484	9489	0	1	1	2	2
89	9494	9499	9504	9509	9513	9518	9523	9528	9533	9538	0	1	1	2	2
90	9542	9547	9552	9557	9562	9566	9571	9576	9581	9586	0	1	1	2	2
91	9590	9595	9600	9605	9609	9614	9619	9624	9628	9633	0	1	1	2	2
92	9638	9643	9647	9652	9657	9661	9666	9671	9675	9680	0	1	1	2	2
93	9685	9689	9694	9699	9703	9708	9713	9717	9722	9727	0	1	1	2	2
94	9731	9736	9741	9745	9750	9754	9759	9763	9768	9773	0	1	1	2	2
95	9777	9782	9786	9791	9795	9800	9805	9809	9814	9818	0	1	1	2	2
96	9823	9827	9832	9836	9841	9845	9850	9854	9859	9863	0	1	1	2	2
98	9868	9872	9877	9881	9886	9890	9894	9899	9903	9908	0	1	1	2	2
98	9912	9917	9921	9926	9930	9934	9939	9943	9948	9952	0	1	1	2	2
99	9956	9961	9965	9969	9974	9978	9983	9987	9991	9996	0	1	1	2	2
N	0	1	2	3	4	5	6	7	8	9	1	2	3	4	5

[1] *An enlarged reproduction of a page from the* Handbook of Chemistry and Physics, Edited by Charles D. Hodgman, *by permission of* Chemical Rubber Publishing Company.

FOUR-PLACE ANTILOGARITHMS[1]

	0	1	2	3	4	5	6	7	8	9	Proportional Parts				
											1	2	3	4	5
.00	1000	1002	1005	1007	1009	1012	1014	1016	1019	1021	0	0	1	1	1
.01	1023	1026	1028	1030	1033	1035	1038	1040	1042	1045	0	0	1	1	1
.02	1047	1050	1052	1054	1057	1059	1062	1064	1067	1069	0	0	1	1	1
.03	1072	1074	1076	1079	1081	1084	1086	1089	1091	1094	0	0	1	1	1
.04	1096	1099	1102	1104	1107	1109	1112	1114	1117	1119	0	1	1	1	1
.05	1122	1125	1127	1130	1132	1135	1138	1140	1143	1146	0	1	1	1	1
.06	1148	1151	1153	1156	1159	1161	1164	1167	1169	1172	0	1	1	1	1
.07	1175	1178	1180	1183	1186	1189	1191	1194	1197	1199	0	1	1	1	1
.08	1202	1205	1208	1211	1213	1216	1219	1222	1225	1227	0	1	1	1	1
.09	1230	1233	1236	1239	1242	1245	1247	1250	1253	1256	0	1	1	1	1
.10	1259	1262	1265	1268	1271	1274	1276	1279	1282	1285	0	1	1	1	1
.11	1288	1291	1294	1297	1300	1303	1306	1309	1312	1315	0	1	1	1	2
.12	1318	1321	1324	1327	1330	1334	1337	1340	1343	1346	0	1	1	1	2
.13	1349	1352	1355	1358	1361	1365	1368	1371	1374	1377	0	1	1	1	2
.14	1380	1384	1387	1390	1393	1396	1400	1403	1406	1409	0	1	1	1	2
.15	1413	1416	1419	1422	1426	1429	1432	1435	1439	1442	0	1	1	1	2
.16	1445	1449	1452	1455	1459	1462	1466	1469	1472	1476	0	1	1	1	2
.17	1479	1483	1486	1489	1493	1496	1500	1503	1507	1510	0	1	1	1	2
.18	1514	1517	1521	1524	1528	1531	1535	1538	1542	1545	0	1	1	1	2
.19	1549	1552	1556	1560	1563	1567	1570	1574	1578	1581	0	1	1	1	2
.20	1585	1589	1592	1596	1600	1603	1607	1611	1614	1618	0	1	1	1	2
.21	1622	1626	1629	1633	1637	1641	1644	1648	1652	1656	0	1	1	1	2
.22	1660	1663	1667	1671	1675	1679	1683	1687	1690	1694	0	1	1	2	2
.23	1698	1702	1706	1710	1714	1718	1722	1726	1730	1734	0	1	1	2	2
.24	1738	1742	1746	1750	1754	1758	1762	1766	1770	1774	0	1	1	2	2
.25	1778	1782	1786	1791	1795	1799	1803	1807	1811	1816	0	1	1	2	2
.26	1820	1824	1828	1832	1837	1841	1845	1849	1854	1858	0	1	1	2	2
.27	1862	1866	1871	1875	1879	1884	1888	1892	1897	1901	0	1	1	2	2
.28	1905	1910	1914	1919	1923	1928	1932	1936	1941	1945	0	1	1	2	2
.29	1950	1954	1959	1963	1968	1972	1977	1982	1986	1991	0	1	1	2	2
.30	1995	2000	2004	2009	2014	2018	2023	2028	2032	2037	0	1	1	2	2
.31	2042	2046	2051	2056	2061	2065	2070	2075	2080	2084	0	1	1	2	2
.32	2089	2094	2099	2104	2109	2113	2118	2123	2128	2133	0	1	1	2	2
.33	2138	2143	2148	2153	2158	2163	2168	2173	2178	2183	0	1	1	2	2
.34	2188	2193	2198	2203	2208	2213	2218	2223	2228	2234	1	1	2	2	3
.35	2239	2244	2249	2254	2259	2265	2270	2275	2280	2286	1	1	2	2	3
.36	2291	2296	2301	2307	2312	2317	2323	2328	2333	2339	1	1	2	2	3
.37	2344	2350	2355	2360	2366	2371	2377	2382	2388	2393	1	1	2	2	3
.38	2399	2404	2410	2415	2421	2427	2432	2438	2443	2449	1	1	2	2	3
.39	2455	2460	2466	2472	2477	2483	2489	2495	2500	2506	1	1	2	2	3
.40	2512	2518	2523	2529	2535	2541	2547	2553	2559	2564	1	1	2	2	3
.41	2570	2576	2582	2588	2594	2600	2606	2612	2618	2624	1	1	2	2	3
.42	2630	2636	2642	2649	2655	2661	2667	2673	2679	2685	1	1	2	2	3
.43	2692	2698	2704	2710	2716	2723	2729	2735	2742	2748	1	1	2	3	3
.44	2754	2761	2767	2773	2780	2786	2793	2799	2805	2812	1	1	2	3	3
.45	2818	2825	2831	2838	2844	2851	2858	2864	2871	2877	1	1	2	3	3
.46	2884	2891	2897	2904	2911	2917	2924	2931	2938	2944	1	1	2	3	3
.47	2951	2958	2965	2972	2979	2985	2992	2999	3006	3013	1	1	2	3	3
.48	3020	3027	3034	3041	3048	3055	3062	3069	3076	3083	1	1	2	3	3
.49	3090	3097	3105	3112	3119	3126	3133	3141	3148	3155	1	1	2	3	4
	0	1	2	3	4	5	6	7	8	9	1	2	3	4	5

[1] *An enlarged reproduction of a page from the* Handbook of Chemistry and Physics, Edited by Charles D. Hodgman, *by permission of* Chemical Rubber Publishing Company.

FOUR-PLACE ANTILOGARITHMS[1]
(continued)

	0	1	2	3	4	5	6	7	8	9	1	2	3	4	5
													Proportional Parts		
.50	3162	3170	3177	3184	3192	3199	3206	3214	3221	3228	1	1	2	3	4
.51	3236	3243	3251	3258	3266	3273	3281	3289	3296	3304	1	1	2	3	4
.52	3311	3319	3327	3334	3342	3350	3357	3365	3373	3381	1	1	2	3	4
.53	3388	3396	3404	3412	3420	3428	3436	3443	3451	3459	1	2	2	3	4
.54	3467	3475	3483	3491	3499	3508	3516	3524	3532	3540	1	2	2	3	4
.55	3548	3556	3565	3573	3581	3589	3597	3606	3614	3622	1	2	2	3	4
.56	3631	3639	3648	3656	3664	3673	3681	3690	3698	3707	1	2	2	3	4
.57	3715	3724	3733	3741	3750	3758	3767	3776	3784	3793	1	2	3	3	4
.58	3802	3811	3819	3828	3837	3846	3855	3864	3873	3882	1	2	3	3	4
.59	3890	3899	3908	3917	3926	3936	3945	3954	3963	3972	1	2	3	4	5
.60	3981	3990	3999	4009	4018	4027	4036	4046	4055	4064	1	2	3	4	5
.61	4074	4083	4093	4102	4111	4121	4130	4140	4150	4159	1	2	3	4	5
.62	4169	4178	4188	4198	4207	4217	4227	4236	4246	4256	1	2	3	4	5
.63	4266	4276	4285	4295	4305	4315	4325	4335	4345	4355	1	2	3	4	5
.64	4365	4375	4385	4395	4406	4416	4426	4436	4446	4457	1	2	3	4	5
.65	4467	4477	4487	4498	4508	4519	4529	4539	4550	4560	1	2	3	4	5
.66	4571	4581	4592	4603	4613	4624	4634	4645	4656	4667	1	2	3	4	5
.67	4677	4688	4699	4710	4721	4732	4742	4753	4764	4775	1	2	3	4	5
.68	4786	4797	4808	4819	4831	4842	4853	4864	4875	4887	1	2	3	5	6
.69	4898	4909	4920	4932	4943	4955	4966	4977	4989	5000	1	2	3	5	6
.70	5012	5023	5035	5047	5058	5070	5082	5093	5105	5117	1	2	3	5	6
.71	5129	5140	5152	5164	5176	5188	5200	5212	5224	5236	1	2	4	5	6
.72	5248	5260	5272	5284	5297	5309	5321	5333	5346	5358	1	2	4	5	6
.73	5370	5383	5395	5408	5420	5433	5445	5458	5470	5483	1	3	4	5	6
.74	5495	5508	5521	5534	5546	5559	5572	5585	5598	5610	1	3	4	5	6
.75	5623	5636	5649	5662	5675	5689	5702	5715	5728	5741	1	3	4	5	7
.76	5754	5768	5781	5794	5808	5821	5834	5848	5861	5875	1	3	4	5	7
.77	5888	5902	5916	5929	5943	5957	5970	5984	5998	6012	1	3	4	5	7
.78	6026	6039	6053	6067	6081	6095	6109	6124	6138	6152	1	3	4	6	7
.79	6166	6180	6194	6209	6223	6237	6252	6266	6281	6295	1	3	4	6	7
.80	6310	6324	6339	6353	6368	6383	6397	6412	6427	6442	1	3	4	6	7
.81	6457	6471	6486	6501	6516	6531	6546	6561	6577	6592	2	3	5	6	8
.82	6607	6622	6637	6653	6668	6683	6699	6714	6730	6745	2	3	5	6	8
.83	6761	6776	6792	6808	6823	6839	6855	6871	6887	6902	2	3	5	6	8
.84	6918	6934	6950	6966	6982	6998	7015	7031	7047	7063	2	3	5	7	8
.85	7079	7096	7112	7129	7145	7161	7178	7194	7211	7228	2	3	5	7	8
.86	7244	7261	7278	7295	7311	7328	7345	7362	7379	7396	2	3	5	7	8
.87	7413	7430	7447	7464	7482	7499	7516	7534	7551	7568	2	4	5	7	9
.88	7586	7603	7621	7638	7656	7674	7691	7709	7727	7745	2	4	5	7	9
.89	7762	7780	7798	7816	7834	7852	7870	7889	7907	7925	2	4	6	7	9
.90	7943	7962	7980	7998	8017	8035	8054	8072	8091	8110	2	4	6	7	9
.91	8128	8147	8166	8185	8204	8222	8241	8260	8279	8299	2	4	6	8	9
.92	8318	8337	8356	8375	8395	8414	8433	8453	8472	8492	2	4	6	8	10
.93	8511	8531	8551	8570	8590	8610	8630	8650	8670	8690	2	4	6	8	10
.94	8710	8730	8750	8770	8790	8810	8831	8851	8872	8892	2	4	6	8	10
.95	8913	8933	8954	8974	8995	9016	9036	9057	9078	9099	2	4	6	8	10
.96	9120	9141	9162	9183	9204	9226	9247	9268	9290	9311	2	4	6	9	11
.97	9333	9354	9376	9397	9419	9441	9462	9484	9506	9528	2	4	6	9	11
.98	9550	9572	9594	9616	9638	9661	9683	9705	9727	9750	2	4	7	9	11
.99	9772	9795	9817	9840	9863	9886	9908	9931	9954	9977	2	5	7	9	11
	0	1	2	3	4	5	6	7	8	9	1	2	3	4	5

[1] *An enlarged reproduction of a page from the* Handbook of Chemistry and Physics, Edited by Charles D. Hodgman, *by permission of* Chemical Rubber Publishing Company.

Index

(References to material in appendixes are preceded by a capital A followed by the page number. Thus: Active metals, A 392.)